The Persistence of Reality II

Science without Unity

The Persistence of Reality
JOSEPH MARGOLIS

The Persistence of Reality

Science without Unity
Reconciling the Human and Natural Sciences

JOSEPH MARGOLIS

Basil Blackwell

Copyright © Joseph Margolis 1987

First published 1987

Basil Blackwell Ltd
108 Cowley Road, Oxford, OX4 1JF, UK

Basil Blackwell Inc.
432 Park Avenue South, Suite 1503
New York, NY 10016, USA

British Library Cataloguing in Publication Data

Margolis, Joseph
    The persistence of reality.
    [2] : Science without unity: reconciling the human and natural
    sciences
    1. Reality 2. Perception
    I. Title
    153.7 BD331

            ISBN 0–631–15173–7

Library of Congress Cataloging in Publication Data

Margolis, Joseph Zalman.
    Science without unity.

    (The persistence of reality; 3)
    Includes index.
    1. Science—Philosophy. 2. Science—Social aspects.
1. Title. II. Series: Margolis, Joseph Zalman.
1924– Persistence of reality; 3.
Q175.M3624 1987 001 87 – 8025
ISBN 0–631–15173–7

Typeset in 10 on 12 point Garamond
by Photo·graphics, Honiton, Devon, England
Printed in Great Britain by Billing & Son Ltd, Worcester

In memory of my mother
1899–1986
when we are, death is not
and when death is, we are not

Contents

Preface

The argument I offer is an unorthodox one. I can't say it is radically novel or simply radical. I hope it is neither, but I'm not sure. The model of science it heuristically sketches in order to attack – the so-called unity of science program – is no longer as ardently or as unequivocally or as explicitly defended as it once proudly was; and there are a great many new speculations about the nature of science that are surely as heterodox as the view I pursue. Still, I think it may be fairly claimed that there has not yet appeared a reasonably detailed countertheory to the various strands of the unity model that answers the natural worries that follow abandoning it – for instance, regarding causality, natural laws, explanation, objectivity. I say only, in as flat a tone as I can command, that I have tried to meet the implied request. There would be no point in saying so, of course, if I thought I had utterly failed in the effort.

I must, however, add a small confession about the manuscript. I distinctly remember teaching some courses in philosophy at a summer institute in Calgary, Alberta in 1968 at the very kind invitation of Terence Penelhum, and discovering (to my great delight), during a paper I attended among the more-or-less regular activities of the University department, that I was surely blessed with a Socratic daimon. Naturally, one does not ignore such a gift. The oddity of it, I found, was that Socrates' voice had been an inner one, and mine was externalized in the voice of the speaker, the man who presented the formal paper. I *knew* at once in an ineluctably reliable way that that man must be completely mistaken, that I had better not follow his lead – ever; and that it was his own voice that betrayed him to me. I actually wrote a paper on the occasion – a successful one I think – with no better clue about the topic than that it should be developed in a way as distant from and as diametrically opposed as possible to the express view of that (nameless) speaker. Since that moment, I have always been on the lookout for other, admittedly more important, external voices that could guide me as reliably along the lines of what I have come to think of as sensible and honest philosophical work. The amazing thing is that I have found several such counterprophetic voices – prominent ones in fact. I'm quite certain they would just as surely remark (if they had a mind to bother) that *I* had strayed from the truth and deluded myself to boot. But there's the confession at least. In bringing to bed the very large argument that follows I could not have found my way without the aid of those corrective voices.

Perhaps to say so is no more than an expression of comic relief, a kind of hysterical epiphenomenon released at the end of a very large labor. But it has its more sober side as well. For I am quite persuaded that serious philosophical debate (and much more than philosophical debate) is profoundly rhetorical – in a non-pejorative sense of that term. Argument entrenches itself in the competing currents of a living society. It convinces because it persuades; and it persuades because it shows us how to organize the flotsam of our discursive society so that favored saliencies yield a clear sense of a non-arbitrary, non-*ad hoc* use of a conceptual strategy that feels as if it won't exhaust itself in trivia. It persuades, because it shows how we may move effectively through familiar, even novel, puzzles that other strategies are not aware of or cannot resolve or cannot resolve as convincingly under the terms of the dialectical contest implied.

I hope whatever else may be thought of the argument that follows that it will signal a way of breaking out of the kind of parochical dispute that has for so long separated Anglo-American and Continental European philosophical practice. For, quite frankly, that is the benign heterodoxy I am convinced we should prefer: we can no longer live within the Maginot lines and Berlin walls and Harvard yards of Western philosophy. No doubt, those epithets themselves harbor a parochialism of their own. I am entirely willing, however, to be introduced to a larger sort of instruction.

I have always felt that the puzzles of the philosophy of art and the philosophy of science were most intimately linked. This certainly runs contrary to the current of thought that prevailed during my own student years. I saw the connection then (and I see it still) both in terms of the actual discipline of theorizing about particular artworks and particular experimental results and in terms of the same intelligence that must be admitted to guide the admirable practices each is concerned to review. Now I find that intuition strengthened, and, thus strengthened, less and less heterodox in the philosophical world. Less and less heterodox but not less contested. The theory of the human sciences may well be the most strategic terrain on which to risk our most sovereign conceptions. Those forces that were, at least at one time, assembled in the service of the unity of science, and now (somewhat scattered) still claim to represent best what is implied by the enviable achievements of the formal and empirical sciences rightly insinuate the need for a responsible adversary to give a fair account of how to replace each and every argumentative strategy belonging to what is still viable within the scattered canon. I acknowledge the legitimacy of the charge. I also admit the often careless way in which many who (in my view) have rightly sensed the inadequacy of the canon have tried to dismantle it. Honor forbids that these matters should be contemptuously conducted. But the sheer volume of the reinforcing drudgery of armies of the unspectacular faithful – notably so well-connected in the way of paper and print and mannerly readers – has surely given a false impression of the true consensus, or even of the true power, of the arguments and strategies the slowly yielding champions of the canon have been able to sustain.

I have found it necessary to improvise a referentially explicit (if dense) and a dialectically gymnastic (if diffuse) way of organizing the counterargument I mount here. I shouldn't want to allow the story to proceed impressionistically. I think the best way to proceed – at least in our own time, possibly *not* thereafter – is to pinpoint what the acknowledged exemplary thinkers have to say, to draw the argument out of their own formulations as those have proved so fatefully influential to others. This is the only way I know to overcome the threat of inaccuracy, of allusion, of misrepresentation, of irrelevance, of arbitrariness, of a lack of charity. The opposing vision – it is after all an opposing vision that is wanted – must be drawn together with care and scruple. But, once assembled, it probably has to be scattered; or, it probably must expect to give way to scatter, since its own canon is that there is and can be no timeless canon. In admitting that, I see I already anticipate the peculiar transience of the effort. But that would have been true in any case. Still, the validity of meeting on its own terms what for convenience I am calling the canon cannot fail to be matched by an admission that the potentially fruitful lines of analysis pursued by the principal movements of contemporary Continental European thought ought not to be denigrated simply because there is no clear way to reconcile them with the canon's own favored projects. What I have sought to do, therefore, is begin to shape a vocabulary that bridges the best of both practices. The theory of the human sciences seems to be a terrain suited to a promising effort of such a sort.

To say anything more on this matter would, effectively, launch us into the issues themselves. I have already said too much, I suppose, since the argument lies not with the preface but with what follows; and there the preface is no longer needed. But it remains to say that, I, personally, have been unable to trade one orthodoxy for another. The good work of our time will surely shape a new idiom better suited to a global discourse than to local geographies. And yet no one can escape his local history.

I have been much affected by the encouragement of many friends. I have the pleasant sense of their waiting to see how the project will turn out. Readers ultimately collect in a natural way; and I feel privileged to have found space enough to begin and finish this particular undertaking, and very probably to be able to share a conversation or two about it thereafter.

Finally, once again, I must thank Grace Stuart for her customary care in putting the manuscript in acceptable order.

Philadelphia, Pennsylvania   J.M.
December 1986

Acknowledgements

With alterations and additions (some extensive) chapter 1 appears in Michael Krausz (ed.), *New Studies in Relativism* (Notre Dame, Ind.: Notre Dame University Press, forthcoming); chapter 2 in Marcelo Descal (ed.), *Cultural Relativism and Philosophy* (forthcoming); chapter 3, in *Topoi* (forthcoming); part of chapter 4 as "Information, Artificial Intelligence, and the Praxical," in Carl Mitchum and Alois Huning (eds), *Philosophy and Technology II* (Dordrecht: D. Reidel, 1986); chapter 7 in *The Monist*, LXIX (1986); chapter 10 as "Emergence," *Philosophical Forum*, XVII (1986). Permission to reprint has been granted.

Prologue: A Sense of the Issue

Our theme is more difficult to hide than to piece together from the local quarrels here collected. We shall be speaking of the human sciences primarily. But the epithet is already a warning. To address the *human* sciences is to confront ourselves with the possible discontinuities between the "human" and the "natural" sciences. We must allow the strange weight of that contrast to sink in: surely, even the natural sciences are the work of human scientists; surely again, the world of human life is no more than the most intriguing part of the natural world. And yet, the sciences do seem different – in the only sense that seriously counts – in the conditions of description and explanation, in the unity or bifurcation of the methods allotted to each in determining what is true regarding the world, and in the converse unity or bifurcation of the substantive sectors of reality they would describe and explain.

To speak of the unity of science is hardly innocent, hardly merely a musing about whatever similarities and dissimilarities may strike the mind in comparing, say, physics or biology and sociology or history. For the unity of science is a well-formed topic (well, a reasonably well-formed topic); also a would-be program or meta-program for any and all sciences, a model by which to gauge the standing of any discipline, a historical movement taking clear form, in the spirit of contest, from the twenties and thirties to our own day, drawing its distinctive sense of mission from older, middle and late nineteenth-century, sources and even more distantly from the very Renaissance beginnings of modern conceptions of science. The "unity of science" is the thesis of a conceptual rationale of what it is to be a *bona fide* science, expressed as a canon and as the abstracted import of the gathering self-discipline of science's own history. Presumably it threads its palpable way from Galileo to Newton to Mach to Carnap to us at least.

The essential question is whether the unity holds – and why or why not. It is a question of method, of knowledge, of reality, of self-discipline, of practice, at one and the same time. There is no hierarchy of privilege that can be assigned any one of these concerns; that itself is a particularly strategic constraint on the would-be force of any would-be answer. For – to risk a decisive conclusion before the argument is even in – what the method of science (or, of any science) is or should be is, rightly, a function of the nature of the domain it would describe and explain; conversely, what we may claim the real properties of a given domain's phenomena are is, rightly, a function of how best, relative to their description and explanation, we may discipline our particular claims. In the heavy jargon of the trade, *methodology and ontology are one* – or symbiotic at least. We cannot treat such inquiries independently or assign them priority one over the other. This is as true of the first-order work of the sciences and the second-order work of philosophy as it is of the work of any of the sub-disciplines of philosophy. In a sense, therefore, queer as it may seem, whether there is a common method or rationale uniting the natural and the human sciences is a perfectly good empirical question. It is a question that presupposes the continuity of science and philosophy, the inseparability of first- and second-order questions, and the eimpirical standing (on those conditions) of philosophy itself. It is also the essential pivot of pertinent contention: whether, that is, the peculiar hegemony of the unity model through the entire expanse of Western thought is or is not conceptually reasonable.

The answer given here is that it is not. It is, in fact, already somewhat heterodox to broach the question seriously. The massive prejudice of the theory of science is hardly an unreasonable one. It insists that the paradigm of scientific rigor and adequacy is exhibited in the work of basic physics, that there is some best way (or some small number of versions of the best) in which to formulate the clearly convergent principle and procedure of rigorous description and explanation in physics, and that with noticeably modest accommodation the paradigm method may be fitted to the entire variety of the more complex sciences, ranging from chemistry and biology to psychology, sociology, history, linguistics, and even perhaps literary criticism. The question of honor among the partisans of the unity canon asks whether this or that version of the comprehensive model gets things exactly right; the disagreements engendered are only family quarrels, the busy disputes of branches that no longer acknowledge one another publicly but (in the sociobiological sense) persist all the same in promoting altruistically their common stock.

It is difficult to demonstrate that the unity model is in jeopardy in the physical sciences. Of course it is. But the history of the quarrel usually takes the form of discarding the exposed infelicities of particular – often, particularly sanguine or monolithic – versions of the canon without ever reaching a compelling sense of betrayal or discontinued membership. The doctrine recovers its composure by retreating to ever cannier, if thinner and more tolerant, manifestos. How much can be given away without abandoning the title altogether is itself a function of the changing tolerances of a changing perception of history.

Certain master themes, however, are noticeably stubborn and central, without yet forming a fixed or fully systematized set of essential tenets. These probably include at least the following: advocacy of a comprehensive physicalism, or tolerance of functional properties reconciled in certain favored ways with a physicalist presumption; insistence on the extensional treatment of causality, or, additionally, the inherent nomological nature of causality, or, additionally, the natural necessity of nomological regularities; elimination of intentional structures in nature, or assurance that intentional structures do not adversely affect the scope of an extensional practice or the assignment of nomological regularities; presumption of closure with regard to the eligible range of causal explanation, or, additonally, an ideally homonomic form of explanation suited to reasonably well-demarcated domains or to the whole of nature; denial of a causal role to history as such, or denial that historical processes have a distinctive nomological structure, or denial that, where it is causally effective, history is intensionally irreducible; insistence that the laws of the complex sub-domains of nature are derivable (strictly or by idealization) from the fundamental laws of nature, or (less strenuously) are linked in favorably systematic ways among themselves through an ordered hierarchy of disciplines, or are at least empirically confirmed by inductive or falsificatory or similar procedures (congruently with such a vision), or are suitable for canonical explanation where empirically supported or confirmed, or are favorably placed (without yet being fully hierarchically linked) within notably comprehensive theories in which nomologically very strong parts of our standing science are pointedly featured.

One sees at once the sense in which our criteria risk becoming quite vague and in which, at one and the same time, it may be claimed and disclaimed that the unity program remains in place through this or that concession; for every concession may be matched by a congruent declension in the defining features of the canon. And, yet, we sometimes do have a definite or reasonable impression that closure may be arbitrary (as with J. B. Watson) or that reduction may be uncompelling (as with B. F. Skinner) or that historical process may not be, or may not be reducible to, mere temporal process (as against Carl Hempel or Adolf

Grünbaum) or that the homonomic structure of the domain supposed to include the physical and the psychological is much too hastily formed (as with Donald Davidson) or that the would-be functional laws of thought and language may not be sufficiently reconciled, in empirical terms, with known physical and biological regularities (as with Noam Chomsky and Jerry Fodor).

Generally, what this means is that the unity program is vestigial when not vigorous; and that, even when it loses its committed champions, its tenets linger on to ensure a ghostly accusation against any outright effort to "bifurcate" the sciences. Even when there is no unity to speak of, an explicit and detailed repudiation of the unity – particularly with regard to the human sciences – invites anathema. One can hardly find any open attempt within the whole of Anglo-American analytic philosophy to shape a reasonably rounded account of the human sciences, that explicitly embraces bifurcation and does so credibly. On the contrary, the stronger the sense of the methodological weakness of a discipline *vis-à-vis* the remembered canon, the more likely its theorists are to insist on the vitality and pertinence of the canon itself. Psychology, psychoanalysis, cognitive science, economics, sociology often produce extraordinarily radical theories upholding the unity program one way or another, whereas theorists of high-level physics are entirely prepared to be baffled by claims of a correspondingly close fit to the achievements of their own domain. The very status of the fundamental laws and fundamental theory of physics is profoundly disputed, as the varieties of so-called anti-realism confirm – just the achievement that should have ensured the victory of the unity program if anything could.

In the human sciences, where there are no canonical achievements at all (though there are indeed tempting achievements), the pressures of conceptual analysis assure us that the systematic attraction of "folk" disciplines and "folk" vocabularies that violate the canon are very much on the wane or should be scuttled as quickly as possible. On the other hand, the attempts of the most prominent Continental European champions of bifurcation – Kantians, Hegelians, Marxists, phenomenologists, hermeneuts, Frankfurt Critical theorists – are embarrassingly easily discounted as simply uninformed about the natural sciences they deny can provide a suitable model for the disciplines they themselves favor. This is surely the fate of Jürgen Habermas, Hans-Georg Gadamer, and Paul Ricoeur at least, who are certainly among the most active Continental discussants of the methodological and ontological difficulties of bringing the human sciences within the ken of something like the unity model. And yet there is a trace of bad faith in that disclosure, because the dismissal of their own fledgling efforts (if that is what they are) cannot really be so easily discounted by merely mentioning their limited grasp of the natural sciences. The issue cuts both ways: for if there is a case to be made *for* the bifurcation thesis (which needn't of course be merely dual – it may well be finely plural), then the advocates of unity may similarly be ignorant of the puzzles of the human sciences. Why not?

What follows, in a word, is a defense of bifurcation and a rejection of unity. It is more than that – or, it is intended to be more. For, even if we should concede that the unity-of-science program no longer really exists as a vital option or has been completely deformed by the concessions that have had to be made since its heyday, even if the question of *method* is already discounted as a coded signal of allegiance to some subset of the "central" tenets collected earlier on (as in rather different ways Paul Feyerabend and Gadamer and Michel Foucault may be supposed to have held), we may deliberately exploit the heuristic advantage of invoking the old debate (if we keep our wits about us) without falling into the trap of the earlier idiom. We may trade on the enviable rigor that was promised (but not delivered) – because assuming the burden of that promise did after all entail a notable measure of discipline; and we may benefit from the sense of history preserved while we strike out in fresh ways. We can hardly escape our history but we need not be blinded by it. Dilthey for instance thought to match the rigor of the natural sciences, since their method seemed crisp and admirably clear to him, however inapt for the disciplines he himself favored. Since matters are no longer thus perceived, we need not, in opposing unity and favoring bifurcation, continue to acknowledge *the* method of science or pursue the *method* of the human sciences. But to say this is certainly not to deny the pertinence of the methodological questions that remain to be considered.

The project, therefore, is quite straightforward. On the methodological side, any effective countering of the central tenets represents a significant increase in the reasonableness of bifurcation. On the substantive or ontological side, any congruent adjustment in the analysis of nature would be similarly supportive. One substantive issue dominates. The strongest charges made by the champions of unity are these: that failure to conform with the central tenets renders science methodologically unintelligible or incoherent; and that their replacement for reasons said to be grounded in the analysis of real phenomena leads at best to a cryptic or unintended restoration of a substantive dualism of mind and matter. Concede the challenge. The project remains entirely clear and may even be caught for advertising purposes in a simple slogan: *bifurcation without dualism.* How may we support it?

There are essentially two constructive inquiries that we can pursue, that complement each other and that are very much on the rise at the present time. One requires following through on the straightforward analysis of a very large cluster of concepts that bear directly on the fate of all versions of the tenets we assigned (above) as "central" to the unity model – the analysis of the mental and the psychological, relations between individuals and the society to which they belong, the relationship that holds between the natural or physical and the cultural, the meaning and structure of history, the historical complexity of human existence, the intentional and the intensional, the nature and structure of language and social practice and traditions and institutions, the causal role of human agents and cultural forces, the process of interpretation and consensus, the significance of ideology and the critical understanding of one's own imputed ideology, the conditions of understanding one's fellows within one's own society and of understanding alien societies, the biological conditions of culture, the neurophysiological conditions of intelligence, the meaning of rationality and the nature of norms and rules and principles, the enculturing power of work and technology, the tacit assimilation of social practices and the acquisition of linguistic and lingual skills, learning in all its forms, the significance of human freedom and choice, the appearance of individual and cultural novelty and creation, the evolution of new cultures, the reality of human purpose, the nature of human behavior and action and conduct, the structure of selves and persons, the linkage between cognition and the aptitudes of persons and selves, the relation between bodies and persons or selves, the sense in which the mental and the cultural are emergent, the structure of consciousness and awareness and sensation and thought and unconscious processes. There is an endless number of such issues – issues we are endlessly puzzled by and realize we have hardly penetrated. Nevertheless, it is fair to say that the fate of what we are calling the bifurcation thesis (and the doctrine that opposes it) hangs in the balance. How can the story be closed when the facts are hardly in?

The second inquiry is distinctly parasitic. It investigates how all disciplined inquiries are affected by programmatic reflections on the relationship (chiefly cognitive) between man and his world. It would be entirely fair to say that the principal inquiries of this sort that currently play a constructive role of any kind include what – with some latitude – may be called *naturalism, phenomenology*, and *deconstruction*. Their increasing if implicit convergence constitutes a concerted effort to legitimate a large, determinable, and systematic body of knowledge (science, preeminently but not exclusively) under conditions that radically preclude the cognitive transparency of the world and preclude any cognitive privilege by which to ensure its validity. Historically, each has contributed in a distinctive way to our grasp of the conditions of our knowledge: each is prone to an excessive and aporetic claim; each is needed to set constraints on the potentially excessive claims of the others. Their relationship in fact functions very much like that of the children's game, scissors/paper/stone. At any rate, that's the best way to understand them. Together, they profoundly affect the pretensions of the unity canon and the advocacy of the canon's tenets. They affect the viability and meaning of any realism in science, as much in the physical sciences as in the human. They affect the fortunes of incommensurabilism, relativism, skepticism, conventionalism, historicism, traditionalism, progressivism, extensionalism; and, in doing that, they challenge the security of *both* unity and bifurcation as stable options.

So there are two senses of the bifurcation thesis to be sorted. On the one, *if* unity is opposed, then bifurcation – that is, opposition to unity – will prevail; and, if unity is already subverted, then bifurcation signifies the reversal of unity's favored tenets *even within the scope of the would-be paradigms of physical science*. There's the larger prize.

If the argument holds, then all the enterprises of science are distinctly *human* achievements – proposals and projections of what the world is like and what we are like in coming to know the world and our own mode of inquiry. Science, after all, is radically human. All of its admirable discipline is secured by *us*, secured by us under conditions that make it more of a marvel and more of a mystery than either the defenders of unity's tenets or the champions of transparency and cognitive privilege (who may well be the same) would have us believe. But that discipline *is* secured – in whatever way it is. "Naturalists" might naively have supposed that the matter was never in doubt. "Deconstructionists" might have insisted that that assurance was the fatal self-deception of the race. "Phenomenologists" might have dialectically exposed the radical contingency of every such presumption while searching forever for a deeper apodictic source. It is only when they come together – merge, in a single undertaking that admits no priorities or privilege among themselves – that the sciences have any chance of being convincingly legitimated. Of course, it is we who must be convinced, and it is we who must convince. So the project harbors its own amusing contingency: the division of labor between naturalism, phenomenology, and deconstruction is itself only a reasonable convenience of the moment; its future history will undoubtedly alter the saliences within which to fix a plausible legitimation of science and philosophy. The symbiosis of method and substance can tolerate no other view.

In our own time, this double puzzle of the human sciences – their decided contrast with the physical or natural sciences and their profound infection of the other (the point of the heuristic use of the canon) – has come into focus in a very large and memorable way. All of the most-debated questions about the standing and discipline of the psychological sciences and of the social and historical sciences that presuppose or incorporate them collect quite marvelously in a single contest of strategies. It is a contest that is only obliquely committed to the old loyalties of the unity canon, for it has been progressively seasoned by all the concessions, deformations, disloyalties, and abandonments of the unity program that the unraveling history of science and the history of the theory of science have forced us to remember. Purist methods are no longer the issue. The principal key lies with the *eliminability of the human*. For the human, if it is a *sui generis* datum for inquiry, is a datum also complexified by its being the precondition of inquiry itself: once that is seriously admitted, the old tenets of unity risk utter eclipse. So the newer pressures tangentially favoring what, in the old days, would have been recognized at once as the canon, are now designed to eliminate the human factor – to regard it frankly as a picturesque vestige of an older "folk" science or a practice that still (slimly) collects a salvageable and important variety of phenomena (permitting us, say, to discard selves and their baggage but to keep "memory" and the "information-processing" of neuronal strata and the like). There is an army of able myrmidons at work on the project, and the issue hangs in the balance. It hardly needs to be christened for it has been naming itself for some time. Call it the contest of *top-down* and *bottom-up* strategies. It is a contest that is at once methodological, ontological, epistemic, critical – also, ideological, rhetorical, political, and moral.

There is a danger, of course, in muddying the waters with such epithets. But there is also the lesson, the unavoidable conceptual linkage between matters that seem at first glance quite disparate. The connection is not our special concern; but it is useful to remark, as an aside, that, within the boundaries of all the busy sciences, the eclipse of the human could hardly fail to affect – for better or worse (who can really say?) – all the familiarly human questions of a normative sort. We must of course resist skewing the resolution of the top-down/bottom-up dispute so as to satisfy ulterior commitments. But it would be foolish to ignore the natural connection, and it would be philosophically inept to suppose that any such quarrel could possibly escape a praxical involvement. It is much better to be open about these things, and it is better philosophy. At any rate, the upshot of the account that follows is the mate of what was promised regarding the unity/bifurcation quarrel: top-down strategies are ampler and more resilient than bottom-up *and* infect the latter. For the rest, in reaching that conclusion, we prepare the ground for a more rewarding study of the essential features of human existence.

Introduction

Preview and Review

1

Truth and Relativism

I

Skepticism is a philosophical threat that *lurks*: that is, no one seriously subscribes to skepticism in those strong forms in which man's confidence in the reality of the world and in his reasonably effective grasp of distributed truths about the world's events are simply rejected or called into perfect question before gaining their first credentials. It is the cognitive achievement itself that makes skepticism parasitically interesting. It lurks in that sense, nagging at the heels of every effort to explain the competence of our science and general inquiry. It is only at a certain advanced level of reflexive doubt about the corpus of what we take our stable knowledge to be that the subversive possibilities of skepticism become manifest; and, in general, there can be no dialogue between science and an independent skepticism, there can only be the internal insinuation of unfathomed uncertainties smuggled into the very fabric of a science that questions itself.

In this regard, relativism is *not* a claim that lurks – or at any rate it need not be. It has of course been so construed, when, by one strategy or another, it is taken to be, or to entail, a version of a deep skepticism or worse. Richard Bernstein has caught one of the standard complaints of this sort – which, let it be said, he does not endorse: "There is still an underlying belief that in the final analysis the only viable alternatives open to us are *either* some form of objectivism, foundationalism, ultimate grounding of knowledge, science, philosophy, and language or that we are ineluctably led to relativism, skepticism, historicism, and nihilism."[1] Bernstein believes there is a third way between objectivism and relativism; but he also holds that relativism *is* essentially the equivalent of skepticism, historicism, and nihilism – and "lurks" with them, in the sense given, on the thesis he sketches. It turns out that, for Bernstein, relativism is tantamount to incommensurabilism – and incommensurabilism, tantamount to skepticism; and that the relativist himself sees his own position thus, hence as a parasitic option that, because of a failed objectivism *and* because of a delusive option between it and incommensurabilism, we are finally driven to.[2] Richard Rorty, on the other hand, quite unceremoniously dismisses relativism as "the view that every belief on a certain topic, or perhaps about *any* topic, is as good as every other." "No one holds this view," he says; but evidently it lurks, as a form of skepticism, incommensurabilism, or, indeed, on Rorty's own choice of epithet, as a form of "irrationalism."[3]

There is no point to contesting these uses of the term "relativism"; but one longs for a sturdier and more interesting antagonist if one be handy. Certainly, Bernstein and Rorty capture a familiar bit of nomenclature. Theirs is a prejudice of *terms*, in the sense that whatever more positive or more powerful form relativism might take remains utterly beyond their ken.

There are, however, substantive prejudices of a deeper sort that relativism is meant to combat – some so profound and ubiquitous as to be well-nigh invisible within the space of familiar philosophical dispute. We must single out one in particular that, by its defect, identifies the *sine qua non* on which any viable relativism depends. A few well-placed examples will bring us quickly to the essential issue.

In Robert Stalnecker's recent account of propositions and of their role in developing an adequate theory of beliefs and of intentionality in general, the following thesis is advanced: "A proposition is a function from possible worlds into truth-values." Stalnecker goes on to say, by way of explanation, "There are just two truth-values – true and false. What are they: mysterious Fregean objects, properties, relations of correspondence and noncorrespondence? The answer is that it does not matter what they are: there is nothing essential to them except that there are exactly two of them."[4] Stalnecker does not pause to link truth-values to the processes of actual inquiry or the resolution of scientific and metaphysical questions: he says, "these questions are concerned less with truth itself than with belief, assertion, and argument, and with the relation between the actual world and other possible worlds."[5] Perhaps. But might it not be a substantive matter regarding truth-values that (1) their number cannot be determined independently of particular inquiries into particular sectors of the world; and (2), relative to particular such inquiries, their number cannot convincingly be fixed as two?

In fact, Stalnecker nearly concedes the point in an extremely candid and telling way, in attempting to strengthen a realism with regard to counterfactuals, context-dependence, and a possible-worlds reading of intentionality – against the encroachments of anti-realism. We shall shortly be able to trade very profitably on his careful remarks in this regard. So let us have them before us:

> The interests, projects, presuppositions, culture, and community of a speaker or writer provide resources for the efficient expression of content, but they also provide something more fundamental: they provide resources which contribute to the construction of content itself. Content, I have been suggesting, can be represented as a subset of some set of possible states of the world. ... But if the space of possible states of the world itself, the way it is possible to distinguish one possible state from another is influenced by the situations and activities of the speakers (or more generally, the agents) doing the distinguishing, then we have a kind of context-dependence that infects content itself and not just the means used to express it. ... I think this is right. ... But it is no retreat from a reasonable realism to admit that *the way we describe the world* – the features and aspects of reality that we choose to focus our attention on – is not entirely dictated by the reality we purport to describe. However we arrive at *the concepts we use to describe the world*, so long as there is something *in* the world in virtue of which our descriptions are true or false, the realist will be vindicated.[6]

Here, we must notice that Stalnecker means to *restrict* the "constructive" feature of descriptions or knowledge to the "content" of such descriptions (or the way we "express" them); but he nowhere frontally considers that the real world as we inquire into it cannot be disjunctively segregated from our would-be descriptions of it. It is true that he criticizes Bas van Fraassen's thesis that "scientific *propositions* are not context-dependent in any essential way." But, at that very point, Stalnecker remarks that he himself assumes "that context-dependence is a matter of the relation between expressions and their content."[7]

Now, this is simply unsupported and quite arbitrary, peculiarly so in the context of attempting to offset Michael Dummett's anti-realism (regardless of the merits of Dummett's position). For, *if* there is no way to support a pertinent realism except *within* the framework of the indissolubility of realist and idealist elements, or of something like a minimal Kantianism (without Kant's "objectivist" claims), or of some form of anti-realism in the sense of admitting concessions in the direction of decidability (though not necessarily Dummett's extreme view of that), or of something like what Hilary Putnam has recently dubbed the "internalist" view of science (again, without subscribing to Putnam's many variant themes), then Stalnecker *could* not restrict context-dependence to descriptive *expressions*: context-dependence would affect *what we take the real world to be*; and, in that case, the naive assurance – "so long as there is something in the world in virtue of which our descriptions are true or false" – would not hold at all.

To concede this much is *not* (necessarily) to retreat from realism or to endorse any extreme version of anti-realism. But it is certainly to concede either that it is not the case that "there are just two truth-values – true and false" or that we are unable to say how many we will *need* without prior attention to what we take *the world we are able to inquire into* to be like. It is essential to relativism, it should be said, that we cannot always reasonably claim that there are just two truth-values and that, in particular sectors of inquiry, it is more reasonable to claim that the truth-values or truth-like values we need are logically weaker than the bipolar pair – but not merely (if we are to hold to realism) because of limitations restricted to the cognitive resources of human investigators.[8]

Michael Dummett's anti-realism is a little surprising, once this distinction is in place. For Dummett opposes what *he* calls the "realist" view: "the belief that for any statement there must be something in virtue of which either it or its negation is true: it is only on the basis of this belief that we can justify the idea that truth and falsity play an essential role in the notion of the meaning of a statement, that the general form of an explanation of meaning is a statement of the truth-conditions."[9] Since the anti-realist interprets "capable of being known" as "capable of being known *by us*," the anti-realist is himself a realist (and the realist may be conceded to hold a defensible position) just insofar as both restrict the "*realist* interpretation" to "those statements [only] which are in principle effectively decidable."[10] The upshot is that Dummett only appears to retreat from a bipolar pair of truth-values, in giving up bivalence; for he gives up bivalence in order to disqualify realism (or what Putnam has dubbed "metaphysical realism"[11]); but he is not at all prepared to give up *tertium non datur*, "the principle that, for no statement, can we ever rule out both the possiblility of its being true and that of its being false, in other words, the principle that there can be no circumstances in which a statement can be recognized as being, irrevocably, neither true nor false."[12] In effect, this is to draw Dummett closer to Stalnecker's sense of realism (though not necessarily to Stalnecker's specific views about the realist import of given kinds of statements – for instance, counterfactuals). In a way, therefore, the same weakness noted in Stalnecker's account reappears in Dummett's – in a different guise. For Dummett never explains *why*, if the realist view of truth (in effect, bivalence) fails because it is "spurious," because it supposes that the truth-values of *any* statement may (for all we know) be decidable by beings whose cognitive capacities exceed our own, the serious *anti-realist* objection might not, relative to a particular domain, extend to bipolar truth-values themselves – to the equally "spurious" assurance that *tertium non datur* will always (may we say "irrevocably"?) obtain. Clearly, Dummett (not unlike Stalnecker) must segregate the question of decidability and the question of reality. But that is an illicit, unmotivated maneuver: illicit, because it must be open to discovery *what* the actual conditions are on which claims *are* decidable; unmotivated, because the anti-realist scruple has not been brought to bear on *tertium non datur* itself.

Finally, and very simply, Dummett had objected to F. P. Ramsey's "redundancy theory of truth" – that is, the thesis that "is true" is "an obviously superfluous addition" to an asserted proposition.[13] Dummett's objection, however, focuses only on the fact that Ramsey has left no room for the anti-realist complaint. But there is a deeper reason why Ramsey's thesis is inadequate: it may just be that, for a particular sector of inquiry, the redundancy theory is actually false; the truth-like values it will support may be logically weaker than the bipolar pair.

There you have the charge. Neither Stalnecker nor Dummett nor Ramsey has actually bothered to consider that the viable claims of determinate sectors of inquiry may not support – *anywhere along a continuum from realism to anti-realism* – bipolar truth-values (or, effectively, "tertium non datur").

A relativism that is not merely a lurking doctrine, in the sense assigned skepticism, obliges us to retreat from bipolar truth-values or *tertium non datur* – but *not* globally, not indiscriminately, not on an all-or-nothing basis. Relativism is a logical thesis, an alethic thesis, applied piecemeal. It could not be a reasonable thesis if it did not resist the kind of disjunctions Stalnecker and Dummett favor (Stalnecker: between "reality" and our expressed descriptions of reality; Dummett: between the decidability of our statements and what, apart from their decidability, decidability cannot fail to reveal about "reality" thus addressed). In a word, a serious relativism fitted to the actual inquiries of the sciences, ontology, methodology, interpretation, appraisal and evaluation and more, *cannot fail to presuppose that, in no respect, is the world, distributively, cognitively transparent.*

II

In effect, there are two very large substantive constraints on all seriously engaged speculations regarding the truth-values (or truth-like values) suited to a given domain of inquiry. They are of rather different sorts, but they must be brought to bear jointly on such strong claims as those of Stalnecker, Ramsey, and Dummett. One stems from Alfred Tarski's *metameta*theoretical discussion of truth; and the other, from the generally convergent finding of recent Western philosophy repudiating all cognitive transparency. Relativism is the beneficiary of the intersection of these two currents; and the attack on transparency is (broadly, perhaps prejudicially, collected as) the upshot of coming to favor (i) a pragmatist reading of holism in science (itself not cognitively confirmed) and (ii) a Nietzschean sense of "originary" metaphors (not skeptically enforced) – which together dismantle every version of the corespondence theory of knowledge (as opposed to the correspondence theory of truth).

The pragmatist theme, for example in the most defensible reading of Quine's "Two Dogmas,"[14] has, in an important sense, just the same effect as has Heidegger's combination (within his theory of *Dasein*) of Husserl's *Lebenswelt*-theme (shorn of pretensions of apodictic phenomenology) and Nietzsche's – the effect of featuring the radically preformative conditions under which an inquiring and active race effectively survives.[15] The heterodoxy of linking these two notions (as well as the larger traditions they subtend) is now no longer as startling as it would have been a few years ago. But in any case that (essentially a piece of gossip) is not our present concern. The point is that the convergence intended yields the protean finding that there is no realism suited to rigorous inquiry that can escape the limitation that the world we inquire into is not *cognitively* accessible in any way that would support a disjunction between *it* and whatever *we* identify as the world we inquire into. That is the idealist face every realist torso must expose if it cannot claim perfect transparency. It is also in a way the master theme of Thomas Kuhn's *Structure of Scientific Revolutions*: we may inquire into an independent world but we cannot state its nature as it is independently of our inquiries.[16]

This is what is implicitly denied or unacknowledged in Stalnecker's and Dummett's disjunctions: we cannot relegate the open, unsystematizable consequences of context-dependence to the mere cultural *expressions* of our science without affecting the demarcation of the structures of the world we examine; and we cannot encumber the claims of our science in the name of decidability, while assuring ourselves in advance that no decidable claims can fail to accord with the formal features of bipolar truth-values.

By a deliberately provocative use of terms, we may dub the mistake in question a form of "logocentrism": the cognitively blind assurance that the real world cannot but accord with certain logical doctrines – largely formal, largely uninterpreted, largely untested – that assure us *a priori* that only research programs congruent with those doctrines could possibly be viable or productive.[17]

In an intriguing sense, these and related claims are due, perhaps a little remotely, to a misreading or misapplication of Tarski's well-known account of truth. Certainly, both Stalnecker's and Dummett's disjunctions are loosely analogous with Tarski's distinction between the concept of logical consequence and the criterion of deducibility.[18] But Tarski addresses the issue only in formal terms suited to the arithmetic of natural numbers, and himself admits the material difficulty of adequately sorting disjunctively logical and extra-logical terms. Without the extension of Tarski's kind of analysis to the whole of that part of natural language that is needed in the pursuit of the empirical sciences (if that makes sense), it would be quite impossible to vindicate either Stalnecker's or Dummett's disjunctions. Tarski quite openly concludes that we may be

> compelled to regard such concepts as "logical consequence," "analytical statement," and "tautology" as relative concepts which must, on each occasion, be related to a definite, although in greater or less degree arbitrary, division of terms into logical and extralogical. The fluctuation in the common usage of the concept of consequence would – in part at least – be quite naturally reflected in such a compulsory situation.[19]

Correspondingly, Tarski was quite convinced that his semantic conception of truth was not in general suited to the "whole" of the language used in the empirical sciences.[20]

It has been said, for instance by Hilary Putnam, that "true," on Tarski's theory, "is, amazingly, a *philosophically neutral notion*. 'True' is just a device for 'semantic ascent': 'raising' assertions from the 'object language' to the 'metalanguage', and the device does not commit one epistemologically or metaphysically."[21] But the fact is that Tarski's complete conception *fits* certain formalized languages only. Either, then, Putnam has prised apart the "bare" concept from its "application" or else has divided Tarski's theory pointlessly and against Tarski's own purpose. One *can* distinguish between Tarski's bare "definitional" notion of "true" and the actual "structural–descriptive" rule or canon for determining the truth of distributed object-language sentences, but doing that would deprive the definitional element itself of any substantive bearing on *the actual sentences* of a functioning science. For instance, one could hold (on Dummett's summary) that, "for any sentence A, A is [materially] equivalent to ⌜It is true that A⌝, or to ⌜S is true⌝, where S is a [metalinguistic] ('structural-descriptive') name for A."[22] But, *if* we separate the two notions, Putnam would be right in treating Tarski's equivalence thesis as "philosophically neutral" only in the undesirably strong sense that it had no philosophical bearing at all in an epistemically pertinent regard; for what would be the point of introducing provision for a structural-descriptive name for A if, in doing so, we had to concede no information about the nature of the relevant descriptions of such distributed sentences? On the other hand, *if* such descriptions did conform with what Tarski offers as the extensionally regimented descriptions suitably fitted to the interpreted formal languages he actually examines, then, first of all, the notion would no longer be philosophically neutral (it obviously would favor a global extensionalism) and, secondly, it (the complete notion) would require, to be theoretically vindicated, an independent argument about the properties of the pertinent language of science – which Tarski, confessedly, nowhere provides. It is Dummett's point that "Rejection of the principle of bivalence [every statement is either true or false] when not accompanied by rejection of *tertium non datur* [no statement is neither true nor false] does not lead to any conflict with the equivalence thesis"; but it is also Dummett's point (against Ramsey's redundancy thesis) that, since the Tarskian truth-definition "is *not* an expansion of the object-language," we cannot *apply* the predicate "true" to the sentences of an object-language (any object-language) – in accord with the truth-definition – "if we do not yet understand the object-language ... [that is, if we do not have] a grasp of the meaning of each such sentence" in virtue of which "true" can be distributively applied.[23]

The point is that *if* Tarski's conception of truth – that is, the entire conception including provision for the metalinguistic structural description of distributed object-language sentences – fits the formalized languages Tarski claims it fits, then it does so on the strength *of an actual analysis of such languages.* It is not construed by Tarski to be an *a priori* matter. Hence, if it is to be extended to natural languages, then, on Tarski's view, it must be suitably justified by an analysis of those languages as well. *If* the Tarskian definition of truth (without the structural-descriptive account) is "philosophically neutral," then it is neutral because it is vacuous, *because it affirms no object-language/*

metalanguage relationship and because it imposes no constraint on any material equivalence between object-language and metalinguistic sentences. It may well capture some profound common intuition about truth, but it could hardly (then) be said to capture such an intuition along the lines of the equivalence thesis. Anyone who supposed it would automatically do that would have to be extraordinarily sanguine about extensionalism – and would in any case be mistaken. Hence, it does not really matter that Tarski's definition invokes bipolar truth-values; for it was meant all along for languages that conformed sufficiently closely to the formal languages that serve as Tarski's paradigms, and, there, it *is* the appropriate choice. There is, therefore, no incompatibility between subscribing to Tarski's conception and insisting that, for particular domains of inquiry, we should (should have to) introduce truth-like values logically weaker than the bipolar pair.

In fact, if theories of truth are viewed empirically, then there are at least two distinct sources of contention *vis-à-vis* Tarski's conception: first, that truth-values logically weaker than bipolar values may be required in particular domains (the formal concession essential to relativism); and, second, that, for either bipolar or logically weaker values or both, the sentences of a given domain of inquiry may not satisfy Tarski's provisions for their "structural" description (an ontological claim that, however minimal in other respects, is not in the least inhospitable to relativism).

Donald Davidson (the effective custodian of all efforts to extend Tarski's account to empirical disciplines) is quite explicit about "treating theories of truth as empirical theories" though he obviously means this more in terms of (what he calls) "Convention T," the "structural-descriptive" account, than in terms of the suitability of restricting the range of bipolar values.[24] Thus he says, answering one of his critics,

> Of course my project does require that all sentences of natural languages can be handled by a T-theory, and so if the intensional idioms resist such treatment, my plan [the grand plan of Davidson's philosophy] has foundered. It seems to be the case, though the matter is not entirely simple or clear, that a theory of truth that satisfies anything like Convention T cannot allow an intensional semantics, and this has prompted me to try to show how an extensional semantics can handle what is special about belief sentences, indirect discourse, and other such sentences.[25]

Davidson's admission shows very nicely the underground linkage between limiting extending the scope of bipolar truth-values by analogy with Tarski's analysis of certain formal languages and supporting an ontology that eschews all cognitive transparency, along lines favoring a pragmatized holism and the indissoluble symbiosis of realist and idealist elements.[26] For there can be no doubt that the prospects of a sturdy and substantive relativism positively depend on such philosophical policies as

(a) rejecting transparency;
(b) advocating scientific holism;
(c) acknowledging the indissolubility of realist and idealist themes within the scope of policies (a) and (b).

If we continued along these lines, for instance

(d) emphasizing the historical contingency of inquiry itself; and
(e) emphasizing the context-dependence of inquiry under the condition of being unable to fix the absolute context of all context-dependence (relative to all possible worlds, say, or to synthetic *a priori* truths)

we should have admitted just the conditions under which "an intensional semantics" would be least likely to be suppressed or eliminated. Projects such as Davidson's would then "founder" (as Davidson admits), and a robust relativism would either be required or strongly favored. The irony is that Davidson *is* pretty well committed to policies (a)−(e), and yet he comes extraordinarily close to advocating on *a priori* grounds the extension of Tarski's strategy to the empirical disciplines (in fact, to the whole of natural language).

But that extension is not easy to defend. There is, in fact, a certain conceptual slippage in Davidson's argument, irresistibly recalling Putnam's hearty claim about the "amazing" philosophical neutrality of Tarski's conception of truth. For consider that, when he sketches the extension of "Convention T" to natural languages, Davidson candidly observes,

> I suggest that a theory of truth for a language [as we want a theory to do] does ... give the meanings of all independently meaningful expressions on the basis of an analysis of their structure. And ... a semantic theory of a natural language cannot be considered adequate unless it provides an account of the concept of truth for that language along the general lines proposed by Tarski for formalized languages.[27]

Here Davidson advocates Convention T as a *criterion* of the adequacy of theories of truth. The point of his proposal is, however, not entirely clear. It definitely risks being equivocal. For there is a sense in which the appeal to Tarski is merely an appeal to *some* (as yet unspecified) metalinguistic description of the structural properties of object-language sentences in virtue of which the truth of such sentences can be shown to depend on the structure and structural relations among their constituent parts. It is perhaps in this sense that Davidson warns us that

> to seek a theory that accords with Convention T is not, in itself at least, to settle for Model T logic or semantics [that is, Tarski's own full strategy]. Convention T, in the skeletal form I have given it, makes no mention of extensionality, truth funtionality, or first-order logic. It invites us to use whatever devices we can contrive appropriately to bridge the gap between sentence mentioned and sentence used. Restrictions on ontology, ideology, or inferential power find favor, from the present point of view, only if they result from adopting Convention T as a touchstone. What I want to defend is the Convention as a criterion of theories, not any particular theories that have been shown to satisfy the Convention in particular cases [Tarski's own application, say], or the resources to which they may have been limited.[28]

On this first line of argument in favor of Convention T, then, the emphasis is placed on structural-descriptive ways of linking the extension of the predicate "true" applied to (object-language) sentences to those very sentences as preferable to any other approach to managing that relationship. Davidson does not (here) deny the power of Tarski's example: it is only that he does not trade on its own particular strategies; on the contrary, here he proposes a more general overview by means of which to demonstrate the superiority of Tarski's full account over all others. But how does he do this?

The apparent argument is simplicity itself:

> Let someone say [Davidson observes], "There are a million stars out tonight" and another reply, "That's true," then nothing could be plainer than that what the first has said is true if and only if what the other has said is true. ... We have learned to represent these facts by sentences of the form "The sentence 'There are a million stars out tonight' is true if and only if there are a million stars out tonight." Because T-sentences (as we may call them) [sentences of the biconditional form just illustrated] are so obviously

true, some philosophers have thought that the concept of truth, at least as applied to sentences, was trivial. But that's not so. T-sentences don't ... show how to live without a truth predicate; but taken together, they do tell what it would be like to have one. For since there is a T-sentence corresponding to each sentence of the language for which truth is in question, the totality of T-sentences exactly fixes the extension, among the sentences, of any predicate that plays the role of the words "is true." From this it is clear that although T-sentences do not define truth, they can be used to define truth predicatehood: any predicate is a truth predicate that makes all T-sentences true.[29]

Look at this a little more carefully, however. When T-sentences are constructed to capture the sense of narratives such as the story of the million stars – when they "are so obviously true" – they are meant to be trivially true; they are *not* meant to affirm a material equivalence between object-language sentences and metalinguistic sentences formed by any (as yet unspecified) structural analysis of those sentences, that thereby gives "the truth conditions of the described sentence[s]."[30] On that reading (the second reading of the equivocation remarked), T-sentences are hardly "obviously true." Putnam's (and Davidson's) claim could only refer to the vacuous thesis ("obviously true") that, *if*, in stories such as the million-stars story, in asserting "There are a million stars out tonight" one is affirming that there are (that it is true that there are), then the corresponding T-sentence is ("obviously") true as well. But that is to say that, in the original irresistible vignette, *nothing is said or intended regarding relations between object-language and metalanguage*, certainly not anything regarding any (specified or unspecified) structural-descriptive analysis of object-language sentences in virtue of which their truth-conditions are given.

No. The irresistible vignette *may* capture a profound intuition about truth. Such a vignette *is* "philosophically neutral." It *may* reasonably be said to be the intuition on which Tarski constructs his account. But the latter (the strong extensionalized account suited to certain formalized languages) and its extension to natural languages (Davidson's project) *and* the import of using Tarski's (so-called) Convention T as a *criterion* of the adequacy of theories of truth (without yet invoking Tarski's own strongly extensionalized program) *are definitely not philosophically neutral*. The statement of these three theses is certainly not obviously true. So Davidson shifts ground considerably when, with all his cautions, he finally affirms,

The reason Convention T is acceptable as a criterion of theories is that (1) T-sentences are clearly true (preanalytically) – something we could recognize only if we already (partly) understood the predicate "is true," and (2) the totality of T-sentences fixes the extension of the truth predicate uniquely. The interest of a theory of truth, viewed as an empirical theory of a natural language, is not that it tells us what truth is in general, but that it reveals how the truth of every sentence of a particular L [a particular language] depends on its structure and constituents.[31]

Here, it is obvious that Davidson must be equivocating on the meaning of "T-sentences" in giving his reasons for supporting Convention T: in (1), "T-sentences" must, to be "clearly true (preanalytically)," take the trivial form of the first reading of Tarski's strategy; and, in (2), "T-sentences" must accord with an extension of Tarski's own strategy (which Tarski never intended and actually believed unworkable) or else must involve some suitably strong analogue of Tarski's own strategy (recalling Davidson's *caveat* about not adopting prematurely "Model T logic or semantics"). In any case, it is certainly not clear that, either in general or by some as yet unspecified structural-descriptive criterion offered in (2) the claim that "the totality of T-sentences fixes the extension of the truth predicate uniquely" could be shown to be true, could be shown to be sufficiently promising to support an empirically responsible claim that it is true, or could be so characterized that we would even have a reasonable sense of *how to proceed to show that it was true*.

This is the sense in which Davidson's claim is effectively, must be, an *a priori* claim, despite all his insistence on treating the theory of truth as an empirical question. Apart from his own frank admission of numerous details that we cannot yet be sure would support the extension of Tarski's strongly extensionalized canon,[32] there is the stunningly plain fact that, to the extent Davidson subscribes (as he apparently does) to Quine's holism, to the extent he admits the historicized and contextualized nature of natural language, to the extent he subscribes to our constraints (a)–(e), it is simply impossible for him to profess (2) *on empirical grounds*.[33]

Davidson actually calls a theory of truth – that is, "a theory that satisfied something like Tarski's Convention T" (where he simply appeals to "a canonical description of a sentence of L" without insisting on Tarski's own canon – but also without refusing its use) – "absolute": "to distinguish [it, he says] from theories that (also) relativize truth to an interpretation, a model, a possible world, or a domain. In a theory of the sort I am describing, the truth predicate is not defined, but must be considered a primitive expression."[34] But, in the *empirical* sense in which the theory proposed is to be fitted to natural language, subscribing to (a)–(e) (or even to a good part of that set) effectively commits one to relativizing determinations of truth-conditions even if the predicate "true" is not itself treated relationally. Davidson himself makes the essential admission: "'absolute' truth goes relative when applied to a natural language."[35] But it does so, surely, in just that sense in which his criterion (2) regarding testing the adequacy of any theory of truth cannot be shown to be effective, cannot eliminate an "intensional semantics," and cannot preclude the need to retreat (in particular domains of inquiry) from bipolar truth-values to logically weaker values. In this sense, no extension of a Tarski-like strategy could, empirically, disallow a substantive relativism. This, then, confirms the sense in which relativism is favored by the intersection of neutralizing Tarskian-like programs for the empirical sciences and exploiting the ramified import of the developing philosophical repudiation of all forms of cognitive transparency. But it also prophesies increasing concessions regarding intensional complications in the analysis of any theory of science and serious inquiry.

III

There is no need for a developed relativism if transparency obtains. On that condition, any relativism signifies merely provisional limitations in the cognitive powers of human investigators. A substantive relativism requires that the scientific description of the "independent" physical world (that is, the real world as it is independently of human inquiry) be an artifact of a cognitive competence that is at once justifiably realist in its claims about that independent world and incapable in principle of being extricated from the preformative conditions of the historical existence of human investigators – whether conceptual, doxastic, conative, institutional, praxical, ideological, tacit, programmatic, or critical conditions. It must, in short, manifest – symbiotically – both realist and idealist aspects. It may be that Kantianism in its strict form preserves a version of transparency (or what effectively functions as a surrogate for transparency) at the same time as it fuses the realist and idealist features of an apt science: that, presumably, is what Husserl objected to in Kant, in charging Kant with "objectivism."[36] An analogous charge has been leveled against Husserl as well, given his obsessive search for the apodictic, his transcendental solipsism, and his sense of the apparent privilege of the phenomenological method.[37] In that spirit, we have adopted Derrida's term of art, "logocentrism," for any theory that fixes, in the absence of or even specifically in opposition to pretensions of transparency, a privileged constraint on any reasonably productive (or realist) science within (at least implicitly) the bounds of a realist – idealist fusion.

Within the analytic tradition, there can be no doubt that Quine's extreme extensionalism – particularly its *a priori* repudiation of intentionality (and consequent intensional complications)[38] must count as a most influential form of logocentrism in the absence of (in opposition to, in fact) all forms of transparency. This is the same sense in which the global, *a priori* insistence on bipolar truth-values, in Stalnecker and Dummett, and the union of that thesis, *via* Tarski, with a subtle version of Quine's extensionalism, as in Davidson, are here construed as forms of logocentrism. The fact is that doctrines of these sorts preclude any substantive relativism. It is an irony, therefore, as well as a clear sign of a profound incoherence, that Quine is the champion both of a draconian extensionalism that precludes an empirically exploratory relativism along intentional lines and of a pragmatized holism that entails a pervasive relativism.

The principal benefit, as far as the fortunes of relativism are concerned, of the realist – idealist symbiosis under conditions precluding transparency and logocentrism is simply that relativism can no longer be restricted to the contingencies of cognitive accident and limitation. That is, the contribution of relativism cannot be exhausted by considerations of probabilizing evidence or of ambiguity or vagueness or of deliberately heuristic or idiosyncratically constructed schemata or interpretations or the like. On the argument, relativism, a robust or substantive relativism, answers to the general finding that *we cannot uniquely fix the structures of reality just at the point at which we are reasonably justified in treating our cognitive claims in realist terms.*

Thus construed, the physical sciences are themselves enterprises that, precisely in positing an "independent physical world," fall within the inquiries of culturally encumbered human investigators. Clearly, the social and historical world of man cannot intelligibly be construed as separable (even in principle) from the investigative aptitudes of reflexive inquirers; although this is not of course to preclude pertinent notions of objective confirmation and validity with regard to the human sciences. It is only to insist that, in the space of human culture, there can be no initial disjunction between the epistemological and the ontological: there is none either in the physical sciences, but there is often said to be one (extensionalism is its most fashionable logocentric expression), and there *is* a systematic function assigned the notion of an "independent physical world" *within* the space of the other, that cannot even in principle (or at least so we are here claiming) be matched in the human sciences.[39] Relativism, so construed, is nothing less than the attempt to recover, within realist terms, whatever forms of objectivity are possible regarding any sustained science or comparable inquiry – where, for reasons affected by the impossibility of uniquely fixing the real structures of this or that sector of things, we are obliged to retreat to truth-like values logically weaker than bipolar values.

Nevertheless, relativism has a bad name. Its ancient form of course, Protagoreanism, is a philosophical scandal: not, let it be said, because Protagoras's thesis cannot be given a coherent reading but because almost no one wishes to give it one. Paul Feyerabend offers these tantalizingly brief sentences, in introducing his collected *Philosophical Papers*: "The reader will notice that some articles [included] defend ideas which are attacked in others. This reflects my belief (which seems to have been held by Protagoras) that good arguments can be found for the opposite sides of any issue."[40] Feyerabend associates this belief with what he calls "democratic relativism," which to many may suggest only a sense of conceptual thrift and an unwillingness to lose potential contributions made by fringe or minority groups. But Feyerabend's view is actually closer to attacking the logocentrism of privileged traditions (under what, notably, as in Putnam's criticisms,[41] are would-be idealist constraints on scientific realism). Hence, when, subverting the hegemony of Western science (Western "rationalism"), Feyerabend observes, "*They simply take it for granted that their own traditions of standard construction and standard rejection are the only traditions that count,*"[42] he is speaking not merely of intellectual tolerance but surely also as a relativist committed to the thesis we have just formulated – that the structures of reality cannot be fixed either in terms of transparency or in terms of logocentric privilege (*a priori*, privileged traditions) – and as one committed as well to the thesis that *that* is a reasonable reading of Protagoras.

This may be a way of understanding sympathetically the ancient report that Protagoras "was the first to say that there were two contradictory arguments about everything."[43] It suggests, regarding his famous maxim, "Of all things the measure is Man, of the things that are, that they are, and of the things that are not, that they are not,"[44] that Protagoras should be construed as anticipating modern forms of incommensurabilism (or a retreat from bipolar truth-values) rather than as merely subscribing to the anciently assigned doctrine that every opinion is true and every opinion false. The latter is usually pressed, even today, as for instance quite straightforwardly by W. H. Newton-Smith.[45] In that form, it is simply a hopelessly stupid thesis. We may then identify "ancient" relativism (*Protagoreanism*) as the stupid doctrine that any and every statement is both true and false, while at the same time we reserve for ourselves the right to construe Protagoras's own view as requiring a retreat from the global adequacy of bipolar values and a shift, under that constraint, toward a form of incommensurabilism.

The pity is we have lost Protagoras's work. There is an ancient report that the Athenians destroyed all the copies of Protagoras's book; Plato's *Protagoras* is certainly at least in part a parody for his own purpose; and Aristotle's refutation of Protagoras, in the *Metaphysics*, on the standard argument regarding the law of contradiction, *could* be offset if Protagoras had meant to retreat from bipolar values. (So read, Protagoras's maxim would afford a pretty counter to that of Parmenides.[46]) In any case, there is a way of attempting to salvage even the ancient form of relativism. (Also, to put matters this way helps, as we shall see, to give form to the notion that relativism is much more flexible than ordinarily admitted.)

The principal modern form is, suprisingly, almost as inexplicit as Protagoreanism. It is usually termed *incommensurabilism* and is formulated chiefly (and disadvantageously) by the opponents of those said to advance the thesis – in particular, against Thomas Kuhn and Paul Feyerabend. Incommensurabilism *begins* at least with the conceptually innocuous truism that not all measures are mutually commensurable (in the sense, for instance, in which the hypotenuse of an isosceles right triangle is incommensurable with its side); although that is no reason for believing that incommensurable measures are not, severally, capable of being successfully applied and of being jointly intelligible, even comparable, to the same rational agent.[47] Incommensurabilism acquires color as a form of relativism only when, one way or another, it extends to what may be called "conceptual relativity" – that is, only when it accords with the thesis that, under the preformational conditions of historical existence or under the conditions of adhering to different theories, different research projects, different paradigms, different modes of training, orientation and the like (or both), different investigators (located either synchronically or diachronically) are, on the available evidence, often unable to incorporate an adequate and coherent picture of one another's conceptions within the terms of reference of their own. The emphasis here is on distinctly finite, real-time constraints in attempting to effect such coherence. But it emphatically is not skeptical about the intelligibility of such divergent conceptions.

So construed, incommensurabilism is not a principled position but an empirical phenomenon that yields a certain relativity of inquiry and results that, over time, might in principle be overtaken. The opponents of incommensurabilism, however, take its contemporary advocates to be advancing a very strong principled thesis – a thesis, in fact, that is either incoherent or unnecessarily (and unconvincingly) skeptical. As with Protagoreanism, therefore, it is the better part of strategy to seek to recover a viable form of incommensurabilism from the excesses of its overzealous opponents.

Two lines of argument have been pursued. One distinctly treats incommensurabilism as the "new relativism" and takes it to be tantamount to "skepticism, historicism, and nihilism" (thus Richard Bernstein[48]); the other treats incommensurabilism as flatly incoherent and self-defeating (thus Donald Davidson[49]). On Bernstein's view, the objectivist is committed to "some permanent, ahistorical matrix or framwork" for resolving all cognitive questions; the relativist denies that there is such a framework and *thereupon* draws the skeptical conclusion.[50] The incommensurabilist thesis is therefore the denial of objectivism construed as disallowing the adjudication of all truth-claims. Bernstein does *not* advocate returning to objectivism. He means merely to urge either that the rejection of that doctrine does not lead to relativism or skepticism or that there may be a form of incommensurabilism that does not lead to the relativist or skeptical conclusion. But he never explains how this is possible; and the relativist, sympathetic with incommensurabilism, *may* simply retreat from the global adequacy of bipolar values *and thereby separate relativism from skepticism*.[51] Bernstein himself apparently believes that the ongoing "tradition" of inquiry somehow preserves the path by means of which unique or very strongly convergent resolutions of contending truth-claims can be counted on to be effected. But this itself is an obvious form of logocentric assurance – one we may christen "traditionalism," one in fact strenuously advocated (but not seriously defended or methodologically specified) by Hans-Georg Gadamer (whom Bernstein professes to follow) and Charles Taylor.[52]

The second counterstrategy (Davidson's) leads us to consider the tenability of "conceptual relativism," the thesis (on Davidson's reading) that *we* are able to individuate plural conceptual schemes different from our own because of *their* partial or total untranslatability into *our* conceptual scheme.[53] The argument is intended as a *reductio* specifically against Kuhn and Feyerabend. But the truth of the matter is that we have no clear idea how to establish untranslat*ability* (nor has Davidson) when we have merely encountered a failure of translation of a suitably stubborn and pervasive kind. So "conceptual relativism" may well be an excessive and incoherent claim, as Davidson avers. But a *moderate* incommensurabilist (certainly Kuhn in his more careful moments, possibly even Feyerabend despite his deliberate provocations) may affirm no more than "conceptual relativity" (as characterized above: sustained failure of translation under real-time conditions). He may then go on to reject "conceptual relativism" (Davidson's impossibly difficult untranslatability thesis) and to lead the first option in the direction of a principled argument, as by

1 denying (what Bernstein terms) the objectivist thesis (which, in an obvious sense, Davidson is logocentrically committed to, by way of his reading of Tarski's semantic conception);
2 affirming some form of the realist–idealist symbiosis; and
3 concluding that, under the actual contingencies of historical inquiry and existence, we can never ensure escaping some range of (moderate) incommensurability.

This is all the relativist needs to ensure the recovery of the incommensurabilist thesis as a valid form of relativism not in the least skewed toward skepticism or nihilism. That thesis is certainly not internally incoherent. Within the analytic tradition of the philosophy of science, perhaps Ian Hacking is as reasonable an advocate of this sort of relativism as any,[54] though it is also fairly assigned to Kuhn and Feyerabend; and in the Continental European tradition it is explicitly favored by Michel Foucault and, more indirectly (even dubiously), by Jacques Derrida – both perhaps as recent champions of the Nietzschean tradition.[55]

Nevertheless, relativism does not achieve its strongest formulation in Protagoreanism and incommensurabilism, though it may avail itself of their resources. What we may now characterize as a *robust* or moderate relativism is a substantive thesis about constraints on truth-values in distributed sectors of inquiry due to (what may be generously construed as) empirical evidence that those sectors (but not necessarily every sector) cannot reasonably support truth-claims in terms of bipolar values. On the argument, we must fall back to logically weaker, many-valued claims (as of plausibility, reasonableness, aptness and the like) if we are to salvage a measure of objectivity with respect to the inquiries of those sectors. Such a retreat may be required, for instance, in advancing claims about explanatory theories and laws in the physical sciences,[56] or about reference to theoretical entities through changing theories,[57] or about ontological schemes of "what there is."[58] If the concession is required here, it is a foregone conclusion that it will be required in historical studies, in interpretation, in the social and human sciences, in criticism, and in appraisal and evaluation as well.

The skeleton of the relativist argument, then, is extraordinarily simple.

It posits the nearly trivial (but not unimportant) thesis that we may introduce by fiat any consistent logical constraints we care to admit on the truth-values or truth-like values particular sets of claims or claims in particular sectors of inquiry may take. We may, by fiat alone, deny to a given sector the power of pertinent claims or judgments to take bipolar truth-values (or, acknowledging an asymmetry between "true" and "false," the power to take the strong value "true"); and then, by introducing logically weaker values, we may admit claims or judgments to be evidentially supported or supportable even where, on a bipolar model of truth-values but not now, admissible judgments would yield incompatible or contradictory claims. On the new model, such judgments could be said to be "incongruent," and a suitable discipline could be provided for their confirmation and disconfirmation; at the same time, on the strength of relevance considerations, certain other logically undesirable possibilities may easily be precluded (in probabilistic contexts, for instance, it would be undesirable to deny that "Nixon probably knew about Watergate in advance and did not know about Watergate in advance" was a contradiction, even though "There is a probability that Nixon knew about Watergate in advance and a probability that Nixon did not know about Watergate in advance" would normally not be construed as self-contradictory).[59]

The essential point is that the formal (relativistic) characterization of truth-like values *appropriate* to particular sectors of inquiry

1 need not be logically incoherent or self-contradictory;
2 effectively captures the salvageable themes of Protagoreanism and incommensurabilism and more; and
3 may be formulated so as to accommodate the salient philosophical convergences of our own time.

The rest of the argument really concerns how to show, in given sectors of inquiry, that a relativistic thesis is as reasonable as, or more reasonable than, any program of scientific *objectivity* – not committed to transparency or logocentric privilege – that still insists (in the manner of Stalnecker, Dummett, and Davidson and so many others) that "there are just two truth-values."[60]

The strongest substantive reason for advocating relativism rests, as we have seen, with the finding that we cannot uniquely fix the structures of reality just where we are justified in insisting on the objective and realist import of our cognitive claims. That finding draws strength from arguments in favor of the realist–idealist symbiosis of science, a pragmatized holism, a moderate incommensurabilism, a historicized and context-dependent sense of human inquiry and existence, the impossibility of totalizing the conditions of cognition, and the general rejection of transparency and logocentrism. It needs to be said as well that, since relativism is to be promoted in a piecemeal and distributed way, there is no incompatibility in managing, within one science, both bipolar claims and the logically weaker claims here advocated; for it is certainly clear that inquiries that promote relativistic values depend on a *theory* of bipolar values even where they retreat from the use of such values. Also, of course, there is no reason to deny that relativistic programs readily accommodate comparative judgments of merit or force, notions of progress, rigorous evidence and the like. In short, relativism is expressly opposed to skepticism and nihilism.

The important question that remain concerns what we may fairly prophesy is the consequence for the human and natural sciences of admitting a robust relativism. The answer is stunningly plain. The admission of relativism under the conditions sketched – not merely the fiat of a certain alethic decision but the substantive grounds for making that decision for particular runs of inquiry – will obviously enhance the import of all the puzzles of intentionality. All notions of objectivity, of stable regularities, of real structures will, *however realist their standing* (and they will be entitled to such standing), be to some extent *artifacts* of a culture that is tacitly and contingently preformed by its developing history, that is capable of a measure of reflexive criticism that still cannot altogether escape from that condition or achieve a complete cognitive penetration of it, and that cannot imagine at any stage of its history how to secure conceptual closure for all possible ways of construing the salient regularities it actually records and dwells upon. Human inquiry is radically context-dependent, and the context-free proposals of the strongest sciences are proposals formed *within* some such diachronized context-dependence. All sciences are human sciences, in terms of their actual achievement. They cannot, therefore, be rightly characterized apart from the features of human culture, human history, human language, human activity, human experience, and human needs and interests. Also, to say that much is to say as well, by a large but hardly mystifying leap, that to admit relativism is, ultimately, to admit deeper intentional and intensional puzzles regarding objectivity than are normally acknowledged in the canonical literature.

In a word, to subscribe to the weakened picture of truth-values is, wherever promising, to press the details of cultural relativity in the direction of a principled relativism.

Notes

1 Richard J. Bernstein, *Beyond Objectivism and Relativism* (Philadelphia: University of Pennsylvania Press, 1893), p. 2.
2 See Joseph Margolis, *Pragmatism without Foundations: Reconciling Realism and Relativism* (Oxford: Basil Blackwell, 1986), ch. 3.
3 Richard Rorty, "Pragmatism, Relativism, and Irrationalism," *Proceedings and Addresses of the American Philosophical Association*, LIII (1980), 727–30.
4 Robert C. Stalnecker, *Inquiry* (Cambridge, Mass.: MIT Press, 1984), p. 2.
5 Ibid.
6 Ibid., pp. 152–3 (italics added).
7 Ibid., p. 151; see Bas C. van Fraassen, *The Scientific Image* (Oxford: Clarendon Press, 1980), pp. 134–7.
8 The full argument regarding realism and anti-realism is given in *Pragmatism without Foundations*, pt II, especially with regard to van Fraassen, Dummett, and Putnam.
9 Michael Dummett, "Truth," *Truth and Other Enigmas* (Cambridge, Mass.: Harvard University Press, 1978), p. 14.
10 Ibid., p. 24 (I have italicized "realist").
11 See Hilary Putnam, *Meaning and the Moral Sciences* (London: Routledge and Kegan Paul, 1978).
12 Dummett, Preface to *Truth and Other Enigmas*, p. xxx.
13 F. P. Ramsey, "Facts and Propositions," *Foundations of Mathematics*, ed. Richard B. Braithewaite (London: Routledge and Kegan Paul, 1931), p. 143.
14 W. V. Quine, "Two Dogmas of Empiricism," *From a Logical Point of View* (Cambridge, Mass.: Harvard University Press, 1953).
15 See Margolis, "A Sense of *Rapprochement* between Analytic and Continental European Philosophy," *Pragmatism without Foundations*, ch. 8.
16 Thomas S. Kuhn, Postscript (1969) to *The Structure of Scientific Revolutions*, 2nd, enlarged edn (Chicago: University of Chicago Press, 1970).
17 The term "logocentrism" is, of course, deliberately borrowed from Derrida. See Jacques Derrida, *Of Grammatology*, tr. Gayatri Spivak Chakravorty (Baltimore: Johns Hopkins University Press, 1976).
18 Alfred Tarski, "On the Concept of Logical Consequence," *Logic, Semantics, Metamathematics*, 2nd edn, tr. J. H. Woodger, ed. John Corcoran (Indianapolis: Hackett, 1983).
19 Ibid., p. 120.
20 See Tarski, "The Concept of Truth in Formalized Languages," *Logic, Semantics, Metamathematics*; also, Alfred Tarski, "The Semantic Conception of Truth," rep. in Leonard Linsky (ed.), *Semantics and the Philosophy of Language* (Urbana: University of Illinois Press, 1952).
21 Hilary Putnam, "Reference and Truth," *Philosophical Papers*, vol. 3 (Cambridge: Cambridge University Press, 1983), p. 76.

22 Dummett, Preface to *Truth and Other Enigmas*, p. xx.

23 Ibid., pp. xx–xxi.

24 Donald Davidson, Introduction to *Inquiries into Truth and Interpretation* (Oxford: Clarendon Press, 1984), p. xiv.

25 Davidson, "Reply to Foster," *Inquiries into Truth and Interpretation*, p. 176.

26 That is, of course, always conceding that "ontologies" need not be committed to transparency – contrary to the fashionable charge (linking, we may say, Quine and Heidegger) in Richard Rorty, *Philosophy and the Mirror of Nature* (Princeton, NJ: Princeton University Press, 1979).

27 Davidson, 'Semantics for Natural Languages' *Inquiries into Truth and Interpretation*, p. 55. Cf. "Truth and Meaning," ibid., p. 35.

28 Davidson, "In Defense of Convention T," ibid., p. 68.

29 Ibid., p. 65.

30 One of the most recent formulations of "Convention T" is given by Davidson in "Reality without Reference," ibid., p. 215.

31 Ibid., p. 218.

32 See for instance Davidson, "Truth and Meaning," ibid., pp. 35–6.

33 See the perceptive remarks of Ian Hacking, *Why Does Language Matter to Philosophy?* (Cambridge: Cambridge University Press, 1975), pp. 154–5.

34 Davidson, "Reality without Reference," *Inquiries into Truth and Interpretation*, pp. 215–16.

35 Davidson, "In Defense of Convention T." ibid., p. 75.

36 See Edmund Husserl, *Phenomenology and the Crisis of Philosophy*, tr. Quentin Lauer (New York: Harper and Row, 1965).

37 This marks a substantial part of the thrust of Derrida's early work, though it is as much addressed to Kant as to Husserl. See Jacques Derrida, *Speech and Phenomena*, tr. David B. Allison (Evanston, Ill.: Northwestern University Press, 1973); also, Irene E. Harvey, *Derrida and the Economy of Différance* (Bloomington: Indiana University Press, 1986), particularly Introduction.

38 W. V. Quine, *Word and Object* (Cambridge, Mass.: MIT Press, 1960), section 45.

39 See Carl G. Hempel, "Studies in the Logic of Explanation," *Aspects of Scientific Explanation* (New York: Free Press, 1965), p. 263.

40 Paul K. Feyerabend, *Philosophical Papers*, vol. 1: *Realism, Rationalism and Scientific Method* (Cambridge: Cambridge University Press, 1981), p. xiv.

41 See Hilary Putnam, "Two Conceptions of Rationality," *Reason, Truth and History* (Cambridge: Cambridge University Press, 1981).

42 Paul K. Feyerabend, "Historical Background: Some Observations on the Decay of the Philosophy of Science," *Philosophical Papers*, vol. 2: *Problems of Empiricism*, pp. 28–9.

43 Kathleen Freeman, *Ancilla to the Pre-Socratic Philosophers* (Oxford: Basil Blackwell, 1948), p. 126.

44 Ibid., p. 125.

45 W. H. Newton-Smith, *The Rationality of Science* (London: Routledge and Kegan Paul, 1981), pp. 34–7.
46 See Mario Untersteiner, *The Sophists*, tr. Kathleen Freeman (Oxford: Basil Blackwell, 1954), p. 50 n. 18; also, Untersteiner's effort at reconstructing Protagoras – ibid., ch. 3, pt 3.
47 Perhaps the clearest expression of this view is given in Thomas S. Kuhn, "Theory-Change as Structure-Change: Comments on the Sneed Formalism," *Erkenntnis* X (1976), where the triangle case is given.
48 Bernstein, *Beyond Objectivism and Relativism*, pp. 79, 2f.
49 Davidson, "On the Very Idea of a Conceptual Scheme," *Inquiries into Truth and Interpretation*.
50 Bernstein, *Beyond Objectivism and Relativism*, p. 8.
51 This is the argument of Margolis, *Pragmatism without Foundations*, ch. 3.
52 See Charles Taylor, "Philosophy and its History," in Richard Rorty et al. (eds), *Philosophy in History* (Cambridge: Cambridge University Press, 1984).
53 Davidson, "On the Very Idea of a Conceptual Scheme," *Inquiries into Truth and Interpretation*, pp. 197–8.
54 Ian Hacking, "Language, Truth and Reason," in Martin Hollis and Steven Lukes (eds), *Rationality and Relativism* (Cambridge, Mass.: MIT Press, 1982).
55 See for instance Michel Foucault, "Truth and Power," *Power/Knowledge: Selected Interviews and Other Writings 1972–1977*, tr. Colin Gordon et al., ed. Colin Gordon (New York: Pantheon, 1980). The theme underlies the conceptual position in Foucault's historical studies, for instance in his *Discipline and Punish*, tr. Alan Sheridan (New York: Random House, 1977).
56 See Nancy Cartwright, *How the Laws of Physics Lie* (Oxford: Clarendon Press, 1983).
57 See Putnam, "Meaning and Knowledge" (The John Locke Lectures 1976), Lecture II, *Meaning and the Moral Sciences*.
58 This is surely the essential point of Quine's notion of "ontological relativity." See W. V. Quine, "Ontological Relativity," *Ontological Relativity and Other Essays* (New York: Columbia University Press, 1969).
59 See further Margolis, *Pragmatism without Foundations*, pt I.
60 It is fashionable but completely unjustified to construe relativism as committed to treating "truth or rational acceptability as *subjective*," as Putnam does ("Two Conceptions of Rationality," *Reason, Truth and History*, p. 123). At best, it is the result of a conventional reading of Protagoras; at worst, a cryptic reference to the arbitrary skepticism of incommensurabilism.

2

Relativism and the *Lebenswelt*

I

The natural home for a robust relativism is the world of human culture. Its principal phenomena will support a relativistic view of judgment if any phenomena will, and any domain of inquiry can be made more hospitable to relativism to the extent that it is itself essentially encumbered in cultural terms. This is why repudiating cognitive transparency, admitting the radical underdetermination of explanatory theories and the theory-laden nature of observation itself, and historicizing the role of particular such theories and categories in all inquiries entitled to be called realist move us inexorably to concede – or to consider conceding – the legitimacy of construing strategic issues even in the philosophy of the physical sciences relativistically.[1] Wherever we examine disciplines addressed to the most characteristic cultural artifacts – for instance, in interpreting literature, in determining the meaning of historical events – it is noticeably difficult to avoid accommodating relativistic claims. But also, wherever we admit that the history of a science substantively infects the findings of that science – as in fixing the reference of theoretical entities and in acknowledging the idealizing and improvisational function of explanatory laws – we enculture science and, in doing that, prepare it for relativistic concessions.

Nevertheless, to funnel the full import of the enculturing of rigorous inquiry into the mere defense of relativism is a considerable extravagance. The viability of relativism is little more than the by-benefit of a more profound revision in our understanding of human existence and human inquiry. It *is* such a by-benefit, however, and there appears to be no convincing way of avoiding its election as a favorable option among the principal alternative theories of objective judgment that, historically, have always crowded relativism out. So there is a point to a certain retaliatory zeal.

The deeper issue concerns what *enculturing* existence and inquiry means (and, of course, why it proves thereupon so easy to defend a ramified relativism). There are many ways in which to attempt an answer. But there is one in particular that threads its way through the relatively recent history of the dominant currents of theorizing about cognition – featuring *naturalism* and *phenomenology* and, therefore, the developing convergence among Anglo-American and Continental European accounts of inquiry and knowledge. We are here primarily interested not simply in tracing that history, but in tracing it well enough to mark the sense in which the complexities of cultural life are seen to affect profoundly the very characterization of objective knowledge: on the argument intended, in doing that, we muster evidence that legitimates the increasingly relativistic drift of contemporary theorizing.

The story to be told obviously trades on Edmund Husserl's decisive critique of the failure of naturalistic theories of knowledge – say, at least through the period from Galileo to Descartes to Kant – to come to terms with the full implications of the mediation of language upon the would-be competence of an objective science. For what Husserl masterfully demonstrates is the dependence of all language-mediated inquiry on the essentially tacit, preformational, incompletely fathomable contingencies of the *Lebenswelt* within which we are first enabled to pursue the objectives of science and do there pursue them. This alone, the attack on naturalism, on the competence of the methods of the natural sciences as extended to the psychological sciences, *and* on the assured "givenness" of nature thus construed (including the givenness of a thus naturalized reason and experience), the attack on the absence of any need to penetrate beneath the perceived, apparently spontaneous fit of our languaged categories to the flux of experience itself – this alone, we must acknowledge, advances a powerful objection to the deep "naiveté" of the principal philosophical tradition up to Husserl's own day.[2] That objection does not, however, specify how actually to remedy the neglect displayed.

Husserl has his solution(s) to be sure, but it is entirely possible to accept the global charge without yet subscribing to (any of) Husserl's own intended solutions. If we divide "phenomenology" between that in the mere critique of naturalism and Husserl's formulation of a would-be foundational science in accord with the *Wende zum Gegenstand* (phenomenology proper), then even the "naturalists" have surely contributed to the critique: Hegel and Marx have certainly insisted on examining the terrain of the *Lebenswelt* in order to expose the conditions under which rational inquiry presumes to move objectively through nature. Here, one can hardly deny a marked convergence among historicized

Kantian accounts of the conditions of understanding (Dilthey's view, for instance), Hegelian and Marxist accounts, the work of Husserl's own mixed progeny (in particular, of Heidegger, of Merleau-Ponty, of Gadamer), the result of mingling radical Nietzschean themes with the phenomenological and/or the Marxist (as in Heidegger, Derrida, and Foucault), the result of mingling hermeneutic themes with the phenomenological and/or the Marxist (as in Gadamer, among the Frankfurt Critical figures, and in Habermas). These theorizing strands all have to do with isolating at least the subterranean, culturally preformative, normally unperceived conditions of every science and every rational inquiry. But they sort themselves as "naturalistic" or "phenomenological" (provisionally) insofar as they do or do not take the positive findings of a corrective science (beyond the initial critique) to be themselves the findings of *a naturalistic discipline*. Husserl's charge is that they are not and cannot be such, that such a reading would merely reinforce the naiveté originally exposed. There you have the dividing line between Hegel and Marx on one side, and Husserl (and, misleadingly put, Nietzsche) on the other. There one finds the ambivalence of Paul Ricoeur, who expressly seeks to reconcile the two tendencies.[3] And there one finds the essential objection Husserl puts to Heidegger's attempted revision of his own *Britannica* article: for Husserl believed that Hiedegger subverted phenomenology by falling back to a disguised form of naturalism.[4]

It is a distortion, of course, to put Husserl and Nietzsche together, although, by an irony of history, many of those strongly influenced by Husserl have been as strongly attracted to Nietzsche (Heidegger and Derrida, preeminently). Nietzsche is the more radical, in the sense that, for Nietzsche (and those he has influenced), there can be no foundational science of any kind;[5] whereas Husserl's charge of the naiveté of naturalism was expressly intended to usher in just such a corrective science. There are, therefore, three possible strategies before us:

1 to treat the corrective discipline as still naturalistic but sensitized (however incompletably) by an understanding of the self-deceptive processes of human history (Hegel, Marx, the Frankfurt Critical movement, possibly Heidegger himself, on Husserl's reading – therefore, also Gadamer);

2 to treat the corrective discipline as invoking foundational sources that, in principle, are *not* encumbered by the enabling (and self-deceiving) currents of the *Lebenswelt* (Husserl); and

3 to deny that the presumed disorder (though we do normally misconstrue the smooth functioning of rational inquiry) *is* a

condition that can be rightly diagnosed and in principle corrected (Nietzsche, Derrida, Foucault).

It is the convergence of these forms of critique that is collected here as a self-styled "phenomenology." With these themes in place, what we may say is this: *if* the foundationalism and cognitive transparency of classical philosophy are unsupportable,[6] *if* the corrective foundationalism Husserl intends cannot be secured despite its deliberate avoidance of the naive error of naturalism, and *if* the validity of both the Husserlian and Nietzschean exposés makes no sense *without some recovery of a viable science and critique*, then

(i)   there is no essential hierarchy of cognitive *profondeurs* that can be assigned any set of naturalistic and phenomenological methods of inquiry; and

(ii)   it must be possible to formulate conditions for the legitimation of inquiry centered in terms sensitive to the shifting constitutive role of the *Lebenswelt* and, thus adjusted, freed from the least pretense of foundational assurances.

The Husserlian maneuver is a double one: first, to disclose the dependence of objective science on conditions affecting the very formation of the discernible self-identity of things under change – that such a science cannot competently confirm or describe; and second, to penetrate to a deeper level of self-evidence *not* thus dependent on language and objective categories or, more generally, not thus dependent on the flux of the *Lebenswelt* – and (there) to determine the essentializing conditions under which the phenomenal objects of our *Lebenswelt* are formed as they are. Robert Sokolowski has put the point very neatly, in Husserl's behalf: "A categorial intention is one in which we intend not a simple perceptual object, but an object *infected with syntax*. A fact or state of affairs, a group, a relation with its relata, are categorial objects." This is the heart of naturalism – "the natural attitude." But

The point of a phenomenological *description* is to show what kind of manifold is involved in *bringing about* the concrete and intuitive presence of the object we are concerned with, which is the identity *within this manifold*. There is no way of explaining why the object appears in such-and-such a manifold, the object simply is like that, and no more can be said. The submission of phenomenology to the way things are means that as philosophy it is *content to describe these entities and manifolds* without trying to explain why they are like this. Husserl would take any such attempt at explanation to be a metaphysical construction.[7]

Sokolowski draws attention to two utterly different respects in which language addresses "objects": the naturalistic, wherein objects taken as given are already aptly "given" *in* linguistically catalogued ways; and the phenomenological, wherein the self-identity of individual objects framed linguistically in whatever ways prove apt is itself a methodical construction, the result of the testing of constructions, *from* a manifold of experience originally not linguistically so structured. Both views are matched with their appropriate schemes of truth and self-evidence; the phenomenological is specifically concerned not with the match between judgment and fact but only with the "punctual," the once-and-for-all, intuition of the presence of objects regarding which normal questions of fact arise.[8]

There is a fair sense in which Husserl has heroically teased out a ubiquitous feature of every science and every conceptual analysis, namely, a sense in which we are persuaded that the schemata we favor enjoy a distinctive aptness over all contending alternatives – a certain irresistibility, perhaps (in the explicit Husserlian spirit) even a necessity, an inevitability, a *ne plus ultra*, a "firstness" or foundational "beginning," an apodictic assurance that our schemes could not (phenomenologically) be otherwise. It is a sense implicitly opposed, say, in Quine's famous doctrine of the indeterminacy of translation,[9] though, there is surely no reason why Quine should have taken up Husserl's concern.

The point of the comparison cuts both ways: Anglo-American speculation, particularly regarding the conceptual resources of the psychological sciences, tends, as Quine's work does, to be only slimly interested in assessing the full cognitive difficulty of replacing concepts that normally service discourse about familiar phenomena; for his part, Husserl disallows, somewhat prematurely and quarrelsomely, large conceptual alternatives. If, of course, Husserl's project were fully vindicated, then (as a by-benefit) relativism would be either precluded or most severely circumscribed. Quine's notion, though essentially relativistic, is relativistic only in naturalistic terms (in a terminological sense Husserl and Quine more or less share).[10] The question remains whether *Husserl's* phenomenological project is also, perhaps even inescapably, open to a relativistic reading. That would be a reading utterly heterodox and inadmissible from Husserl's viewpoint but hardly ineligible along the reversed lines Heidegger favors or along the explicitly hermeneutic lines Gadamer favors (regardless of Gadamer's own resistance to relativism).

The pivotal clue is already to be found in Sokolowski's account. Phenomenology is a *descriptive* discipline. It does not, however, directly describe actual objects: what it seeks to describe – *in a sense free of mere "worldly" (naturalistic) experience* – are the conceptual limits under which objects of given kinds may, through whatever imagined change, continue to be the things they are; and without conforming to which they cease to be "thinkably" such. In this sense, by a method of imaginative "free variations" among conceptual possibilities, we fix the *eidos* of particular things, their essence, their *a priori* structure as the things they are within experience. "In such free variations of a paradigm ...," Husserl asks, "what remains kept as the invariant, the necessary universal form, the essence-form, without which something of this sort ... would be entirely unthinkable as an example of its kind?"[11] So the descriptive effort – relative to what is "first" in subjective experience, relative to what is apodictically discerned as such – essentially relies on a contact with cognitively pertinent sources *not* linguistically, culturally, or historically encumbered (*not* encumbered by our *Lebenswelt*) *and* on a capacity for providing an undistorting record of that original fit *within* the flux of our *Lebenswelt*. These are surely doubtful resources *that no allegedly phenomenological probing can quite ensure.*

These two sources of *first foundation* and *apodicticity* account for Husserl's being driven (correctly: in the sense of grasping what his own theory ultimately requires) toward the extreme solipsism threatened in his transcendental doctrine. We need not follow him in this.[12] The essential point is that, precisely because *phenomenology is descriptive* – albeit in a sense utterly different from that of naturalistic description – it is quite impossible (or, at any rate, Husserl fails to show why it is "unthinkable") to deny that the phenomenological search for foundations, for an apodictic assurance that the foundational is foundational, for an apodictic grasp of what within the range of the foundational is eidetically invariant for given phenomena, is *inextricably infected by the contingencies of naturalistic, languaged practices functioning smoothly within our incompletely fathomed "Lebenswelt."* There is *some* tendency in all thinking that inclines toward the conceptual necessities Husserl pursues; but there is *no* decisive sense in which such a tendency is intuitively grasped *as* originary, *as* apodictically marked *as* originary, or *as* yielding, as such, apodictic truths of an eidetic nature linked to the originary. In a word, Husserl's phenomenological project *is not itself able to be vindicated phenomenologically* – as the logic of the situation would seem to require. And so it must fail. But in failing it strengthens the sense in which all science, all inquiry, all speculation *is* naturalistic (however deepened) – in being so structured that even our most "punctual" intuitions of what could not be otherwise are inseparable from the influence of contingently preformed and preforming conceptual habits with which and within the tacit scope of which we first entertain imaginative possibilities of what might be otherwise. The deeper import of all of this concerns the cultural nature of man (incarnate in some sense in the human animal). The shallower message yields a globalized sense of the inescapability of relativism.

II

Let us turn back, now, to recast the argument with an eye to the larger continuity between naturalism and phenomenology.

There is a certain paradoxical quality that can be informally displayed in our most persistent habits of thought regarding the natural world. On the one hand, we hardly wish to hold that the "real world" (usually we mean the "physical world") depends for its very existence or for its actual nature on those descriptive categories by the use of which we fix what we take the world to be; and, on the other, there is no viable way by which to give determinate form to what (we take it) the world is like except through the use of our conceptual network (which, reflexively, we impute to ourselves under the same presumption). In the refined atmosphere of epistemic questioning, the compelling achievements of human inquiry are characteristically cast in terms of a *fit* or correspondence of some sort; and the implied question has been traditionally articulated in terms of a relationship between *facts* and *propositions*, both of which are singularly abstract, singularly dubious entities, perhaps no more than heuristically invoked to facilitate the discourse of truth-claims itself.

The trouble is: we are unable to replace these referents with more perspicuous ones; and they themselves provide the required match only in a formal sense and only as projected from what looks suspiciously like an already settled, prior, quite independent conviction that the powers of human inquiry can supply the right propositional half. It is certainly the idiom of that perpetually attractive and important disaster, Wittgenstein's *Tractatus*: "The world is the totality of facts [*Tatsachen*], not of things [*Dinge*]"; "The totality of existing states of affairs [*Sachverhalte*] is the world"; "A picture [*Bild*] depicts reality by representing a possibility of existence and non-existence of states of affairs"; "A logical picture of facts is a thought [*Gedanke*]"; "What a picture represents [its sense] it represents independently of its truth or falsity, by means of its pictorial form"; "If a fact is to be a picture, it must have something in common [identical] with what it depicts"; "I call the sign [*Zeichen*] with which we express a thought a propositional sign. – And a proposition is a propositional sign in its projective relation to the world"; "A propositional sign is a fact [a determinate logical possibility of existence and non-existence of states of affairs]."[13]

The same idiom puts in a notorious appearance in a more relaxed but perhaps even more compelling vignette in the well-known quarrel between J. L. Austin and P. F. Strawson: there, mention of the "relationship" between truth and facts positively puts our nose to the glass, forces us to concede the ruinous threat of redundancy in admitting such linguistic or language-like entities as propositions or statements, on the one hand, and facts or states of affairs, on the other. Austin himself confronted so-called correspondence theories (his own for instance) with the essential puzzle: "Either we suppose that there is nothing there but the true statement itself, nothing to which it corresponds, or else we populate the world with linguistic *Doppelgänger*. ..."[14]

Our issue, however, is not *that* quarrel, the quarrel of the fit, but rather that of what eluding its bottomless trap entails regarding cultural contexts. Let it be said at once that correspondence in the sense required in both Wittgenstein's and Austin's accounts has absolutely nothing to do directly with the effective resolution of *any* cognitively determinate empirical question – whatever its "formal" contribution to the semantics of truth-claims may be alleged to be. Statements about the way the world is, independent of the conditions of human inquiry, are themselves (*must be*) artifacts of those very conditions (even when shown to be true, however that may obtain) – and admitting *that* constraint is hardly tantamount to their dismissal: on the contrary, on the argument it is all we have and it is quite enough to permit an escape from the more dire forms of skepticism.

Wherever we suppose that to ensure the required fit we must invoke an independent source of cognitive privilege – a source of assurance that the fit obtains – the philosophical tradition records that two sorts of maneuvers have been particularly favored. Both are colored by the conviction that the flux and endlessness of the world's manifestations may be suitably contained (for purposes of knowledge) even if they cannot be stopped or denied. By means of one, we suppose the world of change to be subject to universal, therefore unchanging, principles – call this *essentialism* (following Karl Popper's usage[15]); by the other, we suppose ourselves capable of reliably fixing certain strategically placed truths that nullify or effectively reduce the threat of cognitive disorder among all other possible beliefs or truth claims – call this *foundational-*

ism.[16] One may be tempted to say, disjunctively, that the first is a claim about the structure of the world and the second a claim about our cognitive aptitudes. But that would be a mistake, because each is obviously implicated in the other. The one holds that there are necessary structures in the real world (*de re*) in virtue of which knowledge is possible; the other holds that the discovery of nontrivial, necessary or undeniable truths (*de dicto*) confirms that the world must be stable enough to be known. Speculation may appear to bifurcate the two, but they are not really separable. The homely fact is that knowledge must have its object and to specify the nature of any such object entails claiming to know enough to be able to do so.

Essentialism and foundationalism need not be thought to exhaust the possible strategies for "stopping the world."[17] In a sense, they are only particularly prominent species of a more generic presumption on which naturalistic accounts appear to converge: the double presumption, namely, *that there is only one world we inhabit and share* (however we describe it) and *that we can describe it objectively if only we adopt the proper initial stance with respect to which the intended fit between fact and proposition* (allowing for error and its correction) *alone obtains.*

The pretty gain is that this double presumption serves to identify as well a common ground between naturalism and phenomenology – the ground, say, that Descartes or Kant and Husserl have in common but exploit quite differently. The opposition between the correspondence notion of naturalistic truth and the intuitive registration of eidetic truth does not adversely affect that benefit. Husserl supposes that the classical tradition is naively confined within the limits of naturalistic theorizing (and he is right about that): the naturalist presupposes that the languaged world is "given" more or less as it is actually catalogued in natural language. But, though Husserl's own would-be improvement escapes the naive limits of that naturalistic stance, it cannot, merely in doing so, escape the deeper common ground it shares with naturalism – *and*, in thus failing (if failing that be), it may be shown to fall inexorably under the constraints of the naturalistic stance it means to displace (the full role of which it therefore fails to assess correctly).

> The life-world [declares Husserl], for us who wakingly live in it, is always already there, existing in advance for us, the "ground" of all praxis whether theoretical or extratheoretical. ... But there exists a fundamental difference between the way we are conscious of the world and the way we are conscious of things or objects ... though taken together the two make up an inseparable unity. Things, objects (always understood purely in the sense of the life-

world), are "given" as being valid for us in each case (in some mode or other of ontic certainty) but in principle only in such a way that we are conscious of them as things or objects *within the world-horizon*. ... On the other hand, we are conscious of this horizon only as a horizon for existing objects; without particular objects of consciousness it cannot be actual [*aktuell*]. ... The world ... does not exist as *an* entity, as an object, but exists with such uniqueness that the plural makes no sense when applied to it. Every plural, and every singular drawn from it, presupposes the world-horizon.[18]

The deep weakness of Husserl's account (the weakness Heidegger seized upon) lies in this: that the discovery of the conceptual features of the "horizon," like the eidetic properties of particular objects, depends on the self-disclosures of consciousness – now, no longer confined within the natural world (a thesis Heidegger could accept) – but the power of consciousness on which it depends is one somehow disconnected from any being that, possessing it, might affirm and legitimate and be responsible for particular claims. Heidegger makes transcendental consciousness dependent on the (mythic) life of *Dasein*: his invention adds that *Dasein* is a being uniquely not found among "worldly," naturalistic beings. For his part, Husserl evidently found this preposterous. But the novelty of Heidegger's maneuver was that it saved the relevance of phenomenological reflection while abandoning the foundational and apodictic concerns of Husserl.[19] The only way in which Heidegger could combine these objectives was by construing phenomenology hermeneutically. The extravagance of both accounts, as well as their incompatibility and their separate coherence, confirms at the very least the less than apodictic force of Husserl's conception of the powers of the would-be science of phenomenology: confirms it in such a way that we are led to see that phenomenology cannot be segregated from naturalistic reflection as a separate, cognitively "deeper" level of certitude and cannot escape the threat of an increasingly radical relativism. Nevertheless, there must surely be a phenomenological-like dimension functioning *within* any respectably self-critical naturalistic stance. There is also little question that Husserl and Heidegger are obliged to revert to "naturalistic" questions in order to pursue and vindicate their respective attacks on naturalism (or objectivism) and on one another.[20]

There is, then, a multiple weakness in Husserl's view of the "world-horizon." Some (notably Nelson Goodman[21]) have implicitly challenged the intuition of the "unity" of the "world" (in spirit and detail oddly not too distant from the theme of Heidegger's account). The "unity" of the "world" may have to be relativized to a particular self at a particular moment, may have to be constructed and reconstructed (for reasons of coherence) over a span of life, may have to be constructed consensually by whole societies over time, and may have to be relativized to particular cultures. (These possibilities may in fact be close to Goodman's intention – they are surely close to Heidegger's and Gadamer's – but Goodman is singularly tight-lipped about the full reasoning behind his views.) Finally, there may be no compelling sense – in a Nietzscheanized spirit – in claiming that there *is* a "uniqueness" to the world in virtue of which "the plural makes no sense when applied to it": it may never, as Roland Barthes and Derrida and Foucault in rather different ways insist (post-Husserl),[22] yield *a sense of the actually unified horizon* within which all variation *is* coherently contained.

If so, then Husserl's *phenomenological* drive for the ground and "beginning" of philosophy and description and for the apodictic with respect to them is doomed – both generically and in one of its essential claims. The putative structure of our unique world may well be a shifting function of the shifting and diverging core experience of its denizens (ourselves); *and* the phenomenological claim that its uniqueness is "given," is given apodictically, and is given in such a way that what holds phenomenologically of particular objects within its horizon is also given apodictically, may rightly be seen to be fatally affected by the contingencies of the "natural attitude" to the world. Paradoxically, the world's uniqueness may be denied or put into more-or-less perpetual doubt *phenomenologically*! That would signify

(a) that phenomenological descriptions need not be apodictic or originary;

(b) that phenomenological descriptions need not be ideally convergent;

(c) that phenomenological descriptions may even be construed relativistically; hence

(d) that phenomenological descriptions are formally distinct from, but not cognitively separable from, naturalistic descriptions; hence, also

(e) that phenomenological descriptions of all sorts are conceptually dependent on and infected by the contingencies of the *Lebenswelt*; and hence, finally,

(f) that there is no meaningful hierarchy of a cognitive reliability between naturalistic and phenomenological descriptions.

These findings (a)–(f), are all radically opposed by Husserl but, to one degree or another, they are pretty well tolerated (even somewhat favored), certainly variously explored, by the most important thinkers Husserl influenced – Heidegger, Merleau-Ponty, and Gadamer – provided only that naturalistic claims of cognitive privilege are entirely abandoned.[23] By an irony of ironies, they may be compatibly added to the exposés of the shallow naturalisms of Anglo-American analysis that characteristically ignore phenomenological challenge.

III

Husserl means (in a double sense) to be the Descartes of phenomenology; to find and fix a suitable replacement for the undeceiving God (the ultimate origin and the supreme referent for the description of the totality of things) and to find and fix a replacement for the geometrized self-evidence of the *cogito* (the apodictic limit of the variability of conceptual options). But, now, it makes no difference whether the project of cognitive certainty is approached naturalistically or phenomenologically; for, on the argument, the two programs are indissolubly linked and fatally affected by the same contingencies of our *Lebenswelt*. The irony is that Husserl affords the most decisive objection to the pretended fixities of all naturalistic undertakings, but in doing that opens himself to the same objection adjusted to his own pretended escape *from the conditions that made the other impossible*. The decisive clue is this (it is at once naturalistic and phenomenological): the flux – of language, of human history, of cultural bias, of fragmentary experience, of variable perspective, of imperfectly cohering phenomenal uniformities, of the presumed unity of memory and of experienced personal identity – *cannot be escaped*; and, being inescapable, that flux, particularly in the face of fundamental challenges to conceptual invariance, cannot but disallow the full (Husserlian or Cartesian) recovery (or even a good "approximation" to the recovery) of the originary and the apodictic. Husserl's findings cannot fix what is constitutive of all rational cognition; nor can it be shown that all cognition is regulated by his determinate claims (or, demonstrably incoherent for failing to do so). The pertinent objectives that might have vindicated his endeavor – most memorably, the search for universal laws of nature, for apodictic laws of thought, for conceptual invariances, for totalized schemata, for all possible worlds – are all idealizations of *genuinely salient regularities within the changing horizon of our "Lebenswelt;"* but that alone hardly makes them demonstrably asymptotic of Husserlian conclusions. There are no context-free sciences in this respect, naturalistic or phenomenological; there is no totalized context of contexts; there is no way to measure the "distance" from any palpable conceptual vector to the true cognitive source from which we gauge the "unthinkability" of would-be conceptual deviances. Our sciences succeed within and only within that flux: their salient regularities are all we have (reflexively disciplined); they must therefore be sufficient for our cognitive purpose (if any conditions are). A context-free science is a context-bound vision. It is simply incoherent to insist on a context-bound assurance that we *are* cognitively capable of escaping our own cultural world, or that we would recognize any such escape if we achieved it, or that there is or must be one and only one ideally convergent way to envision the world ranging over all and future shifting cultures. We are confined, therefore, to inventing a *similitude* of a context-free vision, not (in Popper's sense or, if we may so speak now, in Husserl's related sense) a *verisimilitude*. None is possible.

Consider an object lesson. Nancy Cartwright has provocatively claimed that the explanatory laws of physics must be false in order to serve the very needs of scientific explanation they are invented to serve. Yet they do function effectively: they risk and depend on ineluctable discrepancies between phenomenological and explanatory laws (in a sense of "phenomenological" quite distinct, of course, from that of Husserl's usage); they do not quite capture the full conditions under which perceived phenomena are actually explained; and they invariably require bridge principles to complete the task.[25] It is not essential that Cartwright's reading of physics be uniquely correct but only that it be a genuinely viable alternative – a drastic alternative as it happens, but not easily disconfirmed by way of the more canonical views of testing physical claims from which it obviously departs.[26] The viability of Cartwright's proposal demonstrates at once the inseparability of first- and second-order questions (the inseparability of science and the legitimation of science)[27] and in the process confirms the irresistibility of relativism within naturalistic frames of reference that eschew all foundationalist and originary pretensions. It does so in what is usually regarded as the field of the firmest empirical achievements man can claim.

But Cartwright's account is noticeably inexplicit about the cognitive processing in virtue of which her own option *is* viable as well as credible: she exposes thereby the remarkable informality of naturalistic sources of epistemic confidence on which the precision of advanced physics ultimately rests.[28] She does not, therefore, explicitly provide for the prescientific investigations that Husserl claims the cognitive stability of science itself ultimately depends upon. But what she does show, in subverting the canonical standing of so-called "fundamental laws," is that the description of physical "phenomena" (*not* confined to the observational) is itself an artifact of the historical development of the practice of physics.[29] Cartwright's maneuver, therefore, is a paradigm of the way the most disciplined work of a naturalistic sort alters (however conservatively) our sense of the objects and the conditions of the self-identity of such objects encountered within our *Lebenswelt*: it must, therefore, also affect the linkage between naturalistic and phenomenological inquiry – in the direction of (a) – (f) above – against Husserl or at least in a way that Husserl seems not quite to have addressed. This is also the sense in which Husserl's chief progeny – Heidegger, Merleau-Ponty, and Gadamer – have moved in various ways to diminish or erase the disjunction between the naturalistic and the phenomenological, as well as to favor somewhat a measure of relativism. Our object lesson exposes, therefore, the "naive" presumption underlying Husserl's quite correct exposure of the "naive" presumption of every naturalism. They are one and the same, since the would-be invariances of phenomenological review cannot but be hostage to the potentially radical novelty and contingency of naturalistic inquiry: the extent to which the one is open to profoundly plural, nonconverging, unstable, incompletely collectable, variably salient, ever-productive, hardly systematic, transiently sufficient fragments of conceptual networks must adversely affect not only particular would-be eidetic invariances but our very sense of the prospects of such a project. Put most tellingly, the more one (rightly) emphasizes the deep contingencies and fragmentary uniformities of naturalistic thought, the more impossible it becomes to insist that these can be overcome or corrected by invoking a scientific phenomenology – for the would-be powers of the latter are essentially infected by, in being applied to, the materials of the former.

On Husserl's view, "Objective science asks questions [of the world already codified for scientific study] only on the ground of that world's existing in advance through prescientific life."[30] It does so, however, somnambulistically, claiming (as an "objectivism") to see "what, in this world, is unconditionally valid for every rational being, what it is in itself," moving "upon the ground of the world which is pregiven, taken for granted through experience." On the other hand "transcendentalism," the ultimate phenomenological corrective, says, "the ontic meaning of the pre-given life-world is a *subjective structure*, it is the achievement of experiencing prescientific life."[31]

The momentum of his well-intentioned corrective propels Husserl to a most incredible prophecy. He begins with a scrupulous, self-effacing critique of the unearned legitimation of every objectivist claim and, by stages, moves to an astonishingly sanguine promise of a source of self-conscious certitude – now no longer even human: "in me," proclaims

Husserl, "'another I' achieves ontic validity as copresent [*kompräsent*] with his own ways of being self-evidently verified, which are obviously quite different from those of a 'sense'-perception." (The rest of the statement needs to be scanned to be believed.[32])

"Only when this radical, fundamental science exists ...," Husserl continues, "the universal, a priori, fundamental science for all objective sciences," can we rightly speak of an objective science at all or an "objective logic." Without it, every would-be science

> hangs in mid-air, without support, and is, as it has been up to now, so very naive that it is not even aware of the task which attaches to every objective logic, every a priori science in the usual sense, namely, that of discovering how this logic itself is to be grounded, hence no longer "logically" but by being traced back to the universal prelogical a priori through which everything logical, the total edifice of objective theory in all its methodological forms, demonstrates its legitimate sense and from which, then, all logic itself must receive its norms.

Finally, "this [very] insight surpasses the interest in the life-world which governs us now."[33]

But none of this actually legitimates Husserl's assurance and, on the argument, nothing can. A phenomenologized science may well be deeper than an objectivized one, but its profundity must lie in its dialectical resources, not in its discovery of more and more fundamental cognitive privilege. There are no such sources to be found. We fix our bearings in transit, and we have no inkling of how whatever proves to be stable there is connected with the outmost boundaries of all possible such reflections – similarly fixed in transit. "But where are the bounds of the incidental?" Wittgenstein asks; "in philosophy," he adds, "we often *compare* the use of words with games and calculi which have fixed rules, but cannot say that someone who is using language *must* be playing such a game. – But if you say that our languages only *approximate* to such calculi you are standing on the very brink of a misunderstanding."[34] There is no reason to think that a similar objection cannot be brought against any form of thought said to be cognitively more fundamental than Wittgenstein's target – against Husserl's speculations, in fact.[35]

It would be a mistake to think that the objection to Husserl's project rested only on the charge that neither first foundations nor apodictic certainty can be fixed or approached by any means whatever.[36] That part of the charge still holds, it is true. But the essential point is that the force of Husserl's assessment of competing options is ultimately *not* drawn out dialectically (though it could have been) but depends on privilege: it depends on the presumption that Husserl's own findings rest on sources closer to true "beginnings," *closer* to genuinely apodictic discoveries than those of Husserl's opponents. *Escape* from the *Lebenswelt*, it seems, is the only way to master the flux of the *Lebenswelt*; else, we are little more than its passive creatures.

But there is another possibility – to live *in* the flux and find our bearings there. It is the way favored in all *non*-phenomenological accounts that have (one may almost say) taken Husserl's critique of naturalism to heart. It is also the way variously favored within the later phenomenological tradition, most naughtily by Jacques Derrida, in a Nietzschean voice. For in Derrida the correction is shaped in terms of the charge that Husserl failed to extricate himself from the correspondence claims of the naturalism he attacked. Not that the naturalist could have worked out his own escape: only that no systematic philosophy could. There is a surd, Derrida insists – a gap, an indefinable *brisure*, an inexpressible something that can neither be said to be nor be said not to be – that always obtains, that makes conceptualization possible, that is never overcome, that can never be explicitly fixed within the terms of reference of any conceptual network. Derrida fixes his claim – which cannot consistently be a claim – in a single line suitably subverting both Rousseau (the ancestor of Lévi-Strauss – hence, the emblem of the most extreme presumption of naturalism) and Husserl (whom he elsewhere directly attacks): "The supplement is what neither Nature nor Reason can tolerate."[37] The "supplement" is what can never be completely supplied but is "wordlessly" required in order to ensure that our categories correctly fix the would-be structures of either Nature or Reason – at whatever level either may be analyzed.[38] The stalemate Derrida intends is insinuated in the famous use of that invented sign, said to be "neither word nor concept": *différance*. The plain charge is that, *in* phenomenology as everywhere else, there is an arbitrary presumption of having fixed the relation between words and things as they truly are (returning, by whatever route, "to the things themselves") in virtue of which constative discourse proceeds and prospers. But there is always slippage, an unrecoverable alterity, that a genuinely focused phenomenology would have spotted and duly acknowledged. Husserl failed to recognize the gap. On Derrida's usage, the common presumption of objectivism and phenomenology is the obverse of this lapse: *logocentrism*.[39]

Derrida was clever indeed to mark the common theme. He may already have found it in Heidegger. But he himself is aporetically trapped by his own lesson. He cannot account for the precision of his own exposé or for its sustained import *vis-à-vis* any disciplined inquiry. We have come full circle, therefore. Recall that we began by sorting three possible lines of strategy for accommodating the effect of the *Lebenswelt*: a self-corrective naturalism attentive to the *Lebenswelt* itself, a phenomenology not confined to the *Lebenswelt* but in touch with the conditions through which it is first generated, and a Nietzscheanized critique content merely to expose the self-deceptive practices of all forms of the other two. On the argument, there can be no principled demarcation between naturalism and phenomenology; and again, on the argument, there can be no fall-back to origins or apodictic certainty within either camp or within their union. *But, then, there cannot be any pertinent grasp of determinate failures to meet these two constraints that does not implicitly affirm the ongoing viability of discourse actually guided by them.* The question remains how, plausibly, to characterize what may be recovered of science and disciplined inquiry within such spare terms (what may here be called pragmatism[40]). Still, the force of Derrida's "deconstruction" of every promising theory of competent inquiry presupposes that every such theory entails a logocentric commitment – a commitment to origins, foundations, totalized possibilities, closed systems, all possible worlds, or apodicticity – that cannot be defended and cannot be eluded. That charge, however, is a *petitio*, as it is also in Husserl – a victim of the vestigial attraction of the naive naturalisms Husserl attacked or the equally naive transcendental fixities Husserl offered in their place. Derrida cannot really deny that an effective deconstruction *is* conceptually tied to a formulable charge, even if, for whatever coy reasons he favors,[41] the rhetoric of *différance* and *suppléments* features the theatre of the surd.

The bottom line is this: we *can* reject privilege, objectivism, logocentrism, the transcendentally apodictic, *but we cannot escape our "Lebenswelt."* We cannot, because the admission of the *Lebenswelt* is essentially no more than the reflexive recognition that we are in some way culturally constituted – both with regard to determinate challenges of relatively sessile conceptual schemes and with regard to a tacit or inchoate sense of the preformational forces by which such schemes are endlessly generated and endlessly in need of being similarly monitored. We cannot escape, because we are encultured creatures, unique enough in natural terms, unique enough to accommodate phenomenological scruples, but unable to penetrate to the originary sources of the enabling culture we absorb and, by absorbing, change. It is the absence of cognitive or ontic beginnings, of apodictic sources, of totalized possibilities, of genuinely universal regularities, of essential necessities discovered once and for all, that marks the pathos of the human condition. Appearances suggest that

that absence is hardly disabling. The *démarches* from Descartes to Husserl to Derrida, therefore, trace a somewhat comic trail, for the versions of hyperbolic doubt betray an incapacity on the part of each to release his hold on a deeper privilege.

Human existence is an encultured existence, whatever its other sources. Appeal to the fact is the charm of referring to the cognitive import of the *Lebenswelt*. It sets the realist problem of reinterpreting the whole of human behavior in terms freed from the presumption of privilege and deepened by whatever may be reasonably offered through an analysis of the cultural – particularly an analysis of the linguistic, the praxical, the historical, and the intentional. Here, at most, we have merely set the stage. There is an advantage in having done so. But one sees at once that the prospects of a thoroughgoing relativism are essentially tied to one's theory of cultural life. If there is no escape from the *Lebenswelt*, there can be no escape from relativism. But what form our relativism must take must itself be affected by whatever regular structures may be convincingly assigned the cultural. Ultimately, it is the saliencies of our cultural condition – not the mere by-product of relativism – that are bound to oblige us to recast our vision of what our sciences can accomplish. There is every reason to believe that its full import has been neglected or denied; for that is no more than the lesson entailed in the inseparability of naturalistic, phenomenological, and deconstructive strategies. They are neither separable from one another nor hierarchically ordered in any cognitively pertinent way among themselves. Because of that, the admission of the *Lebenswelt* ineluctably encourages an appetite for relativism. But its deeper lesson lies with recovering the forms of objectivity obscured by speculation keyed more exclusively to the disjoined strategies of one or the other of these three traditions. The natural sciences have been hopelessly elevated beyond the constraints of our *Lebenswelt*; and the human sciences have been cheerfully repudiated wherever they have candidly embraced them.

Notes

1　The background argument appears in Joseph Margolis, *Pragmatism without Foundations: Reconciling Realism and Relativism* (Oxford: Basil Blackwell, 1986).

2　See Edmund Husserl, "Philosophy as Rigorous Science," in *Phenomenology and the Crisis of Philosophy*, tr. Quentin Lauer (New York: Harper and Row, 1965).

3　See Paul Ricoeur, *Hermeneutics and the Human Sciences*, ed. and tr. John B. Thompson (Cambridge: Cambridge University Press, 1981).

4   Cf. the brief account in Timothy J. Stapleton, *Husserl and Heidegger: The Question of a Phenomenological Beginning* (Albany: State University of New York, 1983), ch. 4. See also Herbert Spiegelberg, *The Phenomenological Movement*, 2nd edn, 2 vols (The Hague: Martinus Nijhoff, 1965), vol. I, pt VI.

5   See Friedrich Nietzsche, "On Truth and Lie in a Nonmoral Sense," in *Philosophy and Truth: Selections from Nietzsche's Notebooks of the Early 1870's*, ed. and tr. Daniel Breazeale (New York: Humanities Press, 1979).

6   This is of course the pop theme of Richard Rorty, *Philosophy and the Mirror of Nature* (Princeton, NJ: Princeton University Press, 1979).

7   Robert Sokolowski, *Husserlian Meditations* (Evanston, Ill.: Northwestern University Press, 1974), pp. 31, 103 (italics added). See Edmund Husserl, *Logical Investigations*, tr. J. N. Findlay, 2 vols (New York: Humanities Press, 1970), Sixth Investigation.

8   Sokolowski, *Husserlian Meditations*, pp. 79–80, 235–6.

9   W. V. Quine, *Word and Object* (Cambridge, Mass.: MIT Press, 1960), ch. 2.

10   See W. V. Quine, "Epistemology Naturalized," *Ontological Relativity and Other Essays* (New York: Columbia University Press, 1969).

11   Edmund Husserl, *Phänomenologische Psychologie*, ed. W. Biemel (*Husserliana* IX) (The Hague: Nijhoff, 1962), section 92 (p. 72); cited and translated by Sokolowski.

12   On the central notions of the foundational and the apodictic, see Edmund Husserl, *Cartesian Meditations*, tr. Dorion Cairns (The Hague: Martinus Nijhoff, 1960).

13   Ludwig Wittgenstein, *Tractatus Logico-Philosophicus*, 2nd, corr. edn, tr. D. F. Pears and B. F. McGuinness (London: Routledge and Kegan Paul, 1972), sections 1.1, 2.04; 2.201, 3, 2.22, 2.16, 3.12, 3.14.

14   J. L. Austin, "Truth," repr. in *Philosophical Papers* (Oxford: Clarendon Press, 1961), p. 91. See also his "Unfair to Facts," *Philosophical Papers*; and P. F. Strawson, "Truth," repr. in *Logico-Linguistic Papers* (London: Methuen, 1971).

15   See Karl R. Popper, "The Aim of Sciences," *Objective Knowledge* (Oxford: Clarendon Press, 1972).

16   The most convenient formulaic account is given in Keith Lehrer, *Knowledge* (Oxford: Clarendon Press, 1974), ch. 4.

17   I am taking advantage here of Carlos Castañeda's expression, "stopping the world" – used by his sorcerer Don Juan – in a spirit entirely opposed to Don Juan's purpose. For the sorcerer, it signifies a mystical function arresting the flux of the world according to familiar categories; here, it signifies the hyper-fixity of the ordinary (or of certain successor) categories of the world – so it reverses Don Juan's sense altogether. And yet that hyper-fixity pretends to a privilege that the flux of the world cannot sustain. See Carlos Castañeda, *The Journal to Ixtlan* (New York: Simon and Schuster, 1972).

18   Edmund Husserl, *The Crisis of European Sciences and Transcendental Phenomenology*, tr. David Carr (Evanston, Ill.: Northwestern University Press, 1970), p. 143. See Sokolowski's gloss, *Husserlian Meditations*, pp. 169–72.

19   Heidegger's view is contained in a letter to Husserl (October 22, 1927) regarding the intended revision of Husserl's *Encyclopaedia Britannica* article. The pertinent text is given in Walter Biemel, "Husserl's *Encyclopaedia Britannica* Article and Heidegger's Remarks Thereon," in Frederick Elliston and Peter McCormick (eds), *Husserl: Expositions and Appraisals* (Notre Dame, Ind.: Notre Dame University Press, 1977).

20   The point is very clearly developed in Charles B. Guignon, *Heidegger and the Question of Epistemology* (Indianapolis: Hackett, 1983).

21   Nelson Goodman, *Ways of Worldmaking* (Indianapolis: Hackett, 1978). Goodman's theory is insufficiently explicit on the essential issues.

22   See Roland Barthes, "From Work to Text," tr. Josué V. Harari, in Harari (ed.), *Textual Strategies* (Ithaca, NY: Cornell University Press, 1979); and Michel Foucault, *The Order of Things*, from the French (New York: Random House, 1970).

23   This may be said even of Maurice Merleau-Ponty, *The Visible and the Invisible*, tr. Alphonso Lingis (Evanston, Ill.: Northwestern University Press, 1968), which, though the work of the most "loyal" of the three, nevertheless, by locating the phenomenological in the body itself (flesh, *la chair*) and in acknowledging the idealization of phenomenological description relative to one's history, moves quite far toward erasing the demarcation between the naturalistic and the phenomenological. See particularly chs 3–4. The (Husserlian) idealization of the "invariant" remains (e.g. pp. 111, 114), but it is now an exercise that need neither ensure *nor* approximate in any determinable way the strictly apodictic or foundational; or, if it is made to persist thus, it must do so inconsistently or arbitrarily.

24   One may, not unreasonably, see here an ironic parallel between Popper's well-known doctrine of verisimilitude and Husserl's, in spite of enormous differences in the purpose and nature of their respective claims. See Karl R. Popper, "Two Faces of Commonsense: An Argument for Commonsense Realism and against the Commonsense Theory of Knowledge," *Objective Knowledge*.

25   Nancy Cartwright, *How the Laws of Physics Lie* (Oxford: Clarendon Press, 1980).

26   Contrast for instance Bas C. van Fraassen, *The Scientific Image* (Oxford: Clarendon Press, 1980).

27   This goes decisively contrary to Richard Rorty's now-fashionable disjunction between first-order science and second-order epistemology and metaphysics. Rorty never addresses the issue directly. See Rorty, *Philosophy and the Mirror of Nature*.

28   In fact, one of her principal opponents, van Fraassen, somewhat adopts the phenomenological lingo, but it is clear that van Fraassen means, by

"phenomenology," little more than the naive pronouncements of ordinary folk (what Wilfrid Sellars has dubbed the "manifest image"). Cf. van Fraassen, *The Scientific Image*, p. 74.

29 Cartwright, *How the Laws of Physics Lie*, chs 6, 9. See also van Fraassen, *The Scientific Image*; ch. 6.
30 Husserl, *The Crisis of European Sciences*, p. 110.
31 Ibid., pp. 68–9.
32 It goes on as follows: "Only by starting from the ego and the system of its transcendental functions and accomplishments can we methodically exhibit transcendental intersubjectivity and its transcendental communalization, through which, in the functioning system of ego-poles, the 'world for all' and for each subject *as* world for all, is constituted. Only in this way, in an essential system of forward steps, can ·we gain an ultimate comprehension of the fact that each transcendental 'I' within intersubjectivity (as coconstituting the world in the way indicated) must necessarily be constituted in the world as a human being; in other words, that each human being 'bears within himself a transcendental "I"' – not as a real part or a stratum of his soul (which would be absurd) but rather insofar as he is the self-objectification, as exhibited through phenomenological self-reflection, of the corresponding transcendental 'I'" ibid., pp. 185–6.
33 Ibid., p. 141.
34 Ludwig Wittgenstein, *Philosophical Investigations*, tr. G. E. M. Anscombe (New York: Macmillan, 1953), section 81.
35 See Nicholas F. Gier, *Wittgenstein and Phenomenology* (Albany: State University of New York, 1981); also Laszek Kolakowski, *Husserl and the Search for Certitude* (New Haven: Yale University Press, 1955).
36 An attempt to sort these two themes in Husserl, and to give priority to the originary, is made in Stapleton, *Husserl and Heidegger*, particularly ch. 3.
37 Jacques Derrida, *Of Grammatology*, tr. Gayatri Spivak Chakravorty (Baltimore: Johns Hopkins University Press, 1976), p. 148.
38 The more direct attack on Husserl appears in Jacques Derrida, *Edmund Husserl's Origin of Geometry: An Introduction*, tr. John P. Leavey (New York: Nicolas Hays, 1977); and *Speech and Phenomena and Other Essays on Husserl's Theory of Signs*, tr. David B. Allison (Evanston, Ill.: Northwestern University Press, 1973).
39 Derrida, *Of Grammatology*, pt i.
40 The properties of this newer form of pragmatism are explored in *Pragmatism without Foundations*.
41 See Joseph Margolis, "vs. (Wittgenstein, Derrida)," in Rudolf Haller (ed.), *Aesthetics: Proceedings of the 8th International Wittgenstein Symposium, Part I* (Vienna: Hölder-Pichler-Tempsky, 1984).

Part One

Minds without Substance

3

Minds, Selves, and Persons

I

One of the most celebrated passages in the whole of Western philosophy
– and of course one of the most influential – is that in which David
Hume professes not to be able to perceive or discover "himself"
("*myself* "). But it is a passage not so much obscure as more complex
than its admirers ordinarily concede:

> For my part [he says], when I enter most intimately into what I
> call *myself*, I always stumble on some particular perception or
> other, of heat or cold, light or shade, love or hatred, pain or
> pleasure. I never can catch *myself* at any time without a perception,
> and never can observe anything but the perception.[1]

Hume does not, in an obvious sense, wish to deny the usual – shall we
say pre-philosophical – affirmations in which people suppose that they
do "catch" themselves *in* catching their perceptions, emotions, sensations,
thoughts and the like. He insists only that he never catches "myself" as
a distinct and separable (Lockean) "idea" or (Humean) "perception"; he
catches only "perceptions" as of heat or cold.[2]

Part of the complexity of Hume's thesis rests with the empiricist
notion that perceptions as of heat or cold, love or hatred, are actually
sensory or reflective particulars of some sort distinct from the putative
self to which they "appear"; and part of its complexity rests with Hume's
worries about the propriety of claims of numerical identity, in particular
of personal identity and even of the identity of physical bodies. In fact,
very much the same argument against personal identity is mounted, by
Hume, against the identity of physical bodies – and rests on the alleged
difference between any changing sequence of perceptions (or, more
precisely, of "impressions" and their more-or-less congruent "ideas")

and the attribution of a unified, self-identical object (self or body) that persists through time and change.[3] But if Hume meant, in challenging the perception of ourselves, to challenge for the same reason the ordinary sense in which we perceive physical objects, then the familiar eliminative force of Hume's remark fails at once, since the elimination of physical bodies is known to be a remarkably strenuous, if not an altogether unrewarding, undertaking, certainly not one that rests on experiences thought to be as elusive as catching "myself" reflexively. The usual objections to macroscopic physical objects depend for support on certain forms of scientific realism (for example, on Wilfrid Sellars's views[4]) that claim to be conceptually favored by the rigorous features of the physical sciences; they are not normally thought to depend on anything like the sense in which Hume claims never to perceive "myself ... *without* a perception." Or, of course, the objection comes from the radical empiricist quarter that favors sense data and a constructivist or heuristic theory of macroscopic objects. But extreme empiricism is hardly any longer defended – in cognitive terms.[5]

In drawing attention thus to Hume's wording, we see that we must consider

1   whether not being able to perceive the self "without a perception" is equivalent to not being able to perceive the self;
2   whether "observ[ing a] perception" could ever be said not to entail, or would actually preclude, the existence of the self (an "observing" self), or could ever as such preclude the perceivability of the self; *and*
3   whether observing (or "catching") the (one's) observing (or one's perception) of a "perception" does not effectively entail the perception of the self (which Hume apparently denies occurs).

The beauty of introducing these complications by way of a review of Hume rests in large part with the naive and persistent force of 1–3 as reflections on ordinary experience, without prejudice to empiricist or rationalist proclivities or, for that matter, to any of the extraordinary uses to which Hume has been put in recent discussions of minds and persons. In particular, it rests with our not needing (initially at least) to discuss at all the substantial nature of minds or selves in discussing the logical and cognitive significance of mentioning and referring to the self. It also rests with our not needing (again, initially) to call into question all the sanguine assumptions under which a Humean-like reflection on the nature of selves and persons is so singlemindedly pursued by its numerous aficionados.

The full benefit of abruptly launching into a review of Hume's account of the self remains to be supplied – remains even to be mentioned. But we may anticipate that to treat the sciences as cultural or human achievements, even where (as is so often true) their most extreme champions would eliminate the merely picturesque (perhaps even completely delusive) idiom of the psychological concerns of man, we should have to come to terms somewhere with what we mean by the "self" and with its role in the formation of any science. It seems natural, therefore, to begin by taking account of the most notorious philosophical effort to eliminate the self altogether – and that on the basis of a scrupulous attention to experience alone.

One cannot help noticing that Hume's discussion – and that of much of the Anglo-American philosophy of mind that favors Hume's questions – tends not to introduce in a prominent way (or at all) distinctions regarding the societal, historical, and biological dimensions of a ramified theory of human minds and selves or, for that matter, distinctions regarding possible preconditions structuring our reflexive .theorizing about our own nature – whether critical, phenomenological, existential, ideological, pragmatic, praxical or perspectivist. To remark how narrow the Humean-like speculation must be is to be sufficiently forewarned in this regard. It provides a strategic beginning nevertheless – at least for the reason that that most refined caution known as Humean skepticism threatens (quite amiably) to dismiss, in eliminating selves, the very concern with the *analysis* of selves and persons (if not also of minds). For the Humean undertaking means to replace the analysis of selves with the analysis of certain habitually misleading ways of speaking and thinking of selves – always "skeptically" controlled, in the sense that its upshot is simply that selves or persons (also, physical bodies, on Hume's view) are no more than "fictions" to be explained by attention to certain inveterate linkages among our notions[6]

Surprisingly, countermeasures often nicely match the critical gauge and economy of such subverting maneuvers, without necessarily impoverishing our sense of the very field to be canvassed. On the contrary, their adequacy under such imposed economies may actually help to map our larger options. We enter the contest, then, in the simplest way but not simplistically, for, if the Humeans were right, the need for a certain form of serious analysis (though not of course for others) would be entirely obviated.

One contemporary author, Stephen Stich, takes it that Hume was primarily occupied with the "ontological" question: "What sort of thing or stuff is a mind or mental state? How is it related to matter?"[7] But this is a mistake. It suggests more the eliminative materialist's motive for invoking Hume than it does Hume's own inquiries; it fails to mark the essential pivot of all of Hume's skepticism about minds and selves – namely, the pervasively distorting effect he claims the idea of numerical identity works upon all our pertinent ideas. Stich therefore misses the essential subtlety embedded in the puzzle of identity – namely, that the question of the identity, the numerical identity and reidentifiability, of selves, or of the self-identity of persons, is a question logically quite distinct from that of the "stuff" of minds or selves or persons and may be addressed and resolved without prejudice to Stich's ontological question and may even by answered favorably *if there is no distinctive "stuff" of minds, selves, or persons* to be discerned. Very simply put, this means that the rejection of ontic dualism or the advocacy of some monism (or materialism for that matter) has, as such, absolutely no direct bearing on the elimination of selves, persons, or minds. Stich focuses primarily on the second issue *via* the first, hoping to accommodate what cognitive science may preserve from our "folk psychology"; but his entire venture presupposes an eliminative maneuver along Humean-like lines that he does not actually supply.

The entire fashionable current of eliminative materialism, that favors the outright denial or repudiation of minds, selves, and persons – because it rejects all forms of ontic or Cartesian dualism, all flirtation with nonmaterial substances – is a flat and obvious *non sequitur*. To grasp the point is to grasp the suggestiveness of Hume's way of putting his puzzle.

To press that advantage more argumentatively: Derek Parfit, who may not unreasonably be thought to have attempted the most systematic contemporary extension of Hume's theory of the identity of selves or persons, therefore commits the counterpart mistake to Stich's; for Parfit takes it that:

(i)   questions of the numerical identity of persons can always be dismissed in favor of questions of the physical and psychological continuity of physical and psychological states (or "events");

(ii)   psychological states (or "events") are ascribable without reference to selves or persons or counterpart entities; and

(iii) the denial of (i) and (ii) *is tantamount to* the advocacy of ontological or Cartesian dualism.

Parfit's mistake lies in his having somehow supposed that resisting the reducibility of the question of numerical identity (with respect to persons) to the question of certain psychological and physical continuities is either the same question as, or one that is tantamount to, or one that entails, resisting the reducibility of the nonmaterial "stuff" of minds or selves or persons (with respect to the "stuff" or substance of physical bodies). Doctrines (i)–(ii) are certainly Humean-like, but (iii) is not. In any event, Parfit's is a double mistake; for identity of selves or persons is a matter logically distinct from that of the continuity (or contiguity) of psychological or physical states, and neither of these is the same as that of the "composition" of selves or persons. But it is because of (iii) that Parfit rejects "the Non-Reductionist View, [the view that] a person is a separately existing entity, distinct from his brain and body, and his experiences. On the best-known version of this view [as he goes on], a person is a Cartesian Ego." This is also why Parfit claims that "the Reductionist View is ... the only alternative."[8] Parfit shares Hume's suspicions about identity as well as a fondness for all empiricist-like economies regarding the notion of selves or persons – for which (on Hume's argument) there are no simple or originating reflexive impressions. But conflating the two issues remains a blunder nevertheless. They are simply different matters.

Hume's original passage invites us, in context, to examine the following questions – which Hume himself either does not discuss at all or does not discuss satisfactorily:

(a) what the sense of "self" is in just those circumstances in which "perceptions" are said to "appear" to one;

(b) what the sense of "self" is in just those circumstances in which one is reflexively aware of perceptions appearing to one; and

(c) whether there is a sense of the self-identity of persons or of the unity of self to be drawn from the pertinent cases falling under (b) that is more difficult to eliminate on Humean grounds than that fitting the pertinent cases falling under (a).

It may be reasonably argued that, whatever our sympathies for Humean empiricism or the conceptual economies that Hume's program seems to favor, (b) cannot be managed in quite the same way as (a); Hume fails to grasp (or at least to address) the full force of (b); and (therefore) the question of personal identity – resting primarily on the analysis of (b), though without needing to neglect (a) – cannot be resolved straightforwardly in the eliminative manner: that is, either by replacing (in the logical sense) identity by continuity or by reducing (in the ontological sense) ascriptions of identity (where such ascriptions are meant to resist physicalist reduction as well) to the advocacy of a dualist or Cartesian doctrine. (Of course, it must always be admitted that the precise characterization of Descartes's theory is open to considerable dispute. It is not at all safe to say that Descartes was himself a conventional ontic dualist – though he may have been.⁹) A moment's reflection will confirm that our three questions (a)–(c) are not at all peculiar to Hume's extreme empiricism and may in fact be pressed into service in querying any reasonably responsible enlargement of the theory of sensory experience beyond Hume's strictures. We are in effect preparing the ground for a quite general account of minds, selves, and persons. It seems fair to say, therefore, that we need pay no particular attention here to the various kinds of experience we may wish to include in a ramified theory of mind. The first order of business is to secure the legitimacy of speaking of selves and minds.

Hume begins his discussion of personal identity, remarking: "There are some philosophers, who imagine we are every moment intimately conscious of what we call our SELF; that we feel its existence and its continuance in existence; and are certain, beyond the evidence of a demonstration, both of its perfect identity and simplicity." But he adds at once that "Unluckily all these positive assertions are contrary to that very experience, ... nor have we any idea of self, after the manner it is here explain'd. For from what impression could this idea be deriv'd?"¹⁰ It is very important to understand just how restricted – but pointed – Hume's charge is. He does not deny that we have an idea of self; he does not deny that it has a generative cause; he does not even deny that its cause may involve reference to original impressions. Here is what he very carefully says:

> It must be some one impression, that gives rise to every real idea. But self or person is not any one impression, but that to which our several impressions and ideas are suppos'd to have a reference. If any impression gives rise to the idea of self, that impression must continue invariably the same, thro' the whole course of our lives; since self is suppos'd to exist after that manner. But there is no impression constant and invariable.¹¹

The barest glance shows that what Hume rejects is the notion that the idea of self is due to any one impression, is simple, or is invariable through the whole course of life; and what he means by this is merely that the qualitative experience of any life is constantly changing. Surely, no one – not even one who (*contra* Hume) believed he directly experienced his unchanging self or soul – would deny that his life experiences within the range of heat or cold, love or hate, or the rest of Hume's mentioned items, did constantly change.

The truth is that Hume denies only that we have an idea of self *of a certain sort*: it cannot be an idea that is invariant and it cannot rest on an invariant impression, because "there is no impression constant and invariable" *and* because one never perceives the self "without a perception" (or some other impression or idea) which is one in a chain of perceptions that constantly "succeed each other, and never all exist at the same time."[12] The *usual* argument for the persistent identity of the self requires, on Hume's view, an "impression constant and invariable." There is none because experience is variable *and* what we perceive are various perceptions, which therefore form a "multiplicity" rather than a "unity";[13] *and* because *any* change in a thing "absolutely destroys the identity of the whole, strictly speaking."[14] So the mere inconstancy and multiplicity of experienced life and the uncompromising strictness of numerical identity logically preclude *a suitably generated* (non-fictionalized) idea of personal identity. The idea of identity must cover the multiplicity of experience; hence, *it cannot be due to a simple, original impression unchanging over an entire lifetime*. But that is not to say that we lack an idea of identity over a lifetime. It is only to say that it is a fictional identity.

Now, then, anyone who would build on Hume's argument, or who would claim that Hume's argument is more-or-less paraphrasable (without loss of power) from Hume's original empiricist idiom to a less unwieldy one, would have to claim that Hume's doctrine fairly captures our pre-philosophical sense of experiencing the self. This means that, in countering Hume, *we* need not restrict ourselves to considerations merely internal to Hume's text: for that would hardly explain the attraction of the thesis in an age in which empiricism has been all but effectively dismissed. We may, not unreasonably, introduce ampler notions of reflexive experience, provided that they do not insinuate the offending, the obviously question-begging, doctrine. Once again, the charm of pursuing matters in this way is that it permits us to sort certain common saliencies regarding *minds, selves*, and *persons* in a way that puts considerable pressure on contemporary theories that mean to eliminate utterly (rather than to reduce, say, in physicalist terms) minds, selves, or persons, either all at once or separately. For example, it is plain that Parfit means to eliminate selves and persons (on Humean-like grounds) but not minds; and that, in disputes about artificial intelligence (AI), possibilities are often broached in which machines are said to be intelligent but to lack minds – or to lack minds but to remain, because they are intelligent, artificial persons. Alasdair MacIntyre, for example, remarks that "machines must be mindless" because they do "not possess a genuinely human consciousness" (with which they are, on his account, being compared); although, as automata, they do possess what may be called (humorously following Gilbert Ryle) "Cartesian consciousness and Cartesian intelligence" – some sort of "solipsistic and ghostly consciousness." Perhaps then, on that view, they may also function as ghostly persons.[15]

In any case, we are here canvassing views about minds, selves, and persons in a way that will feature our salient experience as a strongly pertinent (but admittedly modifiable) constraint on what we should admit as an adequate theory regarding familiar phenomena and apparent entities. Hume serves us well, because he professes to stay as close as possible – in certin respects – to that experience, because he means to be as economical as possible in formulating his theory, and because there is a very strong eliminative current in recent work in the philosophy of mind and the cognitive sciences that is distinctly attracted to Hume's way of proceeding (even if it is not in complete accord with Hume's extreme brand of empiricism).

The most strategic pressure applies to the identity of selves – which, for instance, Parfit attacks fully in Hume's manner and spirit, though not as an empiricist. For, in the first place, Parfit favors "Humes's comparison. Persons are like nations, clubs, or political parties."[16] Secondly, in advancing the (his) Reductionist View, Parfit provides an analogue of Hume's empiricist account (without embracing empiricism) by extending the use of "the ordinary sense of the word 'event' so that it will cover the range of use usually covered by "mental state": the reason is that, by that device, psychological phenomena may be redescribed "in impersonal terms" – "Persons would be mentioned here only in the descriptions of the *content* óf many thoughts, desires, memories, and so on. Persons need not be claimed to be thinkers of any of these thoughts."[17] Thirdly, again very much in the spirit of Hume, Parfit holds that, "if we believe that identity is what matters, it is natural to believe that identity is *always* what matters. If we admit one exception, it may be hard to justify rejecting others." Given the difficulties facing this view, Parfit concludes "that *personal identity is not what matters* [to systems in which 'events' of belief and desire and the like obtain, to systems occupied with morality and rationality]. *What matters is Relation R*: psychological connectedness and/or continuity with the right kind of cause" (where "The right kind of cause could be any cause").[18]

Parfit's account shows quite straghtforwardly how Hume's theory can be readily adjusted to a contemporary idiom without subscribing to Hume's own empiricism; it also shows, therefore, the relevance and usefulness of addressing Hume's theory as well as the sense we thereby gain of a certain set of extremely strategic questions regarding the theory of minds, selves, and persons – just those in fact, (a) – (c), that we have already elicited from Hume's account.

One crucial shift, however, moving from Hume to Parfit, isolates a fundamental feature of Hume's account that contributes to its apparent persuasiveness, *that cannot be convincingly retained in any modern, non-empiricist theory – a fortiori*, cannot be retained in Parfit's. The shift, which concerns the notion of personal identity, registers itself in a change of terms. It is a change that, as we shall see, generates a decisive difficulty for *any* eliminative view of selves or persons based on alleged anomalies regarding claims of personal identity. (Other eliminative strategies are thus far untouched.)

In his argument for construing personal identity as a "fiction," Hume emphasizes that to defeat his thesis one would have to identify some one impression that produces the idea of self, that would be "invariably the same, thro' the whole course of our lives; since self is suppos'd to exist after that manner." Since there is none such, Hume rejects the account of his opponents; although, as already remarked, he does not deny that we do have an idea of self. Parfit, however, as we have noticed, insists only that, "if we believe that identity is what matters, it is natural to believe that identity is *always* what matters." One exception will make it difficult if not impossible to refuse others. But, *since Parfit is not an empiricist* (and since hardly anyone is nowadays), Parfit cannot (nor can we) adhere to Hume's impossibly strict (and utterly useless) notion of identity – that is, to the notion that any least change in a thing "absolutely destroys" the identity of that thing. The question remains whether, for any suitably adjusted concept of identity – in particular, for one that concedes the persistence of numerical identity through change in time – there are anomalies or puzzle cases that exhibit the "one exception" Parfit requires. Hence, it is useful to remark that, on Parfit's view, "all Reductionists would accept [the thesis:] A person's existence just consists in the existence of a brain and body, and the occurrence of a series of interrelated physical and mental events."[19] For, by advancing this claim, Parfit makes it quite clear that he does not renounce ordinary numerical identity through change, *with respect to bodies and brains*, and that he reconciles his usage in this respect with the extended use of "event" in the sense already noted.[20]

II

It will prove advantageous to think of Hume and Parfit together, simply because both offer eliminative accounts of selves in similar ways: Hume, by appealing to the most extreme and consistent empiricism; Parfit, by providing contemporary analogues of Hume's arguments within the terms of an adjusted materialism.

Hume does not construe mental states as the states of *some* suitable referent – body or mind or self or person – simply because, on his view, "all distinct ideas are separable from each other" and because, construing ideas as particulars, he is led to pronounce this famous doctrine: "setting aside some metaphysicians ..., I may venture to affirm of the rest of mankind, that they are nothing but a bundle or collection of different perceptions, which succeed each other with an inconceivable rapidity, and are in a perpetual flux and movement."[21] Hume, therefore, does not use a *predicative* idiom in the language of his own philosophical theory, as he quite naturally does when he speaks as an ordinary person or when he seeks to explain the "habits" of thought and speech that are generally favored. Of course, physical bodies are also "collections,"[22] on Hume's view, which shows that Hume supposes he has correctly replaced the predicative idiom *common* to speaking of selves and bodies, and supposes that simple impressions and ideas are simply particulars "such as admit of no distinction nor separation."[23]

However difficult the question and however vague Hume is about it, a simple idea is a particular that may be named for its particular (simple) quality. Ideas, therefore, compress in some inexplicit way what *we* should otherwise distinguish as a denoting and a predicative function. So seen, there *is* a sense in which Hume accommodates the usual predicative idiom in accord with which mental states or experiences are predicated *of some* suitable referent; it is only his extreme empiricism that gives the bundle theory of the mind (as of physical bodies) its apparently radical cast. It is certainly not that Hume holds that mental states (or "events," to use Parfit's term of art) may be identified "impersonally" – on a "no ownership" basis (as Parfit apparently believes);[24] it is rather that the analysis of apparent ownership yields to the compositional strategy Hume pursues everywhere.

When, therefore, Parfit, following G. C. Lichtenberg, says that we could provide "an *impersonal* description" of our experiences, that "we could fully describe our thoughts without claiming that they have thinkers," [25] *he* is not arguing in Hume's manner (since *he* holds to a predicative idiom); also, as it happens, he does not actually demonstrate – for example against P. F. Strawson's well-known objection – that his claim is coherent and viable. Strawson had made a very strong case against the coherence of the no-ownership theory, on conceptual grounds – to the effect that the mere effort to identify and individuate "experiences" logically requires, contrary to the no-ownership thesis, the use of "'*my*', or [an] expression with a similar possessive force."[26] On Strawson's argument, the connection between a putative subject of experience and "that" subject's experiences cannot be a contingent connection at all.[27] Parfit alludes to Strawson's argument but never actually addresses it. "If these arguments are correct," he says, "they might refute my claim that we could redescribe our lives in an impersonal way. *Because these arguments are at a very abstract level, I shall hope to discuss them elsewhere*"![28]

Parfit evidently *takes* it that mental experiences can be coherently predicated of the body (perhaps of the brain), since he has already stated that "A person's existence just consists in the existence of a brain and body, and the occurrence of a series of interrelated physical and mental events." This, clearly, fails to engage Strawson's charge. But it has a further, even more profound, significance; for Parfit never quite manages to explain *what* the relationship is between his "series of interrelated physical and mental events" and "the existence of a brain and body," which, together, is "just" what a person's existence "consists in."

There are only two options available to Parfit; but neither will secure the "impersonal" redescription satisfactorily. On the one, experiences are impersonal because they are ascribed to physical bodies or brains: selves and persons drop out, are eliminated (not reduced). On the other, experiences are simply mental *events*, particulars of a distinctive sort: such that, although selves or persons *may* be implicated in the "content" of particular experiences, experiences themselves need not be said to be "had" or "possessed" by a person or self, and possession is not as such entailed by reference to persons in describing the *content* of a particular experience.

If *physicalism* were true – the thesis (in Thomas Nagel's terms) "that a person, with all his psychological attributes, is nothing over and above his body, with all its physical attributes"[29] – then, of course, *persons would not be eliminated* by merely *reducing* them to bodies. What would result, what results instead, what "repels" Nagel, is that physicalism "leaves out of account the essential subjectivity of psychological states," that "nowhere in the description of the state of a human body could there be room for a physical equivalent of the fact that *I* (or any self), and not just that body, am the subject of those states."[30] Nevertheless, although physicalism "repels" him, Nagel is (at least at the point of this discussion) "persuaded of its truth."[31] *His* puzzle – in effect, Herbert Feigl's before him[32] – is fixed on merely trying to make viable the "dual aspect" idiom through which we speak of "mind" and "body" within "the same universe." In Nagel's own words' "There is something deeply suspect about the whole enterprise of fitting subjective points of view smoothly into a spatiotemporal world of things and processes, and any dual aspect theory is committed to that goal and that picture – that picture of appearances as part of reality."[33]

But to put matters thus is to impoverish quite unnecessarily our grasp of an even deeper puzzle: for it is *not* (or, not obviously or unavoidably) a question of *fitting* "subjectivity" into an "objective" world already more-or-less adequately characterized in physicalist or physicalist-like terms; it is rather a question of providing a fresh characterization of the *world* within which, by various investigative strategies, we are faced with the saliencies of the mental and the physical. Nagel's way of putting the question shows no recognition at all of the quarrel, within Western philosophy, among "objectivist," "phenomenological," and "post-structuralist" or "deconstructive" orientations. Without a grasp of the significance of that historical theme, it would (as, by default, both he and Feigl somewhat ruefully admit) be quite impossible to resolve the mind/body problem.

To remind ourselves how an attack on *that* issue had already been prepared and attempted in recent Western philosophy – and to hint at how it might be restored or pursued within the Anglo-American tradition – we must go to Merleau-Ponty. It is surely at least in his work that one finds vouchsafed the minimal clue to *any* viable solution that would preserve the "subjective." For Merleau-Ponty had attempted to defeat the Cartesian vision – and with it the entire "objectivist" tradition spanning Descartes and Kant most prominently[34] – by enriching the concept of *body* biologically, so that considerations of subjectivity, intentionality, mentality and the like *would be directly ascribable to the "bodily."* The maneuver is what counts: one may, after all, agree or disagree with Merleau-Ponty's systematic efforts. But this surely is the essential clue (if there is any at all) to the resolution of Nagel's bafflement; it must also be, if there is such a thing as a third option to set beside the would-be exhaustive and exclusive pair of options Parfit offers (the Reductionist and the Non-Reductionist), just such an option, and a better than the others.[35] Merleau-Ponty's brilliant suggestion is, of course, that the biological cannot be reduced to the physical; that "body" in the sense of "lived body" (*le corps vécu*) cannot have the same sense it has in the expression "physical body" as used in normal physics; that the Cartesian disjunction between the mental and the physical fatally equivocates and suppresses the distinction between the physical and the biological; that the psychological cannot be separated from the biological; and that, to put the point in its boldest and most provocative form, *the intentional is already a feature of the bodily.*

Without wishing to pursue this theme too much in depth here, we may remark at least two strategic applications. First, on Merleau-Ponty's proposal, it must be the case that, in rejecting Franz Brentano's use of the intentional idiom, W. V. Quine fails to consult the possible need for intentional considerations at the biological level (meant of course to accommodate the humanly psychological as well) and restricts himself – as so many others have done – to the barely mental, to "appearances" (to use Nagel's telltale epithet). There is no reason to think that the elimination of the intentional will work even at the thinnest mental level at which human thought and experience are acknowledged; but the force of Merleau-Ponty's idea would have required Quine to have discussed the relationship between the physical and the biological before repudiating the intentional.[36] It is perfectly clear, even in Quine's most recent statements, however, that he makes no pertinent distinctions in *offering as instances of "physical bodies" and "physical objects," dogs, sticks, stones, mothers, other people, mountains, chairs, and wrecked ships.*[37] Consequently, the authority of Quine's well-known rejection of the intentional hangs on an argument Quine does not himself supply: the psychological may, ultimately, not be separable from the biological, and it may be less easy to deny a real structure of intentionality to a biologized psychology than to a cartoon psychology.

A second application may be drawn from Noam Chomsky's work. Chomsky of course regards himself primarily as a biologically oriented cognitive psychologist, though a scientist strongly disposed to favor the eventual adequacy of the physical sciences. Nevertheless, Chomsky remarks,

> It is perhaps worth stressing ... that the notion of "physical world" is open and evolving. No one believes that bodies are Cartesian automata or that physical systems are subject to the constraints of Cartesian mechanism, or that physics has come to an end. It may be that contemporary natural science already provides principles adequate for the understanding of mind. Or perhaps principles now unknown enter into the functioning of the human or animal minds, in which case the notion of "physical body" must be extended, as has so often happened in the past, to incorporate entities and principles of hitherto unrecognized character. Then much of the so-called "mind – body problem" will be solved in something like the way in which the problem of the motion of the heavenly bodies was solved, by invoking principles that seemed incomprehensible or even abhorrent to the scientific imagination of an earlier generation.[38]

Here, the prospect is offered that the meaning of "physical" may have to be enlarged to accommodate the findings of a scientific linguistics. *If* Chomsky were of a mind to, he might therefore have accommodated an adjustment like Merleau-Ponty's within the scope of the "physical" – as that presumably applies to physics; the reason would have been that, in an anticipative spirit, Chomsky was prepared to treat the genetics of linguistic phenomena as falling well within the scope of physics. Be that as it may, *within* the extension of "physical" as he intends the term, Chomsky introduces an important empirical possibility "that cannot be ruled out a priori; our minds are [perhaps] fixed biological systems with their intrinsic scope and limits."[39] But what if our minds are biological networks of some sort that are not thus fixed? And what if, similarly, selves or persons are specified in terms of biologically grounded functional aptitudes? Chomsky does not discuss the possibility. When (again like Hume) he appears to consider human subjects, he reverts only to the "grammatical function 'subject of'."[40] He therefore mutes all questions of variable context, human history, social experience, improvisation, consensus and interpretation – factors that would enhance our sense of the ineliminable functioning of selves or persons as well as of the implausibility of construing linguistics as an autonomous discipline.

Merleau-Ponty's clue is, however, no more than a promissory note. Even so, it plays a significant role because it exposes at a stroke one of the deepest prejudices of the Anglo-American approach to the analysis of minds, selves, and persons. It is no more than a promissory note, because the fate of the third option (recalling Parfit's schema) ultimately depends on a satisfactory account of the relationship between biology and physics and between biology and the human sciences. For our particular purpose, it may be admitted that we have here been drawn off on a most engaging detour. We must return now to the main track – but we must surely promise ourselves a longer trip along the same road.

To return then: Hume eliminates selves and persons because of empiricist strictures on numerical identity, not because he favors the no-ownership view; and Parfit favors the no-ownership view, but not because of any logical or empiricist strictures on identity or because he has any doubts about the satisfactory functioning of identity with regard to changes over time (for instance, with regard to physical bodies). Strawson claims that mental predicables – "experiences," if you like – must, to be identified at all, be indexed (must necessarily be indexed) by the use of "my" or some suitable surrogate. Parfit does not meet this objection, which at least signifies a possible function for selves and persons if there are independent reasons (as, even on Humean grounds, there are) for speaking of such. (There are, incidentally, complications that Strawson does not consider – regarding referents that exhibit "ownership" but are inferior to selves.) For his part, Nagel is nonplussed by the difficulty of showing how the physicalist reduction fails; he seems to be convinced that the eliminative maneuver (Parfit's, for instance) can be shown to fail if he can show why physicalism fails. But the missing reason for the latter is already implicitly afforded by Strawson – namely, that the purely logical role of predication (presumably common to discourse about bodies and selves) has nothing as such to do with conceptual restrictions on discourse about "experience" (call it the question of "possession" as opposed to "predication"[41]): we may predicate having experiences of some person; but the "having" of an experience is (on a view sympathetic to Strawson's) a necessary logical encumbrance on an experience's being an experience and on its being identifiable as the particular experience it is.

Applying the distinction to Parfit's argument, we may say not only that Parfit has not met Strawson's objection, but also that, if he had managed to show that experiences could be predicated of bodies or brains, he would still have had to show that, in eliminating reference to persons or selves, *he would have thereby eliminated persons or selves.* Notice that, however uncomfortable he may have been about physicalism, Nagel does not conclude that there are no persons or selves on the physicalist thesis; persons and their properties (on that view) simply are bodies and their properties. Generally speaking, the same thesis appears in such different versions of the physicalist thesis as J. J. C. Smart's and Donald Davidson's.[42]

Hume does not predicate mental states of physical bodies – in whatever sense he accommodates predication (or what would have been predication, absenting the stringencies of empiricism). He *does* predicate mental states of selves or persons, provided we understand selves or persons not to be (qualitatively) unchanging over an entire lifetime and provided we understand selves or persons to be mere fictions (if supposed unchanging). For, in the first place, it is essential to Hume's position that "all sensations [impressions] are felt by the mind, such as they really are"; that is, they have no "existence DISTINCT from the mind and perception" and we have no grounds for attributing "a CONTINU'D existence to objects ... specifically different from our perceptions."[43] On this reading, Hume actually opposes the no-ownership thesis – this is his answer to question (a) in the preceding section. Secondly, in discussing the passions, Hume explicitly says that "pride and humility, tho' directly contrary, have yet the same OBJECT. This object is self, or that succession of related ideas and impressions, of which we have an intimate memory and consciousness."[44] Hume does not deny memory.

He "merely" insists that it is "impossible to recall the past impressions, in order to compare them with our present ideas, and see whether their arrangement be exactly similar"; so the difference between memory and imagination (normally) "lies in its [memory's] superior force and vivacity."[45] Within the limits of empiricism, therefore, Hume admits that one can be aware of comparing two ideas and that one can be aware of remembering an idea: one can be aware of being one and the same self that remembers an idea and has (now) the idea of remembering that idea. But this, however truncated, *is* part of the very notion of the numerical identity of selves.

There is no sense, in Hume, in which memory is a mere fiction. On the contrary, "impressions or ideas of the memory" are as entitled to the designation "*reality*" or "*realities*" every bit as much as are the impressions of the senses.[46] Hence, impressions and ideas of "reflexion" not only require the receptivity of the mind in the same sense as do impressions of "sensation";[47] they also require, because of the iterative nature of "reflexion" itself, some minimal notion of what would otherwise have been called self-identity – were it not the case that Hume reserves that term for the fictional identity of a self "thro' the whole course of our lives." Here, then, is Hume's answer to our questions (b) and (c).

Consequently, *if* we replaced Hume's useless notion of identity with a normal one that accommodated identity through change, there would be no question that Hume's concessions would amount to an admission of a persisting self – *whatever its substance and whatever our analysis of its nature*. The truth is we have no separable or distinct idea of self (on Hume's view) because *no* impression or idea, however separable and distinct from any other impression or idea, *is* separable from the recipient mind: apparently, even the simplicity of simple impressions accommodates this complication, for the "linkage" is definitely not a *relationship* of any sort. Hume does not discuss the matter, but the very opening line of the *Treatise* states that "all the perceptions of the human mind resolve themselves into two distinct kinds, which I shall call IMPRESSIONS and IDEAS."[48] In fact, reminding ourselves of Parfit's use of Hume, even if "the soul [is] properly [compared] to a republic or commonwealth," Hume admits that "the true idea of the human mind, is to consider it as a system of different perceptions or different existences, which are link'd together by the relation of cause and effect, and mutually produce, destroy, influence, and modify each other." Furthermore, "Had we no memory, we never shou'd have any notion of causation, nor consequently of that chain of causes and effects, which constitute our self or person." Quite naturally, therefore, Hume concludes "memory does not so much *produce* as *discover* personal identity, by shewing us the relation of cause and effect among our different perceptions." This is certainly Hume's persuasion, though he retreats at once to his special brand of skepticism regarding "all the disputes concerning the identity of connected objects" – which, he says, are "merely verbal" (or fictional).[49]

The point of this extended review of Hume's position is a double one: first, it confirms that the most extreme empiricism, which appears to eliminate the self, really does no such thing – it rather attenuates every pertinent admission by applying impossibly (unworkably) heavy strictures on the language of numerical identity; secondly, it confirms that the most extreme eliminative strategy (Parfit's) cannot avail itself of any analogues of Hume's argument about the fictitious identity of the self – it cannot by such means demonstrate that the putative numerical identity of the self can be viably replaced by reference to the continuity of psychological states (or "events"). We may reasonably conclude, therefore, that there is no known way to eliminate selves *by attention to the formal features of identity applied to the barest notion of "experience."* The strength of our argument lies precisely in its *not* resting on any ramified analysis whatever of the nature of mental life, the substance of minds or selves, dialectical quarrels abut problematic attributes such as intetionality, or the systematic unity or bifurcation of the sciences.

Parfit, as already remarked, has only two options befre him when he commends eliminating selves. The first fails because, if experience is attributed to physical bodies, it does not follow that selves have been eliminated – they may merely have been reduced to bodies (in the physicalist's manner, say). The point is that numerical identity would still apply to *whatever it was* that selves would have been reduced to. Parfit is not altogether explicit on this matter. Unlike the physicalist, he says (as we have already remarked) that "all Reductionists would accept [the thesis:] A person's existence just consists in the existence of a brain and body, and the occurrence of a series of interrelated physical and mental events." He does not reduce mental events to physical events, and he does not explain the relationship between mental events and the existence of a brain and body. By analogy at least, if physical events are, in some sense, "parts" of a functioning body (or brain, itself a functioning "part" of a body), then mental events are also "parts"; or perhaps both mental and physical events are "parts" of larger systems in which body and brain function – and thus are not narrowly "parts" of a body or brain. But it is easy to see that the idiom of "events," which Parfit introduces to forestall rejection of the no-ownership thesis, does not as such preclude or even directly affect the issue of the numerical identity of physical bodies *or* of selves or persons, and does not actually address the no-ownership thesis. It is merely a place-holder for that thesis; the convenience of the "events" idiom presupposes an argument favoring no-ownership. None, however, is supplied. This means that Parfit's second option either fails (given Strawson's objection) or is never really pursued at all. For our purposes, it is a reasonably forceful conclusion to draw to affirm that *the elimination of selves or persons, on identity considerations, requires an independent defense of the no-ownership thesis.*

This finding may be considerably strengthened. When Parfit says "that personal identity is not what matters. What matters is Relation R: psychological connectedness and/or continuity with the right kind of cause," he means us to take R in the no-ownership sense. That would lead at best to a stalemate. As it stands, it is an *obiter dictum*. But now, strangely, Parfit offers the following strong disjunction: either we should adopt the Reductionist View (which entails the no-ownership doctrine and which presupposes its independent defense) or we should adopt the Non-Reductionist View. The Non-Reductionist apparently holds (is made to hold, by Parfit) to the doctrine of the "Cartesian Ego" that "a person is a separately existing entity, distinct from his brain and body, and his experiences."[50] No doubt this *is* a possible version of a position opposing the Reductionist's. But it is certainly not clear that Parfit has (or why he supposes he has) offered us a pair of exclusive and exhaustive alternatives. Why should he not have admitted that a Non-Reductionist *merely opposes the no-ownership thesis or does so without (or without necessarily) insisting that a person is a separately existing entity, distinct from his brain and body?*[51] Why should he not have anticipated something like Merleau-Ponty's option?

Parfit assures us that his argument applies "to all peoples, at all times."[52] He goes on to say, "We have sufficient evidence to reject the Non-Reductionist View. The Reductionist View is, I claim, the only alternative. I considered possible third views, and found none that was both non-Reductionist and a view that we had sufficient reasons to accept."[53] Effectively, then, the two views exhaust the possibilities; all other seeming options either reduce to the one or the other or are aborted at the very start. Since, however, Parfit has not actually shown the no-ownership thesis to be viable, it is clear that he cannot have shown the Reductionist View to be "the only alternative" and he cannot yet have shown that the view (call it V) which holds (a) *that no-ownership is incoherent* and (b) *that the self is separable from the body and brain is not a viable position* is not itself a viable position. The only possible way in which he could have shown V to fail would have been to show that *no* pertinently sustained appeal to the numerical identity of selves or persons could be supported without contradiction. But *if* no-ownership is incoherent, *if* Parfit has failed to explain how to specify psychological "events" or "experiences" without reference to selves or persons, *if* all his would-be counterinstances to a coherent use of identity (*not* to the defense of what he calls the Non-Reductionist View) presuppose no-ownership, *if* the "events" idiom is simply an idiom of art for handling predication without prejudging the entitative theory of selves, then Parfit cannot have shown that V can be ruled out. We need not rush to subscribe to Merleau-Ponty's thesis, but at the very least a biologized account of selves or persons (not Cartesian Egos, not entities separable from body and brain) may well be a viable option. The point is that *all* of Parfit's marvelously intricate counterinstances to any appeal to the numerical identity of selves invoke or are meant to illustrate R ("psychological connectedness and/or continuity with the right kind of cause"); *but R entails no-ownership*. Hence, there is no reason to suppose that Parfit has satisfactorily discharged all versions of V.

III

We need to tidy up this inevitably sprawling account. Certainly, the complexity of attempting to fix acceptable notions of minds, selves, and persons can hardly be ignored; and, certainly, we have collected only a very small part of the required inquiry. But it is a most strategic part. On the argument provided, we may reasonably conclude that, if we admit experience, mental states or mental events or mental processes, consciousness, reflexive consciousness or self-consciousness – in the familiar sense in which human beings report, affirm, avow or similarly express or manifest what they experience or have experienced or are experiencing – then there is no known conceptual strategy by which to support the no-ownership thesis; and, if there is none, then *some* form of admission of selves or persons (or, possibly, living bodies or animals) cannot be refused, and personal identity (among humans) *is* "what matters." The bottom line is simply that predication and reference are conceptually coordinate and that exerience must be "possessively" indexed.

To say this much, however, is not to say how we should construe personal identity over an entire lifetime, so-called cases of multiple personality, reincarnation, hemispheric transplants, science-fiction fission cases, and the like. It is only to say, insofar as such considerations involve numerical identity (because no-ownership fails), *some* admission of self or person cannot be avoided. To reinstate identity is also not to neglect to distinguish between the *numerical identity of selves* and the *unity of selves numerically identified*; nor is it to prejudge the relationship and difference between *individuating selves* and *individuating human bodies*. These are entirely separate matters. It is not inconceiveable – as forgetting and remembering, tacit learning, habit, hypnotic suggestion, schizophrenia, dreams, parapraxes, akrasia, self-deception, quasi-memory, borderline ego identity, and the rest of the familiar supply of puzzling cases attest – that the conceptual need to invoke numerical identity could be reconciled with empirically available evidence that the "self" is or may be hierarchically and in other ways functionally compartmentalized and compartmentalizable. All we are claiming is the ineliminability of some suitably supple notion of self or person, once experience is conceded to be real. We need not yet analyze the substance or nature of selves, or say whether referents less complex than selves are compatible with ascriptions of the mental, or say how far the ascription of the mental may be attenuated in the direction of the primitively animate or the artifactually more and more complex; and *we have not even precluded the elimination of selves*. We have precluded it only on the condition of admitting *experience*. Construe the mental as MacIntyre and Putnam somewhat generously do – as not involving experience – and you may or may not opt for eliminating selves.[54]

This is why Parfit cannot avail himself of eliminative strategies that try to salvage – not by reductive or physicalist means – what (so far) has been (said to be) only "picturesquely" characterized in the experiential terms of our "folk psychology": that is, by replacing the predicative schema in which (Strawsonian) considerations of "possession" or "ownership" obtain by another, by any other, schema that preserves predication all right but *eliminates experience* and the need for such "possession" *and*, with that, eliminates the need to posit selves or persons.[55] Parfit *admits* experience. He favors "folk psychology."

So, presumably, does Daniel Dennett – at least up to a point. But, if so, then Dennett simply misjudges the complexity of his own undertaking when he affirms "The personal story [that is, descriptive remarks that treat things or systems as persons] has a relatively vulnerable and impermanent place in our conceptual scheme, and could in principle be rendered 'obsolete' if some day we ceased to *treat* anything (any mobile body or system or device) as an Intentional system – by reasoning with it, communicating with it, etc."[56] Dennett confuses the reductive and the eliminative projects. He conflates the project of "eliminating" persons, *having admitted* experience and mental states (with all their puzzles), with the entirely different project of *eliminating experience* altogether (though not necessarily the "metal") – which would have laid a foundation for eliminating selves or persons. The first is not possible, as we have argued; and the second is nowhere or not yet provided (though it must be taken seriously).

Paul Feyerabend once recommended a large measure of patience in considering the formulation of an adequate materialism (construed as the view, sufficient for Feyerabend's purpose, holding that "the only entities existing in the world are atoms, aggregates of atoms and ... the only properties and relations are the properties of, and the relations between such aggregates"). On that account, "experience, thoughts etc. are not material processes" at all and do not enter into casual relations of any kind. They are simply not real, and there are no facts regarding them. We are urged by Feyerabend to consult the "facts," not mere "beliefs" (apparently linked to our conceptual addiction to experiences and thoughts), which, somehow, are to be isolated by attention to real casual relations but not to mere meanings.[57] So Feyerabend is definitely not subject to the same sort of objection that dogs Parfit and Dennett; although, for his part, Feyerabend has failed to bring his project to completion. In any event, we are here merely fixing the scope and force of the argument so far supplied.

Perhaps, in all fairness, we should collect a final piece of evidence to assure ourselves that Parfit has genuinely misgauged the weakness of his strategy. We cannot hope to track down all of his most intricate cases; but we need not, either.[58] What's needed is easily afforded. Parfit asks us to consider that a new technology of artificial eyes (involving electrical patterns sent through the optic nerve), replacing blind natural eyes, would yield results that would be *just as good as* seeing."[59] Yes, of course. But when he goes on, on the basis of that example, to say, "Some people would regard division [as of one's brain, distributed to other living bodies] as being as bad, or nearly as bad, as ordinary death [but I would say] his reaction is irrational. We ought to regard division as being about as good as ordinary survival,"[60] he effectively cheats: he shifts from predication (the sight case) to numerical identity (the continued identity of the self); in context, "survival" can only mean "survival of one and the same person," unless (what is contrary to fact or at least contrary to what Parfit has established regarding the facts) the no-ownership thesis is true. If, then, all of Parfit's cases are damaged in this way – as they appear to be – Parfit *cannot* undermine doctrine V (as already sketched). *V, therefore, represents the least concession regarding selves or persons that it seems possible to make.* Only the utter elimination of experience could possibly vindicate the elimination of selves.

But the admisssion of V is entirely unencumbered by particular philosophical persuasions: it is entirely accessible to materialists and idealists, objectivists and phenomenologists, pragmatists and deconstructivists, and to those of any other conviction it may be thought important to consider. One may for instance explore it along the lines recommended in Julian Jaynes's extraordinary theory of the bicameral mind.[61] There need be no *a priori* prejudice against that. But, when Marcel Mauss declares that "the idea of the 'person,' the idea of the 'self' (*moi*) ... one of those ideas we think of as innate – was slowly born and grew through many centuries and many vicissitudes, to the extent that even today it is still hesitant, delicate, precious and requires further elaboration,"[61] we cannot be entirely sure (on the argument supplied) that Mauss's thesis is actually coherent. Mauss does add, it is true, "In no sense do I maintain that there has ever been a tribe or a language in which the word *'je – moi'* (I – me or self ...) did not exist and did not express something clearly represented."[63] But that may not be the same point as the first. Mauss may have been referring to the evolution of the self from sub-human animal sources, or he may have meant that the self is a fresh cultural achievement for each individual human animal, or he may have meant that the very *nature* of human selves continues to change through cultural epochs (which is Jaynes's thesis), or he may have simply contradicted himself. Furthermore, when Michel Foucault, in a well-known paper that construes the "author-function" of literary ascriptions as an artifact of certain seventeenth- and eighteenth-century notions of property, wishes us to understand that changes in the notion of an author are not necessarily entirely different from changes in the notion of a person, we are not told what the limits of such a concession are or what risk of incoherence it may be forced to face: "the author," Foucault says, "is an ideological product ... the author does not precede the works, he is a certain functional principle by which, in our culture, one limits, excludes, and chooses, in short, by which one impedes the free circulation, the free manipulation, the free composition, decomposition, and recomposition of fiction."[64] True, Foucault admits that "we can find through the ages certain constants in the rules of author-construction," that we can detect the author's "historically real function"; but it is certainly not clear that Foucault is himself conceptually entitled to make such a saving distinction in accord with his own more radically Nietzscheanized views of history and truth.[65] We are in danger here again of following a fascinating detour. But we have managed to suggest, nevertheless, the critical difference between the bare principle of *the identity of self and its inseparability from the admission of experience and from the biological, evolutionary, developmental, and cultural history of the self..* We have been concerned here with the first and not with the second.[66]

Having said all this, what, minimally, may be established in a conceptual way about minds, selves, and persons? We shall not explore the enormous possibilities that present themselves at once upon the strong installation of selves or persons – for example, along the lines of socializing and historicizing psychology or along the lines of characterizing mental phenomena as linguistically informed in various ways. It is perfectly clear that the entire movement of Anglo-American conceptions of mind, of the science of psychology, of the cognitive and information sciences, has tended to ignore – even to dismiss – these dimensions of the mental, in formulating its most characteristic theories. But we must bear these possibilities in mind in shaping our minimal conceptions.

There is no reasonable sense in which we may claim straghtforwardly to discover what we should correctly posit as *mind*, *self*, or *person*. We seek a reasonable and manageable convenience, hospitable to what we suppose to be the most fuitful lines of inquiry regarding very large, promisingly coherent accounts of the central questions of epistemology, the methodology of science, psychology, the cognitive sciences, the human and social sciences, linguistics, and the various interpretive disciplines. But we do not need to pursue such advantages directly; we need only to avoid characterizations that distinctly impoverish or distort our approach to what such inquiries may require or wish to entertain. One obvious recommendation suggests itself: *mind* may be taken to be the abstracted and aggregated nominalization (mythically, as a nominalization) of whatever distinct particulars would instantiate in instantiating being "minded," as by exhibiting, or by behaving in ways that involve, or by being capable of, what normally falls within the mental life of humans; and *self* may be taken to be the functional capacity (hence, itself a "minded" capacity) of whatever may be said to be thus minded, with particular attention to the self-identity of *what* is thus minded as well as to its "possession" or "ownership" of minded powers and attributes thus instantiated; or, else, self may be taken to be *that* which is so minded and which is, self-identically, the "owner" (in Strawson's sense) of such powers and attributes.

The great convenience of this suggestion is easily missed because it is so obvious and so simple. "Self" is defined only functionally and as a conceptually symbiotic notion linked (with reference to humans) with "mind"; and "mind" is specified in a completely uncontroversial way in terms only of its usual extension, entirely unencumbered by special theories regarding intentionality, privacy, *qualia*, the structure of thought and perception, or anything of the kind, treated (grammatically) in predicative terms only. (We had, we must remember, found that Parfit had failed to explain how his psychological "events" were to be related to body and brain – which, taken "together," he somehow construed to be what a person "consists" in.)

When "self" is treated entitatively but assigned only functionally (abstractly) specified minded powers, or when the "minded" is itself treated only functionally – as in accord with MacIntyre's suggestion – it is entirely suited to the anthropomorphized vocabulary usual in the description of certain computers. It would be a point of minor quarrel only whether such usage was an extension of the "standard" usage of those terms or whether it was not.[67] Certainly, as artifactual, computers (like artworks) invite such usage. The important point is that, although "self" is defined functionally, human selves cannot function except as the complex creatures they are – biologically, psychologically, societally, linguistically endowed and developed. What the relationship is between the human creature, the individual member of *Homo sapiens*, and the functional self we acknowledge within the space of human existence is, as yet, entirely open to dispute – but it is *not* open to the option of a detached, or detachable, purely abstract entity (such as Parfit had entertained and rejected and such as Merleau-Ponty had in a particularly powerful way shown us how to avoid).

It may be reasonable to identify, as one and the same, being a member of *Homo sapiens* and being (or functioning as) a self, but then again it may not. For example, it would not be obviously incoherent to hold (though it may be false) that persons or selves may be reincarnated from life to life; and it is similarly possible, as in extremely dissociative multiple-personality cases, to concede that there are functionally distinct multiple "selves" that may be concurrently assigned (in some complex, distributed way) to the same biological organism.[68]

By defining self functionally, we gain three advantages at least. First, we do not have to deny that actual selves are manifested or realized in and only in whatever is the "matter" or "substance" or "stuff" of the world; and we do not have to specify what the sciences will eventually decide is the nature of the natural world that supports the existence of selves (for example, what the relationship is between the "physical" processes of physics and astronomy and the "biophysical" and "biochemical" processes of the life sciences) or what, physically or biochemically, is necessary or sufficient for the admission of selves. (This points to the profound inadequacy of what has been termed *functionalism*, though *not* to any grounds for rejecting a functional account of self.[69]) In its central usage, then, mind or being minded must be *incarnate* – indissolubly so – and its mention as *merely functional* (as in speaking of computers) must be peripheral even if it does not entail a change of sense.[70] Secondly, we do not have to decide at this point whether or why the principle "one body, one self" should be sustained or abandoned or merely thought to be normal but not necessary.[71] And, thirdly, we do not have to decide at this point whether or why selves or persons are numerically identical with (certain) bodies or organisms or systems or are related to such entities in more complex ways – for instance, by some inseparably emergent development involving distinct embodying entities.[72]

The third question concerns the conceptual matching (the "adequation") of the nature and powers of selves and the minded attributes ascribed to them. In this connection, Daniel Dennett usefully mentions the following "danger" confronting "AI and other styles of top–down psychology" (psychologies that "factor" selves or molar systems that function similarly to selves):

> designing a system with component subsystems whose stipulated capacities are *miraculous* given the constraints one is accepting (e.g., positing more information-processing in a component than the relevant time and matter will allow, or, at a more abstract level of engineering incoherence, positing a subsystem whose duties would require it to be more "intelligent" or "knowledgeable" than the supersystem of which it is to be a part).[73]

On Dennett's view, both Freud's and J. J. Gibson's psychologies may rightly be charged with "miraculous capacities."[74] The problem is a real one, but the charge is difficult to make stick in cases such as Freud's and Gibson's. It may be much easier to charge "miracle" – *on* Dennett's criterion – for systems (such as Feyerabend's or Feigl's or Parfit's or Chomsky's or Stich's or Dennett's own) that confine the pertinent referents to physical attributes (in terms of physics or biology) and then, without adequate or convincing analysis, ascribe full-blooded mental or psychological powers and attributes *to* such systems. If what are putatively miraculous powers are ascribed to the functioning *sub-systems* of some functioning self, then, in principle at least, they may be ascribed to the (molar) self as well, even if not to other sub-systems of it – for instance, to conscious capacities. In this regard, Freud's notion of the unconscious is hardly miraculous even if it proves untenable. The essential question remains whether and in what way mental attributes can be consistently ascribed to purely physical systems – in a sense similar to that of Parfit's (and Dennett's) proposals.

The way we have put matters permits us to sort, though not to treat

altogether separately, the analysis of the nature or composition of selves
and the analysis of the distributed functional powers of selves. We must
admit, however, that selves, in the strong sense in which humans are so
characterized, may well be uniquely human. There is of course the
tantalizing fact that chimpanzees can be shown to be capable of a measure
of self-recognition (as in certain famous mirror experiments), and there
is the tantalizing fact or claim that, in achieving whatever mastery they
manage with regard to language, chimpanzees have been said to lie –
which certainly would have to count as more than the mere incipience
of self.[75]

The origins of self are clearly obscure. One is tempted to assign them
already to nativist or biological sources. There is a sense in which this
may be assumed in the views of language and conceptual competence
favored by Chomsky and Fodor. Chomsky, however, explicitly notes
the frontal question and turns it aside (quite reasonably, given his nativist
thesis) by refusing a disjunctive use between "knowing" English and (by
a term of art) "cognizing" (at some level at which we could not be said
to "know" in the ordinary sense) "the innate grammar that constitutes
the current state of our language faculty and the rules of this system
as well as the principles that govern their operation."[76] Also, in Jerome
Bruner's studies of the cognitive development of the child, the question
is turned aside. Bruner stresses that "pre-linguistic communicative acts
precede lexico-grammatical speech in their appearance," that there is a
certain biologically prompted predispostion to sociality and a need for
social communication in the human infant that actually facilitates the
acquisition of language.[77] Bruner's essential theme straddles the question
just as Chomsky's does, though perhaps in an ampler way:

> While the *capacity* for intelligent action has deep biological roots
> and a discernible evolutionary history, the *exercise* of that capacity
> depends upon man appropriating to himself modes of acting and
> thinking that exist not in his genes but in his culture. There is
> obviously something in "mind" or in "human nature" that mediates
> between the genes and the culture that makes it possible for the
> latter to be a prosthetic device for the realization of the former.[78]

What these considerations show, what we have been edging towards
anyway, is that there must be a sense in which minded attributes are
rightly ascribed to psychologically apt creatures below the level at which
selves appear or are recognizable. This hardly settles the famous question
pursued by G. H. Mead and Jean Piaget and Lev Vygotsky; but it does
show the reasonableness of admitting that – both with respect to human

infants and with respect to sub-human creatures of intelligence – the conceptually symbiotic linkage between "mind" and "self" must be only one among an eligible array of such linkages, in which the relevantly endowed entity lacks (in a graded or evolutionary sense) certain of the crucial functional powers associated with self. For every such pairing, however, there will need to be an answer regarding whether the intended match escapes Dennett's charge of the miraculous.

It should now be clear that, moving in the reverse direction, *persons are selves*. The term is bound to be elastically used, however, just as "self" and "mind" are. But the principal point is that "person" is used to identify those entities that, functioning as selves, exhibit as well aptitudes beyond the mere incipience of self – associated with the mastery of one's cultural world, with linguistic aptitude, with the capacity for reflection and reason and choice, with certain productive and performative skills, with a sense of social membership and responsibility. Again, there is no need to dispute the correct restriction of the term. But what we must see is that its central use requires a notion of numerical identity more robust than Hume was ever willing to concede officially – which, obliquely, *he* nevertheless admits (in yielding to Locke) to belong to our irresistible (though fictionalizing) habits of mind.

Put more compendiously, even atomistic (Humean) experiences involve selves nonrelationally; and the conscious comparison of experiences, as well as the reflexive awareness of memory, extends, however minimally, the scope of the self-identity of the self. More liberally put, the concept of self serves three distinct but interlocking functions at least:

1   to identify the nonrelationally specified entity with regard to which (paradigmatically) experiences, thought, cognizing powers may be referred to and individuated;
2   to fix the numerically accessible referents of reflexive experience, incorporating function 1; and
3   to fix the theoretically posited entity normally taken as self-identical over an entire lifetime, necessarily accommodating functions 1 and 2.

Enrichment of the powers assigned in accord with function 2, particularly along linguistic and cultural (public) lines, leads directly to a fuller theory (3) of persons; and impoverishment of the powers thus accorded leads to the replacement of selves by inferior but related referents, still keyed to 1 – as for instance in speaking of animals and merely sentient or irritable organisms. Minds (or minded attributes) are normally predicated through the range 1–3, but may (as by functional abstraction – still within the range of 1 and some impoverished analogue of 2 at least) – be predicated of machines. Whether selves are entirely cultural artifacts or biologically incipient "entities"(among humans at least) culturally and only culturally realized is either an empirically inaccessible question at the moment or a matter of terminological taste.[79] Finally, these distinctions do not in any way prejudge what should enter into the analysis of mental states or attributes.

The appeal of all this is simply that it commits us to very little that is quarrelsome, once we succeed in resolving that most strategic question, the eliminability of self by way of puzzles about numerical identity. Once we have done that, the advantage speeds us through a great many other quarrels. Ineliminability of selves counts as a strategic and salient invariance; but invariances of this sort are not so much discoveries of the fixed and final structure of reality as they are confessions of relatively persistent limitations regarding what we are able to think.

This accords rather nicely with a noteworthy dictum of Gaston Bachelard's: "In general, material reality may be defined in terms of anything that remains invariant over a sufficiently broad range of applications. The same is true of mathematical reality."[80] The same, we may add, is true of human reality. As Bachelard also makes clear, such invariances would be finally fixed if we could (as we cannot) totalize over all possible uses of all possible conceptualizations. Since, however, the would-be system of such applications is itself a function of given concepts, a fortiori of the (eventually) new concepts that will replace them, closure (in a creative sense) is only (but at least) evidence of a persisting – and therefore significant – failure of conceptual imagination.[81] To admit the ineliminability of selves is not to oppose the eligibility of the future prospects of a radical cognitive science, but it is also not to anticipate now what it may yet accomplish. We have proceeded by testing (within our ken, of course) the dialectical power of the most radical conception of experience (empiricism) and the most radical disregard of self under the condition of acknowledging experience (the no-ownership thesis). In proceeding thus, we have not favored in any apparent way any particular philosophical program (say, a Cartesian or a Kantian or a Husserlian or a Foucauldian). Hence, our reflection may rightly be regarded as transcendental, as yielding, that is, a plausible invariance of a fundamental kind – for the time being at least. Philosophical psychology, like science and philosophy in general, advances always and only by dint of its most strenuous failures. Alternatively put, we have fixed the apparent ineliminability of a numerically *unitary* self (or selves) with respect to experience; but we have intruded absolutely no presumptions regarding the *unity* (if any) of the internal structure of the self or its composition. That is as it should be, drawing as much as we can from the least quarrelsome admission and leaving the field of substantive theories as little altered in its geography as possible. Even the eliminationists can claim another inning; but they must do so on other grounds than those that attack the nonrelational "possession" of experience and the ineliminable use of the notion of numerical identity in predicative contexts.

IV

That said, we may step back to the point at which we first entered the argumentative lists. The Humean question with which we began is viewed in more than a suspicious way by the phenomenologically minded. The reason of course is that it is introduced by Hume in an empiricist setting. It is not, however, a merely empiricist question – and it cannot be escaped by the phenomenologists (or the deconstructionists, for that matter) any more than by the empiricists or the rationalists or the Kantians or the post-Kantian idealists. The implied misunderstanding is due to a deep equivocation regarding reference to the self. On a much-favored reading (which we shall call reading 1), to speak of the self as a cognizing power, as the "possessor" of experience, is to install the so-called "Cartesian model," to give unquestioned priority to "epistemological" orientations in virtue of which it is supposed that a cognizing *subject* is capable of discerning without irremediable distortion the structure of the real (independent) *objects* of the world – including its own states. The *self*, then, is that *substance* that validly authorizes itself to disclose, distributively, by statement, the way the world is. The natural criticism of this stance argues that how the self comes to affirm the way the world is is undoubtedly a function (a function it cannot in principle fully fathom by the use of the competence presumed in its original stance) of the existential circumstances under which it emerges as it does – circumstances that preform its cognitive powers, that determine those powers constitutively, hence that blind it to its own limitations at the very moment that they enable it to perform in the cognitively productive way it does.

This complaint informally captures the generic feature of what has been called "critique" or "critical" philosophy – neutrally, as among Kantians, Hegelians, Marxists, Husserlians, Heideggerians, existentialists, Frankfurt Critical theorists, deconstructionists, and others. The intended corrective is, without a doubt, the single most important theme of contemporary Western philosophy. On an argument barely adumbrated,[82] any presumption that that corrective can be supplied by appeal to a privileged cognitive source of any kind would vitiate the essential point of the corrective and would hardly escape being hopelessly vulnerable to the same objection. Not only that: the very effort to be "critical" in this sense cannot possibly be acknowledged or called into play without invoking the reflexive cognitive powers of the very self it means to restrain.

There you have the essential paradox of critical philosophy – once we deny all forms of privilege. For, the point of the corrective is to disallow any first-order cognitive privilege (which, on the foundationalist's claim – classically, Descartes's; contemporaneously, Roderick Chisholm's[83] – entails a second-order privilege as well); and yet, in doing that, it must allow for the play of the cognizing self in discerning and specifying just that corrective. So, *if* the corrective is cognitively reasonable, it must presuppose *the cognitive functioning of the self in a way that escapes the presumption of reading 1.* Furthermore, the argument shows that this reading (reading 2) must restore the cognizing power of the self acknowledged but too zealously endowed in reading 1 – that is, the power of affirming, distributively, by statement, the way the world is. Unless we opt unconditionally for the skeptic's position, which reading 1 classically provokes, particularly in its representationalist forms,[84] the critical corrective cannot be meant to disallow our first-order sciences but only to qualify their realist achievement in a scrupulous way; in doing that, it already entails an admission in accord with reading 2.

Reading 2, then, catches up what should be the minimal concession regarding the self: that it is a cognizing entity, the center or focus of *whatever* cognitive power may reasonably be taken to subtend and generate (under critical constraint) the body of our science, our knowledge, our self-knowledge, our knowledge of other selves, our initiative as intentional agents, our know-how, and our second-order grasp of the contingencies of these. Whatever the detailed puzzles such cognizing capacity may confront us with – for example, regarding self-consciousness[85] – there is no reason to suppose that we could ever coherently repudiate this minimal concession: that the self is an entity capable of cognition, that it "possesses" experience (nonrelationally), that its cognitive powers depend on its having experience, and that it manifests this capacity saliently at least (not necessarily essentially, though the matter is disputed) by its linguistic utterances (and linguistically informed central states), normally by the use of a predicative idiom. The supposition that we *could* do without the cognizing self (without yeilding to skepticism) unaccountably requires turning a blind eye to the plain fact that that pronouncement itself (and its supporting arguments) presuppose and engage the self-same power.

The concession is precisely what confirms the inseparability and the equality (in the absence of privilege) of the contribution of naturalistic, phenomenological, deconstructive, and (*a fortiori*) critical reflections on the cognitive powers of the self. Naturalism, ranging classically from Galileo to Kant, affirms the minimal cognitive function of the self, once its presumption (along the line of reading 1) is repudiated. Phenomenology affirms the dependence of naturalistic claims on the preformative functions of the *Lebenswelt* or the conditions of existence, once *its* presumption (along the lines of apodicticity and constructive origins) is repudiated. Deconstruction merely affirms the ultimate indefensibility of any presumption of privilege either along the naturalist's or the phenomenologist's lines (or any other's), but it fails to make explicit provision for the recovery of the self's function along the lines of reading 2. Finally, critical philosophy (which may be naturalistic, phenomenological, or deconstructive) must, consistently, recover a viable – as by way of transcendental arguments[86] – version of reading 2 without falling victim once again to the presumptions of reading 1. There is no reason to think this is an impossible project, even if it is true that it is an unending one.

In any case, it is the confusion of readings 1 and 2, the *non sequitur* of supposing that, in rejecting 1 we must, or can coherently, reject 2, that has led to the repudiation of (second-order) practices of legitimation[87] and to the sheer chaos (*not* anarchism) of seeming claims that refuse to distinguish carefully between science and fiction.[88] As far as the project of eliminating the self is concerned, the confusion offers no advantage at all – although the world it would appear to "inherit" would have become thereupon hopelessly truncated or disordered. We must see that to restrain the cognitive pretensions of human agents is hardly to eliminate them – or the world they engage. This is the clear (but not quite intended) message of those redoubtable critics of naturalism (and of one another), Husserl and Heidegger.

At the risk of a too-easy summary, we may claim that, regarding the self (whatever their respective temptations may have been), neither Husserl nor Heidegger denies a substantive self but both hold "only" that it cannot be captured by "the subject–object relation," cannot be captured by the Cartesian model, cannot be adequately described in terms of a timeless present in which cognizing self and cognized world are perpetually fitted to one another so that the independent second is always perspicuously disclosed to the first, cannot count on such a preestablished harmony. In Heidegger's terms, this imposes on us the need to reinterpret the Greek ontology of Being or *ousia* – of *presence* (*Anwesenheit*) – in terms of the inseparable problematic of the "temporality" of *Dasein*.[89] Hence, the Cartesian fixity of the self's cognitive powers is totally undermined *by* virtue of the essential fixity of *Dasein*'s existential nature. There's a pretty paradox.

Heidegger, it is true, appears to have been moved by the exfoliating implications of this finding – with which he begins his work – to realize that his *own* attempt at determining the essential existentialia of *Dasein* cannot but be similarly doomed; hence, that "Philosophy is ending in the present age."[90] Still, in the context of that very remark (that relatively late remark), Heidegger clearly means to prophesy (gloomily but not utterly without resistance) the replacement of philosophy by "the independent sciences": "The sciences," he says, "will interpret everything in their structure that is still reminiscent of the origin from philosophy in accordance with the rules of science, that is, technologically."[91] This could not, then, possibly have been meant to eliminate the self. Indeed, Heidegger's counsel is to turn "the task of thinking" to determining the task of thinking itself.[92] That turning (*Kehre*) would undermine the presumption of both metaphysics and cybernetics (in Heidegger's quaint idiom), but it would not preclude the reflective self. It only means that the peculiar persistence of the self as an entity sufficiently substantial to function as a center of cognizing power (the false interpretation of which Heidegger worked so hard to forestall, as by attenuating the account of *Dasein* along "historized" lines, by construing *Dasein* relationally, projectively, "stretched" between birth and death, by insisting on authenticity) ultimately resisted his own best efforts. The *aporia* resisted his efforts, in the dual sense that they (*his* efforts) proved to be touched by the same Cartesian presumption as he meant to counter, and that the cognizing self he tried to dissolve *need* not be assigned any such offending "nature":

> *If the "I" is an Essential characteristic of Dasein, then it is one which must be interpreted existentially.* In that case the "Who?" is to be answered only by exhibiting phenomenally a definite kind of Being which Dasein possesses. If in each case Dasein is its Self only in *existing*, then the constancy of the Self no less than the possibility of its "failure to stand by itself" requires that we formulate the question existentially and ontologically as the sole appropriate way of access to its problematic.[93]

Heidegger, therefore, clings to his thesis – even at the price of admitting his failure to vindicate it philosophically *and* of admitting the impossibility of doing so in his own way.

For his part, Husserl had no similar temptations, though a corresponding and opposite irony infects his findings. The following extraordinary passage may serve to fix the sense in which Husserl both overcomes the Cartesian stance and instantiates it in a more profound way then ever Descartes could have been charged with doing:

> The existence of this world is self-evident for me because it is self-evident only in my own experience and consciousness. This consciousness is the source of the meaning of the world and of any worldly objective facts. But, thanks to the transcendental epoché, I perceive that whatever is in the world, including my existence as a human being, exists for me only as the content of a certain experiential apperception in the mode of certitude of existence. As a transcendental ego I am the ego which apprecieves actively and passively. This happens in me although it is concealed from reflection. In this apperception the world and the human being are first constituted as existing. Any evidence gained for worldly things, any method of verification, whether pre-scientific or scientific, lies primarily in me as transcendental ego. I may owe much, perhaps almost everything, to others, but even they are, first of all, others for me who receive from me whatever meaning or validity they may have for me. They can be of assistance to me as fellow subjects only after they have received their meaning and validity from me. As transcendental ego I am thus the absolutely responsible subject of whatever has existential validity for me. Aware of myself as this ego, thanks to the transcendental reduction, I stand now above all worldly existence, above my own human life and existence as man. ... Is it not self-evident that the world, which for the natural attitude is the universe of all that exists without qualification, possesses only transcendentally relative truth, and that only transcendental subjectivity exists without qualification?[94]

The upshot is that the eliminationist maneuver is conceptually hopeless. If we admit experience of any kind, we require an entity capable of possessing it (nonrelationally); and, if we attempt to qualify critically the cognitive achievements that such experience may enable, no attenuation can be expected to permit its actual dismissal. Our finding, therefore, is a very slim one – but also a decisive one. For, without intruding *any* substantive view regarding particular "critical" constraints on the self's cognitive role or on its very nature, we see that every theory of the sciences cannot but be affected by the self's ineliminability and the conceptual encumbrances that that will entail.

Effectively, an interesting conception of the self must consider three matters at least: knowledge of the natural world, self-knowledge and self-direction as an active agent in the world, and knowledge of other selves – *and* the interconnections among these three. Broadly speaking, naturalism and phenomenology have confirmed in their respective ways the symbiosis of the first two sorts of knowledge. It is quite remarkable how consistently the best work of these two traditions resists the conceptual symbiosis between the first two sorts of knowledge and the third. It is really principally in the Hegelian tradition (if, in so speaking, we may include, without intending any distortion, the Marxist and Frankfurt Critical movements as well as much that is currently call Nietzschean or post-structuralist – Foucault's work, for instance) that the full symbiosis of the three is, or at least could be, developed.[95] To confirm the point in a full way would decisively affect our theory of the relationship between the natural and the human sciences. One sees, therefore, the strategic importance of making or resisting the eliminationist's maneuver. But also, in pursuing that issue, one sees how, at the same time, the ground is ineluctably being prepared for a comprehensive account of the complexities of human existence itself.

V

We are of course in danger here of opening an inquiry we mean to postpone, namely, an inquiry into the *nature* of the self. This is still not the place to attempt that inquiry; but there is an obvious advantage afforded by the progress we have already made that need not be squandered without a care to the further complications we know we shall have to face. What we have managed to establish is slim enough: (a) that "experience," distributed or attenuated in any way that the empirical saliences of the familiar world oblige or invite us to admit, requires, in a conceputally symbiotic sense, some form of "possession" (*not* merely predication) on the part of a "self" (or suitable surrogate) adequated to the particular attributes and aptitudes thus ascribed; (b) that the "self," correspondingly parsed in whatever way the empirical evidence favors in speculating about the lowest sensitive organisms, the most sophisticated machines, and the acual variety of humans, is rightly accounted an entity in whatever we take the formal sense to be in which we justify speaking of physical bodies as entities – without prejudice to the possible variety or economy or provisionality of such entitative posits; (c) that the *unicity* of the self, thus construed, serves a multiple function in managing coherently the predication of pertinent mental or psychological attributes and the reidentification of the referents of such predication; (d) that the question of the nature – in particular, of the internal *unity* – of the self or subject or agent of such ascriptions – remains a completely open matter, at least as far as the foregoing tally is concerned; and (e) that the formal questions here considered cannot themselves be operatively resolved without attention to what we should regard as a fair answer to the further question of the nature of the self. Having collected this much regarding our discourse about selves, we may venture two further themes without yet attempting a full account of the nature of selves: (f) that the "minded" attributes assigned the self must be *incarnate* attributes (and that selves or persons or their surrogate must be congruently qualified); and (g) that the functional unity of the self is best construed as an empirically posited normative model, hardly essential, that may (on the pertinent evidence) be breached or qualified in a large variety of ways attuned to historically variant and changing forms of life, alternative careers favored or tolerated within particular societies, intra-psychological functional divisions of labor, clinical experience and the like. Perhaps a brief word about these last two themes would be in order here, without encouraging the expectation that a rounded theory of selves or persons is actually being offered.

What is understandably misleading about mental attributes is due to the most distinctive ability humans manifest, the ability to be reflexively aware of the exercise or instantiation of a number of mental abilities. If, for instance, the occurrence of pain or the exercise of the various powers of sensory perception justifies attributing a measure of *consciousness* to particular organisms, then it is reasonable to argue that consciousness does not entail *reflexive awareness* and such awareness need not be equivalent to *self-consciousness*. These distinctions are intended and are efficiently pressed into service in an extremely informal way in ordinary discourse. There is an advantage in resisting turning them into technical terms. But even within the range of these rather gross-grained distinctions, it is easy to see that several puzzles would need to be resolved in order to sort the difference between human capacities and the capacities of other animals: for one, we should need to know whether and in what regard sub-human animals are rightly ascribed conscious powers in a realist rather than a heuristic spirit; and for another, we should need to know whether and in what regard the conscious powers ascribed in a realist spirit to man are or are not invariably intentional in structure. The first poses the Cartesian question as to whether animals may be ascribed "minded" attributes at all; the second, Brentano's question regarding the problematic extension of intentionality to "sensations."[96] On Brentano's view, where sensations are taken to be "effects of physical stimuli" only[97] or where they are merely "that which is presented" (in consciousness), sensations (including the effects of stimulating the *sensory* systems) are not as such mental phenomena. "By presentation," Brentano says, "I do not mean that which is presented, but rather the act of presentation" (the "act" in which "something appears to us").[98] But then, colors, like pains, have as such no "reference" or "content" or "directed-ness" (though the "interpretation" of them will constitute mental phenomena).[99]

The complexities are plain enough: we are not concerned here to sort them satisfactorily.[100] But we may take note of the fact that, on the grounds supplied (and in Brentano's own view), since "consciousness," specified intentionally, may be "unconscious," consciousness and (what we are calling) reflexive awareness (awareness *of* intentional structure) need not be coextensive. Nevertheless, on Brentano's view and in criticism of Hume (that is, in criticism of the usual reading of Hume – which Brentano himself allows), self-consciousness (consciousness of self, of oneself) can be "derive[d] from any impression [presentation] whatever."[101]

These distinctions, important as they are, do not yet bear directly on what is crucial to resisting ontic dualism: namely, the consideration that, because within the range of reflexive awareness we discriminate intentional structures or "presentations," we are (as Brentano is) inclined to oppose the mental and the physical rather than to construe the mental (or psychological) as a range of *indissolubly complex* phenomena of which only a certain "feature" – the intentional or presentational – is apt for awareness or for deliberate reporting and which (for that reason) may be injudiciously posited as the "essentially" or exclusively mental.

There is only one way to identify the *mental* – if we are to preserve (1) its reality, (2) its causal efficacy, (3) its characteristic intentional distinction (*à la* Brentano, though in a way open to Husserlian and further elaboration), *and* (4) its being not susceptible to ontic dualism. That way requires that the mental *not* be opposed to the physical (or biological), while still being distinct from and not reducible to the physical: it must be (5) an *incarnate* phenomenon, which is to say it must be (5') *emergent* with respect to the physical or biological and (5") *monadic* with respect to its internal intentional structure. In this sense, the notion of the mental accommodates the *abstraction* but not the *separation* of the mental from the physical or neurophysiological. Thinking, for instance, incarnately involves the activation of certain neural circuits even though, reflexively, we are able to report only the intentional structure of our thoughts (or, better, what we are able to report reflexively is what we take our "thoughts" to be).

On this reading, the *functionalist* treatment of the mental is an entirely avoidable extravagance: for it construes the mental as purely abstract and compromises both its reality and its causal efficacy.[102] When, therefore, Merleau-Ponty fixes the locus of intentionality phenomenologically – so that it is pre-predicative, pre-recognitional, pre-cognitive – by assigning it to the "lived body," we may accept his very bold economy provided we neither depreciate its cognitive complexities nor the complexities (as yet here unmentioned) of the societal conditions (also intentional) within which the other arises. "What distinguishes intentionality from the Kantian relation to a possible object," Merleau-Ponty affirms, "is that the unity of the world, before being posited by knowledge in a specific act of identification, is 'lived' as ready-made or already there. ... It is a question of recognizing consciousness itself as a project of the world, meant for the world which it neither embraces nor possesses, but towards which it is perpetually directed – and the world as this pre-objective individual whose imperious unity decrees what knowledge shall take as its goal. This is why Husserl distinguishes between intentionality of act, which is that of our judgments and of those occasions which we voluntarily take up a position – the only intentionality discussed in the *Critique of Pure Reason* – and operative intentionality (*fungierende Intentionalität*), or that which produces the natural and antepredicative unity of the world and of our life, being apparent in our desires, our evaluations and in the landscape we see, more clearly than in objective knowledge, and furnishing the text which our knowledge tries to translate into precise knowledge."[103] We may take Merleau-Ponty's account to entail, relevantly though minimally for the present argument, that the intentional is incarnate in the "mental." We need not press that distinction further here.

Insistence on the physical or biological incarnation of the intentional improves our grasp of another, dual puzzle of the greatest importance: on the one hand, texts (in the commonplace hermeneutic sense) and, on the other hand, selves (in the sense here advanced) are phenomena or entities that exhibit incarnate attributes. For this reason, the question of the *unity* of texts and selves or persons – both culturally emergent entities[104] – is standardly open to certain useful constraints that support the presumption of objectivity in relevant descriptions and interpretations. The point is implicit at least in part in Merleau-Ponty's speaking of the world of "operative intentionality [as] *furnishing the text which our knowledge tries to translate into precise knowledge.*" This is not quite accurate, of course – and is actually much too sanguine (if taken on its face) – because it wrongly conveys the impression that our conscious, linguistic, cultural, historical projects are best construed as interpretations of the "ready-made," "antepredicative unity" that is "*already there*" in the world. But it does suggest that a theory of *texts* too much confined to the abstract, intentional dimension of the cultural (not sufficiently attentive to the incarnate) or a theory of *selves* or *persons* (again too much confined to the abstract) will be tempted to exaggerate how problematic the unity or individuation of texts and selves really is and how radical is the sense in which both texts and selves are social or cultural constructions (possibly even conventionally or arbitrarily demarcated as such).

This surely fixes the glee with which Roland Barthes offers his famous piece of purple: "The Text ... practices the infinite deferral of the signified. ... Every text, being itself the intertext of another text, belongs to the intertextual, which must not be confused with a text's origins: to search for the 'sources of' and 'influence upon' a work is to satisfy the myth of filiation."[105] In the context of selves, on the other hand, the essential benefit of insisting on the incarnate rests with tempering the social constructionist theory of selves: for an emphasis on biological incarnation permits us to identify the precultural incipiencies of functioning selves that a culture "grooms" into such functional aptitudes; and it also encourages us to concede the symbiosis of the psychological and the societal in virtue of which selves cannot be merely social constructions or fictions due to societal or collective forces somehow detached from or insufficiently assigned to the real activity of persons. For one thing, the collective forces of cultural institutions operate *only* through the activity of aggregated persons or selves; for another, there *are no* actual collective agents.[106]

Hence, with all due appreciation for Rom Harré's deliberate "myth" designed to defeat the reduction of psychology to the "metaphysical scheme of the physical sciences," there may be too much conceded by him in the opposite direction, to hold (with him) that "In conversations, basic entities are neither people nor their thoughts, but speech-acts": "It might, then, be a useful fiction," Harré continues, "to base research programs on the assumption of a universe of speech-acts located in a people space. A people-space, as a manifold of discrete points, makes speakers into places-locations for speech acts. The array of people is to be thought as the space in which the forces of speech-acts as entities interact and move."[107] In fact, Harré confesses, "I believe [solving Hume's problem] that it is the people makers of a culture who are responsible for the synthesis of self. By teaching a suitable grammar and by inculcating the practices of self-assessment favored in that culture,

they make possible the kind of unity found in the midst of their culture.
... The *concept* of 'self' as the bearer of 'inner unity' should therefore
be studied as an analogue of the social concept of 'person', since persons
are the public loci of clusters of publicly accountable action and
speeches." We assume we are "selves," he adds, by subscribing to a
theory "modelled on that of the public person, as that concept is
understood in [our] culture." The entire picture is no more than a
convenient "proposal" but it is meant nevertheless to "account for the
persistent but culturally necessary illusion of selfhood."[108]

Harré does not sufficiently distinguish the need for entitative selves
as the numerical referents of mental and moral ascriptions – as servicing
the need for *unicity* rather than merely for internal *unity*: in discounting
an essential and ramified unity (reasonably enough), he does not quite
provide for the incarnate standing of what must be ascribed to unitary
selves (persons). The very idea of constructing selves presupposes the
effective agency of selves themselves; and the unicity of selves requires
at least a certain measure of unity within selves – thought not of that
kind or measure (of rather different sorts) associated with the views of
Descartes, Kant, Brentano, and Husserl. Selves or persons are culturally
emergent entities; persons cannot fail to possess the minimal functions
of selves – both with regard to unicity and unity. But there must already
be in place biologically functional dispositions toward such minimal
selves (formed in culturally diverse ways) that would justify the ascription
of a large range of the usual mental predications, even if it is true that
the peculiarly private and introspective unity characteristic of the
contemporary Western world cannot have been universal for all cultures.
The reasonable presumption that it cannot have been is what links the
otherwise marvellously preposterous speculation of Julian Jaynes and
the culturally perceptive bifurcation of societies of tradition and societies
of ideology so admirably formulated by Alvin Gouldner.[109]

Selves (persons) are socially constituted, then – not "constructed" –
even if it is true that the full Western belief in an essential, noumenal,
private, even disembodied self *is* an illusion and a construction. But
there is a fair sense in which humans are biologically "prepared" for
such cultural formations – by way of a social process distantly analogous
to (but of course much looser, culturally more open than) the societally
triggered, temporally deployed patterns of animal imprinting.[110] Also,
paralleling imprinting in their distinctive way, cultural traits (societally
emergent) must be biologically incarnate – modifications of some
genetically *limited* dispositions.

This of course is precisely what is meant by speaking of culture as
man's "second nature." The point has been put in a most economical

way by Marjorie Grene – uniting and correcting the visions of man of Heidegger, Merleau-Ponty, Wittgenstein, and Helmuth Plessner:

> Human being, as Being-in-a-world ... is possible only as the achievement of a certain kind of living being, with certain organic endowments and a certain kind of biological as well as social environment. *Animalia* are a necessary presupposition of *existentialia* [a criticism of Heidegger's view of *Dasein* by way of Plessner's philosophical anthropology]. Note, however, I am not now proposing to replace the principle of the primacy of historicity by a principle of the primacy of life. What we have to recognize is the place cleared *within* nature for the possibility of the human, that is, historical, or historicizing-historicized, nature.[111]

Following Plessner, Grene stresses "the natural artificiality of man": "a human being is an individual member of the species *homo sapiens*, to whose development, even as a living thing, inherence in an artifactual medium is necessary. It is our nature to need the artificial; we come to ourselves not only as users of, but as dwellers within, a tightly woven net of artifacts."[112] Our adaptation, as selves, to the artifactual world of culture is distributed among an indefinitely varied set of historical options *not* fitted to but emergent within the *sui generis* biological adaptability of the species. Possibly, Harré's constructionism is meant to accommodate the peculiar continuity of man's "first" and "second" natures – more in terms of the theory of persons than of selves (since Harré tends to favor Strawson's account of persons[113]); but it has its difficulties.

Two critical difficulties may be mentioned. For one thing, *if* selves are "constructed," they are constructions (on Harré's view) *made by persons* – who cannot thus be constructed in the same sense. Selves, in particular the deeply illusive notion of unified selves favored in the Western world, are really fictions about certain would-be abstract entities. The notion, therefore, entrenches the functionalist idiom. But persons (on Harré's view) are characterized, in effect, in terms of incarnate attributes: the constructionist theory does not directly bear on the conditions under which they evolve or emerge. Here, the truth is that persons appear with the appearance of human societies – with the emergence of culture itself. So a constructionist view of persons would require a constructionist view of culture – which is impossible. Secondly, *if* persons are the "public loci" of certain culturally formed aptitudes, then *persons cannot but exhibit a measure of internal unity as well as the unicity assigned selves*; so the two notions cannot be treated as entirely conceptually distinct. Persons emerge as functioning selves. Selves exhibit incarnate properties, though they need not exhibit (and may never exhibit) the profound sort of internal unity characteristically assumed in Western theories. "[A]nimate beings," Harré says, "are persons if they are in possession of a theory – a theory about themselves. It is a theory in terms of which a being orders, partitions and reflects on its own experience and becomes capable of self-interventions and control."[114] Harré means to disallow, therefore, any merely biological "maturation theory" of *persons*: persons require a cultural space in order to appear (speech acts serve metonymically in this regard). Of course, he is right. But, by the same token, *they* cannot then be "constructed." And, in the argument just given, *selves cannot be constructed*, even if the "selves" we favor in our ideologies and theories and moral doctrines and religions *are* fictional constructions imposed on actual persons (actual selves).

Harré explicitly links his analysis of selves (of the illusive, interior, unified centers of personal actions and control) with the same inability to "discover" the self that is announced by Hume (and Husserl).[115] What he means to oppose is the discovery of a noumenal entity. A small adjustment would bring Harré's theory into accord with our own. The following considerations are essential to both (if we ignore Harré's special use of "self"): (i) persons must be construed entitatively, to facilitate reference and predication; (ii) their natures must be adequated to their attributes being incarnate (call that feature their being "embodied"[116]); (iii) the postulation of persons as entities need not be intended in essentialist or cognitivist terms, though the aptitudes attributed as characteristic of them presuppose and entail the cognitively reflexive powers *of* such entities; (iv) the formational powers of societies to "constitute" persons are the powers of aggregates of actual, already "constituted" persons; (v) there is no ontological priority to be assigned to the "constructed" or "constituted" nature of cultural attributes and culturally emergent entities *vis-à-vis* one another *or* to the "constituted" or "constructed" nature of human societies and human persons *vis-à-vis* one another, though there can certainly be historically divergent forms of life, divergent forms of persons, and societies lacking the sensibilities of Western "selves"; (vi) the "constitution" of cultural phenomena (whether attributes, persons, selves, institutions, societies, histories, artifacts, art, language, actions or the like) can only be understood in terms of the conceptual linkage holding between culturally *emergent* and *incarnated* phenomena and the physical and biological order in which they are embedded – so the "construction" of selves can only be of "local" significance, that is, it must be a generational phenomenon of some sort that alredy presupposes the generation of the cultural *from* the merely biological; and (vii) within the space of the cultural, persons (*a fortiori*, selves) and societies are symbiotic, in the sense that persons possess attributes that exhibit both "mental" features narrowly linked to their incarnating biological natures and "societal" features more broadly linked to their sharing institutions and practices.

The truth is that the constructionist theory of selves is remarkably widespread and yet not easily recognized. Its principal weakness – and charm – lies in analogy with the usual theory of minds that favors an abstract and functionalist view of the mental and the intentional. This is also a key to the parallelism of the prevalent theories of texts and selves in our own day. Hence, it is particularly rife in those forms of contemporary hermeneutics, post-structuralism, and deconstruction that pretty well ignore the biological, the incarnating, dimension of human existence. For instance, this surely explains the ease with which – both with respect to texts and selves – Gadamer emphasizes the constructionist unity of all historical phenomena, the strong denial of "substance" (which is not quite the same thing as the denial of entities). It is in a sense the deeper meaning of his doctrine of the "fusion of horizons" (*Horizontverschmelzung*). Thus Gadamer says:

> We say ... that understanding and misunderstanding take place between I and thou. But the formulation "I and thou" already betrays an enormous alienation. There is nothing like an "I and thou" at all – there is neither the I nor the thou as isolated, substantial realities. I may say "thou" and I may refer to myself over against a thou, but a common understanding [*Verständigung*] always precedes these situations. We all know that to say "thou" to someone presupposes a deep common accord [*tiefes Einver-ständnis*]. Something enduring is already present when this word is spoken. When we try to reach agreement on a matter on which we have different opinions, this deeper factor always comes into play, even if we are seldom aware of it. ... It is not so much our judgments as it is our prejudices that constitute our being.[117]

Ultimately, then, we are first formed by some self-determining process moving through language and culture. But the implicit arbitrariness of the process is betrayed by Gadamer's restricting his attention to the sheer play of abstract histories: for it proves impossible to *account* for the actual structures and causal influences of human history, and it proves the barest *obiter dictum* to have announced the conserving tradition of the classical – somehow mysteriously abiding in its reliable and reassuring way through all the shifting currents of cultural variety.[118]

In a very odd and totally unexpected way, then, the dominant tendencies of our age to recover the primacy of the historical and cultural nature of man may sometimes be read as the progeny of an incorrectly read Hume.

Notes

1   David Hume, *A Treatise of Human Nature*, ed. L. A. Selby-Bigge (London: Oxford University Press, 1958), p. 252.
2   See Barry Stroud, *Hume* (London: Routledge and Kegan Paul, 1977), ch. 2, for a clear introduction to Locke's "ideas" and Hume's "perceptions."
3   Cf. Hume, *A Treatise of Human Nature*, pp. 219, 253–4.
4   Wilfrid Sellars, "The Language of Theories," *Science, Perception and Reality* (London: Routledge and Kegan Paul, 1963), particularly p. 126.
5   See Sellars, "Empiricism and the Philosophy of Mind," *Science, Perception and Reality*.
6   Hume, *A Treatise of Human Nature*, pp. 254, 259.
7   Stephen P. Stich, *From Folk Psychology to Cognitive Science* (Cambridge, Mass.: MIT Press, 1983), p. 30.
8   See Derek Parfit, *Reasons and Persons* (Oxford: Clarendon Press, 1984), pt III, particularly pp. 275–6. For a sustained examination of Parfit's theory, see Joseph Margolis, critical notice of Derek Parfit, *Reasons and Persons*, in *Philosophy and Phenomenological Research*, XLVII (1986).
9   See for instance Lilli Alanen: "Studies in Cartesian Epistemology and Philosophy of Mind," *Acta Philosophica Fennica*, XXXIII (1982); "On Descartes' Argument for Dualism and the Distinction between Different Kinds of Beings," in S. Knuuttila and J. Hintikka (eds), *The Logic of Being* (Dordrecht: D. Reidel, 1986).
10 Hume, *A Treatise of Human Nature*, p. 251.
11 Ibid.
12 Ibid., p. 252.
13 Ibid., p. 201.
14 Ibid., p. 256.
15 Alasdair MacIntyre, "The Intelligibility of Action," in Joseph Margolis et al. (eds), *Rationality, Relativism, and the Human Sciences* (Dordrecht: D. Reidel, 1986). See also Hilary Putnam, "Brains in a Vat," *Reason, Truth and History* (Cambridge: Cambridge University Press, 1981), mentioned by MacIntyre.
16 Parfit, *Reasons and Persons*, p. 277; cf. Hume, *A Treatise of Human Nature*, p. 261.
17 Parfit, *Reasons and Persons*, pp. 251, 211.
18 Ibid., p. 215.
19 Ibid., p. 211.

20   For attempts to analyze the notion of numerical identity under conditions of change, see David Wiggins, *Sameness and Substance* (Cambridge, Mass.: Harvard University Press, 1980); Eli Hirsch, *The Concept of Identity* (New York: Oxford University Press, 1982).

21   Hume, *A Treatise of Human Nature*, pp. 79, 252.

22   Ibid., p. 219.

23   Ibid., p. 2.

24   Hume regularly acknowledges the irresistibility of the persistence and identity of the self (ibid., sections II, VI *passim*).

25   Parfit, *Reasons and Persons*, pp. 223–6, particularly p. 225. See Bernard Williams, *Descartes* (Harmondsworth, Middx: Penguin, 1978), pp. 95–100.

26   P. F. Strawson, *Individuals* (London: Methuen, 1959), pp. 95–7.

27   It may be claimed that Strawson's point is obliquely anticipated by Sellars's "Empiricism and the Philosophy of Mind" (*Science, Perception and Reality*), particularly in its criticism of sense-datum languages.

28   Parfit, *Reasons and Persons*, p. 225 (italics added). Cf. Sidney Shoemaker, *Self-Knowledge and Self-Identity* (Ithaca, NY: Cornell University Press, 1963), particularly chs 3, 5.

29   Thomas Nagel, "Physicalism" and Postscript (1968), repr. in David M. Rosenthal (ed.), *Materialism and the Mind–Body Problem* (Englewood Cliffs, NJ: Prentice-Hall, 1971), p. 96.

30   Ibid., pp. 110, 108.

31   Ibid., p. 110. Nagel's later book *The View from Nowhere* (New York: Oxford University Press, 1986) repudiates physicalism but does not actually offer any sustained argument against it – except an enlargement of what Nagel had originally found to be repellent about it. For example, he announces that "The subjectivity of consciousness is an irreducible feature of reality – without which we couldn't do physics or anything else – and it must occupy as fundamental a place in any credible world view as matter, energy, space, time, and numbers"; but he admits that the question how to account for the relationship between mind and body "cannot now be settled" (pp. 7–8). He declares physicalism "unacceptable" (p. 16), actually remarks that the no-ownership view is not "intelligible" ("*Something* must be there in advance, with the potential of being affected with mental manifestations, if lighting a match is to produce a visual experience in a perceiver" – p. 30), and objects to Parfit's account in terms of a so-called "dual aspect theory" (p. 45). But the truth is that he really registers a civilized complaint at not being able to defeat physicalism and the no-ownership theory. He advances no effective argument. See also, Geoffrey Madell, *The Identity of the Self* (Edinburgh: Edinburgh University Press, 1983).

32   See Herbert Feigl, *The "Mental" and the "Physical": The Essay and a Postscript* (Minneapolis: University of Minnesota Press, 1958, 1967). Feigl, however, all but repudiates his own "double-knowledge, double-designation view" in the Postscript (pp. 138, 141).

33 Nagel, *The View from Nowhere*, pp. 28, 31.
34 The theme somewhat explains the tangential connection between recent Continental European currents and Richard Rorty's *Philosophy and the Mirror of Nature* (Princeton, NJ.: Princeton University Press, 1979), which has had such an immense popularity in the United States. Oddly, Rorty's book resonates in a way with the discoveries of the Continental philosophers, but he does not actually mention Merleau-Ponty. In fact, he does not actually accommodate a distinctly biological orientation; and where he addresses the mind/body problem, he shows himself (one can only suppose, inconsistently) to favor something like a strong physicalism: see especially pp. 387–9.
35 Perhaps the central theme in Merleau-Ponty may be most perspicuously isolated by comparing Maurice Merleau-Ponty, *Phenomenology of Perception*, tr. Colin Smith (London: Routledge and Kegan Paul, 1962), and *The Visible and the Invisible*, tr. Alphonso Lingis, ed. Claude Lefort, (Evanston, Ill.: Northwestern University Press, 1968).
36 See W. V. Quine, *Word and Object* (Cambridge, Mass.: MIT Press, 1960), pp. 219–21.
37 W. V. Quine, "Things and their Places in Theories," *Theories and Things* (Cambridge: Harvard University Press, 1981).
38 Noam Chomsky, *Rules and Representations* (New York: Columbia University Press, 1980), p. 6.
39 Ibid.
40 See for instance Noam Chomsky, *Knowledge of Language: Its Nature, Origin, and Use* (New York: Praeger, 1986), pp. 59ff.
41 See Joseph Margolis, "The Perils of Physicalism," *Mind*, LXXXII (1973), for an extended discussion of Nagel's failure to draw out the distinction.
42 J. J. C. Smart, "Sensations and Brain Processes," in V. C. Chappell (ed.), *The Philosophy of Mind* (Englewood Cliffs, NJ: Prentice-Hall, 1962); Donald Davidson, "Mental Events," *Essays on Actions and Events* (Oxford: Clarendon Press, 1980). Davidson would deny that he is a physicalist or reductionist, but that is largely because he rejects psychophysical laws.
43 Hume, *A Treatise of Human Nature*, pp. 188–9.
44 Ibid., p. 277.
45 Ibid., p. 85.
46 Ibid., p. 108.
47 Ibid., pp. 7–8.
48 Ibid., p. 1.
49 Ibid., pp. 261–2.
50 Parfit, *Reasons and Persons*, p. 275.
51 For counterproposals to Parfit's, see Bernard Williams, "Personal Identity and Individuation," *Problems of the Self* (Cambridge: Cambridge University Press, 1973); also, David Wiggins, "Personal Identity," *Sameness and Substance*.
52 Parfit, *Reasons and Persons*, p. 273.

53 Ibid., p. 276.
54 See for example Herbert A. Simon, *Models of Discovery* (Dordrecht: D. Reidel, 1977), ch. 5.1.
55 This for instance is the bold undertaking of Stich's *From Folk Psychology to Cognitive Science*.
56 Daniel C. Dennett, *Content and Consciousness* (London: Routledge and Kegan Paul, 1969), p. 190.
57 Paul K. Feyerabend, "Materialism and the Mind/Body Problem," *Review of Metaphysics*, XVII (1963).
58 The most developed example is explored in Margolis, critical notice of *Reasons and Persons*, in *Philosophy and Phenomenological Research*, XLVII.
59 Parfit, *Reasons and Persons*, pp. 200–9.
60 Ibid., p. 261.
61 Julian Jaynes, *The Origin of Consciousness in the Breakdown of the Bicameral Mind* (Boston, Mass.: Houghton Mifflin, 1976).
62 Marcel Mauss, *Sociology and Psychology*, tr. Ben Brewster (London: Routledge and Kegan Paul, 1979), pt III: "A Category of the Human Mind: The Notion of Person, and Notion of 'Self,'" p. 59.
63 Ibid., p. 61.
64 Michel Foucault, "What is an Author?" tr. Josué V. Harari, in Harari (ed.), *Textual Strategies* (Ithaca, NY: Cornell University Press, 1979), pp. 148–59.
65 Ibid., pp. 150, 159. See also, for instance, Michel Foucault, "Truth and Power," *Power/Knowledge: Selected Interviews and Other Writings 1972–1977*, ed. Colin Gordon, tr. Colin Gordon et at. (New York: Pantheon, 1980).
66 Mention may be made here – to give a further impression of the amplitude of the unity and developmental question – of relatively recent work in the loosely Freudian literature termed ego psychology, in particular in the analysis of the phenomena of the so-called "split ego" and the "borderline syndrome." Theorists such as James Masterson offer convincing narrative sketches that favor the occasional failure of a child's developing ego to resolve the critical dynamic challenge of "separating" from the mother ("functioning as an auxiliary ego for the child" – between, say, three and eighteen months of life) and of gathering sufficient momentum to achieve "autonomy." Regardless of quarrels about the precision of ego psychology, the difference between the conceptual roles of the *identity* and the *unity* of self can hardly be confused. See James F. Masterson and Jacinta Lu Costello, *From Borderline Adolescent to Functioning Adult: The Test of Time* (New York: Bruner/Mazel, 1980), particularly chs 1–2; also D. W. Winnicott, *The Maturational Process and the Facilitating Environment* (New York: International Universities Press, 1965), and James F. Masterson, *The Narcissistic and Borderline Disorders* (New York: Bruner/Mazel, 1981).
67 See Hilary Putnam, "The Meaning of 'meaning'," Philosophical Papers, vol. 2 (Cambridge: Cambridge University Press, 1975).
68 For a sample of the recent literature, see *The Psychiatric Clinics of North*

America, vol. 7, no. 1, ed. Bennett G. Brown (Philadelphia: N. B. Sanders, 1984).

69 See particularly Jerry A. Fodor, *Psychological Explanation* (New York: Random House, 1968); also Hilary Putnam, "The Nature of Mental States," *Philosophical Papers*, vol. 2. (Putnam has retreated from the extreme foundationalism offered here.) See Joseph Margolis, *Philosophy of Psychology* (Englewood Cliffs, NJ.: Prentice-Hall, 1984), ch. 4.

70 See Hubert L. Dreyfus, Introduction to *What Computers Can't Do*, rev. edn (New York: Harper and Row, 1979), for a sense of the history of mental descriptions of computers.

71 See particularly Williams, *Problems of the Self*; and Strawson, *Individuals*.

72 See particularly Joseph Margolis, *Culture and Cultural Entities* (Dordrecht: D. Reidel, 1984), ch. 1.

73 Daniel C. Dennett, "Artificial Intelligence as Philosophy and as Psychology," *Brainstorms* (Montgomery, Vt: Bradford Books, 1978), p. 112.

74 Ibid., p. 113.

75 For a convenient review of chimpanzee achievements, see David Premack and Ann James Premack, *The Mind of an Ape* (New York: W. W. Norton, 1983).

76 Chomsky, *Rules and Representations*, pp. 69–70. See also Massimo Piattelli–Palmarini (ed.), *Language and Learning: The Debate between Jean Piaget and Noam Chomsky* (Cambridge, Mass.: Harvard University Press, 1980).

77 Jerome S. Bruner, *Child's Talk* (New York: W. W. Norton, 1983), p. 38. See also Bruner, *Beyond the Information Given*, ed. Jeremy M. Anglin (New York: W. W. Norton, 1973).

78 Bruner, *Child's Talk*, p. 23.

79 See for example the pertinent disagreement between R. Harré, "Social Sources of Mental Content and Order," and Paul F. Secord, "Social Psychology as a Science," in Joseph Margolis et al., *Psychology: Designing the Discipline* (Oxford: Basil Blackwell, 1986).

80 Gaston Bachelard, *The New Scientific Spirit*, tr. Arthur Goldhammer (Boston, Mass.: Beacon Press, 1984), p. 25.

81 Cf. ibid., 31, 40, 53.

82 See ch. 2, above.

83 See Joseph Margolis, *Pragmatism without Foundations: Reconciling Realism and Relativism* (Oxford: Basil Blackwell, 1986), ch. 10.

84 There is a perspective discussion of this linkage in the context of our present issue in Charles B. Guignon, *Heidegger and the Problem of Knowledge* (Indianapolis: Hackett, 1983), chs 1–2.

85 See for example the recent form of the standard quarrel as it appears in Ernst Tugendhat, *Self-Consciousness and Self-Determination*, tr. Paul Stern (Cambridge, Mass.: MIT Press, 1986), Lecture III; and Dieter Henrich, "Selbstbewusstsein: Kritische Einleitung in eine Theorie," in Rüdiger Bubner (ed.), *Hermeneutik und Dialektik*, vol. I (Tubingen: Mohr, 1970), cited by Tugendhat.

86   See Margolis, *Pragmatism without Foundations*, ch. 11.
87   Unlikely as it seems, this may well be the source of Rorty's well-known thesis, in *Philosophy and the Mirror of Nature*.
88   A clear specimen is afforded by Paul de Man, *Blindness and Insight*, 2nd rev. edn (Minneapolis: University of Minnesota Press, 1983), ch. 1. See ch. 11, below.
89   Martin Heidegger, *Being and Time*, tr. John Macquarrie and Edward Robinson from 7th German edn (New York: Harper and Row, 1962), p. 25 (in the German edn); p. 47 (in the English).
90   Martin Heidegger, "The End of Philosophy and the Task of Thinking," *On Time and Being*, tr. Joan Stambaugh (New York: Harper and Row, 1972), p. 58.
91   Ibid. Rorty's debt to Heidegger is reasonably clear.
92   Ibid., pp. 72–3.
93   Heidegger, *Being and Time*, p. 117 (in the German edn); pp. 152–3 (in the English).
94   Edmund Husserl, "Phenomenology and Anthropology," tr. Richard G. Schmitt, in Roderick M. Chisholm (ed.), *Realism and the Background of Phenomenology* (Glencoe, Ill.: Free Press, 1960), p. 138. It needs to be said that this paper by Husserl and Heidegger's "The End of Philosophy" are certainly meant by the authors, at least in large part, as attacks on each other's mode of philsophy – after the unpleasantness regarding Heidegger's revision of Husserl's *Encyclopaedia Britannica* article. There is justice on both sides.
95   A most revealing example of the inherent limitation of·the most recent literature attentive to the best naturalistic and phenomenological analyses of human history and human existence is afforded by David Carr's *Time, Narrative, and History* (Bloomington: Indiana University Press, 1986). Carr's essential concern is to explicate, against the currents of naturalism and phenomenology, Hegel's intention "to surpass the first-person-singular standpoint": not merely "to bridge the gap between consciousness and the external world but to resolve the deeper problem of transcending the I, not toward the external world, or for that matter toward the illusory Universal Subject, but toward the other *I*" (p. 139). Carr takes as his central theme Hegel's excellent motto of the *Phenomenology*: "an I that is We, a We that is I"; but unaccountably he construes it as "a concern for the first person plural which is not only substantive but also methodological" (p. 138). The truth is – unfortunately, Carr both alters Hegel's intention and loses an extraordinary opportunity to correct the philosophical record – that Carr *never* grasps the difference between an *aggregative* use of "we" and a *collective* use of social, societal, cultural, historical, institutional, traditional predicates and attributes. We are of course running ahead of our story here; but the master linkage between self-knowledge and knowledge of the natural world is – will, as we shall see – prove to be inextricably encumbered by the complexities of our knowledge of other selves, which presupposes our knowledge of the collective life in which we live. See below, ch. 12.

96 On the general analysis of consciousness and the problem of attributing minds to animals, see Margolis, *Cultural and Cultural Entities*, chs 2 and 3. On Brentano's treatment of perception and sensation, in particular his treatment of consciousness, see Franz Brentano, *Psychology from an Empirical Standpoint*, ed. Oskar Kraus, English ed. Linda L. McAlister, tr. Antos C. Rancurello et al. (New York: Humanities Press, 1973). In "Inner Consciousness" for instance Brentano clearly reserves the term "consciousness" for *all* mental states properly so characterized: "no mental phenomenon exists which is not ... consciousness of an object"; "the term 'consciousness,' since it refers to an object which consciousness is conscious of [by which Brentano means to permit us to speak of 'unconscious consciousness'] seems to be appropriate to characterize mental phenomena precisely in terms of its distinguishing characteristic, i.e., the property of the intentional in-existence of an object ..."; "All mental phenomena are states of consciousness" (p. 102).

97 "Further Investigations concerning Psychological Method. Induction of the Fundamental Laws of Psychology," ibid., p. 46.

98 "The Distinction between the Mental and the Physical," ibid., p. 79; "Classification of Mental Activities into Presentations, Judgments, and Phenomena of Love and Hate," ibid., p. 198.

99 "The Distinction between the Mental and the Physical," pp. 79-80; cf. the editor's note 2, pp. 79–80. See also, E. B. Titchener, "Brentano and Wundt: Empirical and Experimental Psychology," in Linda L. McAlister (ed.), *The Philosophy of Brentano* (Atlantic Highlands, N.J.: Humanities Press, 1976).

100 We shall return to the intentional more pointedly, in chs 7 and 9.

101 Roderick M. Chisholm, "Brentano's Descriptive Psychology," in McAlister, *The Philosophy of Brentano*, p. 99. Oskar Kraus cites a letter from Brentano to Carl Stumpf bearing on this matter (which Chisholm mentions), dated June 16, 1906, in which the following appears: "any observation is concerned with ourselves in a way. Anyone who analyzes a musical chord is really *apperceiving* components of himself considered as someone who is hearing. He finds that, as someone who is hearing the chord, he is simultaneously someone who hears the different notes," Kraus's "Introduction to the 1924 Edition," in *Psychology from an Empirical Standpoint*, p. 405.

102 See, below, ch. 9.

103 Merleau-Ponty, *Phenomenology of Perception*, pp. xvii–xviii.

104 See further, Joseph Margolis, *Art and Philosophy* (Atlantic Highlands, N.J.: Humanities Press, 1980), chs 1-3; and *Culture and Cultural Entities*, ch. 1.

105 Roland Barthes, "From Work to Text," tr. Josué V. Harari, in Josué V. Harari (ed), *Textual Strategies* (Ithaca: Cornell University Press, 1979), pp. 76–7. See, also, Harold Bloom, *The Anxiety of Influence* (London: Oxford University Press, 1973).

106   See below, ch. 12.
107   Harré, "Social Sources of Mental Content and Order," p. 100. See also,
      Rom Harré, *Personal Being* (Cambridge, Mass.: Harvard University Press,
      1984); *Social Being* (Totowa, N.J.: Rowman and Littlefield, 1979); and
      John Shotter, *Social Accountability and Selfhood* (Oxford: Basil Blackwell,
      1984).
108   Harré, "Social Sources of Mental Content and Order," pp. 105-6.
109   Jaynes, *The Origin of Consciousness in the Breakdown of the Bicameral
      Mind,* and Alvin W. Gouldner, *The Dialectic of Ideology and Technology*
      (New York: Oxford University Press, 1979), particularly ch. 2.
110   Konrad Lorenz, "Companions as Factors in the Bird's Environment; The
      Conspecific as the Eliciting Factor for Social Behavior Patterns," *Studies
      in Animal and Human Behavior*, tr. Robert Martin, vol. 1 (Cambridge,
      Mass.: Harvard Univesity Press, 1970).
111   Marjorie Grene, "The Paradoxes of Historicity," in Brice R. Wachterhauser
      (ed.), *Hermeneutics and Modern Philosophy* (Albany: SUNY Press, 1986),
      p. 185.
112   Ibid., p. 169. See also, Marjorie Grene, *The Understanding of Nature*
      (Dordrecht: D. Reidel, 1974), chs 3 and 18; note particularly the summary
      of Plessner's view of adaptation and adaptability, p. 40.
113   Harré, *Personal Being*, pp. 26-7.
114   Ibid., p. 95.
115   Ibid., p. 94; see, further, the whole of ch. 4.
116   The idiom of "embodied" and "incarnate" is elaborated in Joseph Margolis,
      Culture and Cultural Entities, ch. 1; and Joseph Margolis, *Art and
      Philosophy* (Atlantic Highlands, N.J.: Humanities Press, 1980), chs 1-3.
      In the present context, we need not pursue the full ontology of embodied
      entities; we need only take note of the importance of providing a suitable
      account.
117   Hans-Georg Gadamer, "The Universality of the Hermeneutic Problem,"
      in *Philosophical Hermeneutics*, tr. David E. Linge (Berkeley: University
      of California Press, 1976), pp. 7-9. See also, Brice R. Wachterhauser,
      "Must We Be What We Say? Gadamer on Truth in the Human Sciences,"
      in Wachterhauser, *Hermeneutics and Modern Philosophy*, *passim*.
118   See Hans-Georg Gadamer, *Truth and Method*, tr. (from 2nd edn) Garrett
      Barden and John Cumming (New York: Seabury Press, 1975), pp. 254–8.
      See also, Jürgen Habermas, "A Review of Gadamer's, *Truth and Method*"
      in Fred R. Dallmayr and Thomas A. McCarthy (eds) *Understanding and
      Social Inquiry*, (Notre Dame: University of Notre Dame Press, 1977).

4

Self and World

I

To admit the ineliminability of self is to encumber, correspondingly, our reflections on the world disclosed to self. For the first is the result of attempting in every possible way (and failing in every way) to acknowledge experience as the beginning of science without at the same time attributing such experience to some entity minimally aware of (or nonrelationally "possessing") it; and the second is the result of whatever, on the basis of such experience, we cannot reasonably refuse to regard, again minimally, as a fair inkling of the structure and properties of the real world. Naive experience certainly favors the salience of self and world; but the positing of self and world relative for any and every conceivable cognitive understanding rests on the accumulating (but hardly completed) record of the apparent impossibility of separating self and experience. *Selves* are, paradigmatically, cognizing entities or cognitive agents, and the *world* is, paradigmatically, the ordered ensemble of whatever they may be said to know.

Clearly, there must be some proportionality (or adequation) between how we regard the self as structured and how we regard the world as structured. But, in merely conceding the ineliminability of self, we are conceding no more than the apparent necessity of the salient, *not* anything absolute or essential or necessary *de re* in a deeper sense. We are conceding the self to be a cognizing entity but we are not (here) assigning any cognitive feat in particular to the self thus admitted – nor for that matter any particular model of the structures of such an entity. So, for example, we cannot (in the sense of the salient) treat the "world" as a mere fiction or as the object of an accurate representation furnished on the strength of experience "interior" to the self, without risking the classic forms of representationalism and skepticism.[1] On the other hand, we cannot treat the "world" as apparently disclosed to self, or "self" as apparently disclosed to itself, as assuredly such as it naively appears to be, without risking the classic forms of cognitive privilege and transparency. Realism in science, it seems prudent to say, will avoid these polar extremes and embrace the symbiosis of self and world. But the history of the topic does not entirely confirm this expectation.

There is nevertheless a certain stability and effective viability to standard pictures of self and world, within the margins of the life of the human species, that, on the best guess, depends for its survival on its distinctive cognitive prowess and on a relatively undistorted sense of the powers of self and the structure and properties of world manifesting that prowess. This, very roughly, is what *pragmatism* comes to: or, better, what the admissibility of objective claims comes to, or of *objectivism* construed non-pejoratively *within* the holist context of pragmatism. The fact that the ineliminability of self is sustained only with respect to that entity that (nonrelationally) possesses experience signifies that, in the dialectical sense intended, selves must be treated as whatever remarkably endowed entities they may prove to be *within* the very world that we theorize is cognitively disclosed to them. Of course, to speak of entities and their attributes, in speaking of selves, will seem to some somehow to privilege an idiom that means to penetrate to the independent structures of an actual and independent world – to "correspond," fortunately, to the true order of things.[2] But there are two ready answers to this insinuation: first, the linguistic hegemony of a grammar of subjects and predicates is at least as ineliminable as the entitative treatment of selves, regardless of the uses to which we may wish to put language; and, second, inasmuch as the deconstruction of the correspondence picture employs the same grammar, it must be a *petitio* to insist that, whenever not used to subvert the offending (the privileging) picture, the predicative idiom cannot fail to be committed to it (the correspondence view).

Admittedly, we need to keep these paired distinctions as supple as possible. To concede the near-invariance of the predicative idiom is not to favor essentialism or a correspondentism or a certain characteristic Aristotelian privileging of the verb *to be*;[3] it is only to identify a functional habit of the grammars of natural language and natural thought, not any particular structures that the forms of thought and language must exhibit. We may resist (against Noam Chomsky, for instance) the notion of universal, determinate grammars for natural languages, but it is quite another matter to resist (and to show the reasonableness of resisting) the functional universality of the predicative nature of thought. Similarly, we may yield as much as possible in the direction of W. V. Quine's tolerance for alternative ontological parsings of any array of posited data, but it is quite another matter to assume that the testing of such alternatives will actually confirm (either in first -or in second-order terms) Quine's indeterminacy thesis or an evidentially impenetrable holism or the indefeasible extensional equivalence of alternative entitative parsings when confronted by empirically reasonable hypotheses regarding the biological and social interests of the species.[4]

Also, to treat self, somehow, as prior to the world, as the precondition of the very world, as originally constructing or constituting the world itself, is a hopeless extravagance hardly required by the mere ineliminability of self. That extreme doctrine produces utterly unnecessary and insoluble paradoxes: as by installing an inescapable solipsism, or by risking the permanently fictional nature of the world (or worlds) we suppose we inhabit, or by making a mystery of the relationship between the plurality of human selves and "that" original constituting self and between each individual self and the society of aggregated such selves. It is, of course, just the doctrine that lies at the heart of so much (perhaps all) of Edmund Husserl's phenomenology. It cannot possibly be sustained, regardless of whatever use we may be able to make of phenomenological reflections freed from such a needless *aporia*.[5] In effect, it is a doctrine that risks the most extreme forms of idealism – ironically, without ever having intended to. The decisive correction is this: Husserl is quire right, in resisting "objectivist" privilege, to deny that the *Sinn* (or sense) of the world, assignable to the world's objects, is assignable independently of the initiating (intentional) "acts" of cognizing subjects; but the respect in which a thus-meaningful "world" is "constituted" cannot be equivalent to the self's constituting the actual world. Husserl himself is quite clear about this when he emphasizes that the perception of a physical object *is* the perception of an actual physical object, though it is (also) the "intended object" of the perceptual "act."[6] Nevertheless, there can be little doubt that, in developing his mature account of the "intended" *Sinn* of the world, Husserl does not escape (or does not clearly escape) the extreme idealist formulation. The symbiotic thesis is more balanced.

It is similarly impossible (as impossible as subscribing to an extreme idealism) to deny that contending accounts of the "world" (and of the many cognizing "selves" within it) *are* provisional constructions of the way the world is taken to be by inquiring selves – within the constraints of a holism that respects the realist import of species survival and, therefore, tolerates (in *that* holist or pragmatist sense) the objective (or objectivist) standing of what remains noticebly difficult to avoid acknowledging within the determinate inquiries of human *praxis* and human science. What is *objective* in the way of knowledge is, primarily, what is both salient and difficult to deny or eliminate, not merely in terms of our first-order inquiries but also at every level of critical or transcendental reflection upon whatever appears to be thus disclosed. (It should be added at once, of course, that nothing so far conceded precludes a relativistic account of our sciences.)

Hence, the positing of self does not entail any privileged claim regarding the actual structure of self or of world: *self is simply the principled locus of cognitive power itself.* It is not an arbitrary posit, but it is not a flat discovery either. *What we discover, rather, is the continual failure of efforts to posit effective and compelling alternatives to that posit* – as, indeed, we also do in entertaining alternatives to the predicative idiom. To specify what a "self" knows is to specify what the "world" is like; the privisionality with which the one is formulated is the very provisionality of the other – we cannot escape tendering matched claims about both. So the objectivism here espoused is an objectivism of salience, not of cognitive privilege. For by *salience* we mean some selective consensual recognition, tacit as well as explicit, central to and persistent regarding the experience of particular historical societies, possibly relativized within and across societies, second-order as well as first-order, capable of surviving disciplined challenge diachronically as well as synchronically, fitting and responding to accepted norms of systematic adequacy, scope, coherence and the like (norms that are themselves subject to similar challenge in terms of shifting salience). In effect, this is to espouse a strong convergence between naturalistic and phenomenological inquiry – both deprived of their characteristic claims of cognitive privilege.

What this means is that every form of truth-claim – logical, factual, explanatory, methodological, interpretive, ontological, phenomenological, critical, transcendental – is eligible (recoverable) in terms of the idiom of salience without at all risking a presumption of privilege. The idea that *every* such claims betrays (or necessarily betrays) the presumptions of cognitive privilege is utterly gratuitous, an abuse of ordinary distinctions. Perhaps a fair way of putting the point is simply to affirm that claims within an idiom of salience are *not*, as (within limits) are most of the idioms of privilege, committed to strong forms of correspondence, logocentrism, the "mirror" thesis or the like.[7] Once the entirely reasonable warning is in hand, we may take ourselves to have bracketed every such claim as a claim meant to fall entirely within the limits or the "givens" of salience. The ease of such an adjustment exposes just how inflated must be the charge of the aporetic nature of every science and every philosophy, or the claim of salvageable division between every proper science and every merely vacant and self-defeating philosophy.[8]

When, therefore, Richard Rorty, apparently siding with the "political utopians since the French Revolution," affirms (or at any rate appears to affirm) that they "sensed ... not that an enduring, substratal, human nature has been suppressed or repressed by 'unnatural' or 'irrational' social institutions but that changing languages and other social practices can produce human beings with quite new natures, people of a sort who had never before existed," we must notice the obvious uneasiness of the immediately added *caveat*:

> The difficulty faced by a philosopher who, like myself, is sympathetic to this suggestion, one who thinks of himself as auxiliary to the poet rather than to the physicist, is to avoid hinting that this suggestion gets something right, that my sort of philosophy corresponds to the way things really are. For this talk of correspondence brings back just the idea which my sort of philosopher wants to get rid of, the idea that the world or the self has an intrinsic nature.[9]

To admit that we *do* "get things right" – though only within the precincts of salience and not of privilege or transparency – is to risk conceding a full role to truth and truth-claims and deflating utterly the original deconstructive intent.

Understandably, Rorty opts for the charming (but quite fatal) extreme of recommending that "we should drop the idea of truth as out there waiting to be discovered," that "our purposes would be best served by ceasing to see truth as a deep matter, as a topic of philosphical interest, or 'true' as a term which repays 'analysis'." Rorty goes so far as to add, "To say that Freud's vocabulary gets at the truth about human nature, or Newton's at the truth about the heavens, is not an explanation of anything. It is just an empty metaphysical compliment which we pay to writers whose novel jargon we have found useful."[10] But to say this is to miss or ignore a very large option. It is simply, perhaps discontentedly, to impoverish the whole of our conceptual resources: *for it fails to accommodate the "discovery," within the contingencies of the salient, of whatever proves noticeably difficult or (thus far) impossible to deny or eliminate or alter significantly by way of new conceptual posits.* Truth, therefore, remains a "deep matter" – deeper now that it is freed of the correspondence obsession, never so thin, however, as to obviate all talk of discovering the structure of reality. The lesson to be learned is that the ineliminability of self is and must be matched by the ineliminability of world. But *what*, cognitively, may be said about either is certainly an artifact and a contingency of actual inquiry. Rorty's remark is, at bottom, a deconstructively flowered adherence to (and distortion of) Quine's indeterminacy thesis – unjustified (and unable to be justified) simply because Rorty conflates the "external" objection to all forms of cognitive transparency with the "internal" question of the weight of evidence for or against this or that scientific claim. Quine does not face the same difficulty (but another), because his naturalizing of science and philosophy entrenches the symbiosis of self and world; although, in doing that, Quine fatally fails to loosen his holism in the direction of the "saliencies" he himself acknowledges:

> Episodes leave traces. ... Such *traces*, whatever their physiological nature, are essential to all learning. The trace or an episode must preserve, in some form, enough information to show perceptual similarity between that episode and later ones. ... Each episode ... is a brief time in the life of the subject in his bodily entirety. All impingement is included, sparing no bodily surface. ... Salience shows itself in behavior through the behavioral evidence for perceptual similarity.[11]

The very notion, furthermore, of the symbiotic linkage between world and self, as the mythically "originary" condition from which human *praxis* and science devolve as the contingent or historicized "career" of *that* mythically hypothesized encounter, is essentially what, by an admittedly extravagant idiom, Martin Heidegger offers as the corrective to Husserl's inescapable idealism, the meaning of the primacy of (what in the Continental European tradition has come to be understood as) *subjectivity*. The theme is in part an ingenious and spacious warning: that every objectivism is provisional (not that there is no objectivity), that there is no cognitive privilege (not that knowledge is impossible), that there is no originary beginning of science or *praxis* (not that there is no science or competent *praxis*), that "self" (*Dasein*, in Heidegger's idiom, not to insist on too fine an analysis) has (merely as such) no cognitively pertinent weighting to offer in assessing determinate and opposing claims (not that nothing can be objectively assessed as true or false by suitably endowed selves). Self, or *Dasein*, is the abstracted power of human selves (or of other selves, if there be any) answering to the bare cognizability of whatever is real; but science and effective *praxis* are the actual achievements of earthly selves, structured in some way that inquiry has always to posit in positing any particular body of truths. The symbiosis of *Dasein* and Being, in Heidegger, corresponds to the leaner holism of the pragmatist conception of inquiry – in at least the sense that both provide for the radical provisionality of historically phased claims, for the eligibility of the sciences nonetheless, for a reflexive assurance regarding the competence of cognizing selves, for the symbiosis between our knowledge of selves and our knowledge of the world, and for the denial of any privilege entailed in the mere admission of selves as essentially cognizing entities. In Heidegger's oddly inflated form of compression,

> Dasein is an entity which does not just occur among other entities. Rather it is ontically distinguished by the fact that, in its very Being, that Being is an *issue* for it. But in that case, this is a constitutive state of Dasein's Being, and this implies that Dasein, in its Being, has a relationship towards that Being – a relationship which itself is one of Being. ... *Understanding of Being is itself a definite characteristic of Dasein's Being.* Dasein is ontically distinctive in that it *is* ontological. ... Sciences are ways of Being in which Dasein comports itself towards entities which it need not be itself.[12]

Slimly put, cognizing selves are uniquely reflexive entities.

The posit of an actual structured world is the posit *of* a cognizing power (is intentional in that sense), even if it is the posit of a world that, on a thesis, does not depend *for* its structure on that entity's intervention. Subjectivity, in the Heideggerean idiom (as opposed to the Husserlian) is not an expression of idealism or solipsism: it is rather, by default, by default of the ineliminability of self, the admission of the symbiosis of self and world (or, rather, of *Dasein* and Being) in the "originary" sense of the symbiosis of cognition and the cognizable; so that, at the level of any determinate inquiry, we find that we cannot overcome the symbiosis of the realist and idealist of any human science. In particular, we cannot escape that symbiosis in attempting to assign determinately the structures of our science to the separate roles of self and world – which is precisely the claim Husserl believed himself to have exposed in Kant preeminently: the projects, pejoratively construed, of "objectivism."[13] The symbiosis of self and world is, we may say, the single theme that (effectively) commits us at one and the same time to (1) an inescapable holism, against transparency, *and* (2) the need to rely (*within* the space of that holism) on salience rather than privilege. Nearly all the ontological myths of Anglo-American and Continental European philosophy agree on (1) but many of them (witness the convergence of such otherwise oddly linked discussants as Quine and Rorty and Heidegger and Derrida) fail to grasp or admit the full import of (2). ((1) and (2), taken together, form a most economical summary of what the new pragmatism comes to.

II

There is another important theme linked to this symbiosis. Concede the historicity of self – for instance, in the naturalized way Foucault has favored,[14] or as in Heidegger's account (speaking of the "temporality" and "historicity" of *Dasein* rather than directly of human selves): we are then encouraged to shape a theory of self that profoundly accommodates the salient historicity featured in all current views of science and *praxis*.[15] The point of pressing this theme is to indicate how naturally, once the ineliminability of self is admitted, our very sense of the objectivity of science and of the validity of scientific realism positively requires for their support some conceptual linkage with the symbiosis of self and world. In fact, in pressing the theme, we are helped to formulate a very clear contrast between the viability of an objectivism restricted to the salient and any version of the excessive, impoverishing zeal of deconstructive substitutes for it. In this sense, the *objectivism of the salient* (as we may term it) falls between the objectivism of the privileged and the deconstruction of both; and that thesis cannot be satisfactorily formulated without attention to the ineliminability of self and the symbiosis of self and world. Nevertheless, to place the prospects of cognition thus is not yet to attempt their assessment – for instance, along the lines of transient contingencies, the forestructuring of inquiry, historical horizons, tacit ideologies, pluralism and relativism, continuity and discontinuity of theory and method, or transcendental criticism.

It is both amusing and sobering, however, to be able to draw, quite straightforwardly, the most telling consideration regarding the historicity of self from an early "contemporary" text in the philosophy of science: Bachelard's *The New Scientific Spirit* – certainly the contemporary of the Vienna Circle's program, very nearly the contemporary of Heidegger's *Being and Time*, also the contemporary of Husserl's *Crisis*. For the themes of Bachelard's extraordinary anticipation of the gathering import of the historicizing of science and of the philosophy of science, in Popper, Kuhn, Feyerabend, and Lakatos particularly, are noticeably absent (or very weakly represented) in the principal currents of Anglo-American analytic philosophy from the positivist period very nearly down to our own day. Bachelard very simply observes (in a chapter tellingly titled "Non-Cartesian Epistemology"), "Concepts and methods alike depend on empirical results. A new experiment may lead to a fundamental change in scientific thinking. In science, any 'discourse on method' can only be provisional; it can never hope to describe the definitive complexion of the scientific spirit. The fact that sound methods are constantly changing in science is a fundamental feature of *scientific psychology*."[16] Bachelard's theme is the "constructive" nature of scientific knowledge: "For science ... the qualities of reality are functions of our rational methods. In order to establish a scientific fact, it is necessary to implement a coherent technique. Scientific work is essentially complex. Science is a discipline of active empiricism. ..."[17] One remarkably suggestive application of Bachelard's thesis (to which we shall return[18]) points to the difficulty of denying that what we impute to ourselves as our cognizing capacities is a function of our changing science *and* that those capacities themselves may actually change because of our changing experience. There is some correspondence, therefore, between our theory of the methodology of science and our theory of the capacities of mind: fixity regarding the first tends to be matched with a kind of nativism regarding the second; conversely, to construe methodology in historical terms puts considerable pressure on the plausibility of insisting (say, in Noam Chomsky's or Jerry Fodor's way) on the biologically determined, modular, unchanging cognitive competences of the human mind. (The details of nativism will occupy us toward the end of this account.)

Bachelard is not an idealist, however, in the extreme sense threatened by Husserl. On the contrary, the insistent realism that the history of the physical sciences makes almost irresistible is seen to be defenseless without a strong concession to the constructive function of orienting theories that are forever empirically underdetermined anyway. That perception is itself unfathomable without coming to terms, along the lines already sketched, with the symbiosis of self and world. This is the sense, all right, in which *an adequate science must account for the very achievement of science*: science is itself a phenomenon that falls within the space of nature.

So it is not merely that science is an artifact of the symbiosis of self and world, or that its imputed structure (and the structure of the world) cannot but conform to the historicized saliences that contingently dominate human experience; it is also that science is a datum for science, it is also that our analysis of the imputed structure of the world cannot be segregated from our analysis of the datum of an historicized science. The critical theme, from at least Descartes to Husserl, cannot vindicate business as usual regarding the "best" account of the natural or physical world – now that we have admitted the contingency of all such accounts on the enabling powers of human investigators. For, to construe science as a datum for science is to concede as well the indissolubility of first-

and second-order questions. Now, the symbiosis of self and world is seen to infect (but never to decide), distributively, the weighting of every detail internal to our science. It is not merely that science depends on human experience: it is rather that the posited regularities of the world may be (and almost always are) inadequately indexed to the fragile saliences of a labile theorizing science.[19]

At any rate, the notion of the "constructive" theorizing function of scientific inquiry, under the very condition of disallowing cognitive privilege, both ensures an ineliminably idealist component in *any* orderly *realist* science and risks a radically idealist vision in which the world is simply "constituted" whole cloth by an exercise of that same theorizing function. The *only* way to offset the latter and to make room for the former is to opt for a pragmatist reading of scientific holism; for such a holism (1) ensures the global realism of human inquiry on non-cognitive grounds, on grounds linked to the survival of the species; and (2) favors, as such, no determinate or distributed truth claims, no particular scientific program. Consequently, (3) it is *only* by a strategy of pursuing the salient within such a holism that a realist reading of the determinate work of the particular sciences can possibly be vindicated. *But that work can be vindicated thus.* Broadly speaking, this is the (unperceived – even resisted) point of the profound agreement between W. V. Quine's well–known holism and Heidegger's conception of subjectivity – with respect to science.[20] It marks an agreement, however, of which Quine himself fails to see the implications, even from his side of the equation. For, although he naturalizes epistemology and science (which is to say, he acknowledges that both science and philosophy must be coordinately constrained by the conditions of holism), he does not admit or even perceive or investigate the implications of the symbiosis of *self* and world. There is no discussion, in Quine, of the complications that this entails, unless (too sanguinely for Quine) reading back in terms of the competence of a cognizing or developing self whatever, on independent grounds, Quine believes favors his extreme extensionalism itself constitutes a genuine inquiry into the question.[21] Paradoxically, Quine's very account of holism and the analytic/synthetic dogma depends on the symbiosis of self and world; also, the plausibility of Quine's extensionalism cannot but be adversely affected by his own neglect of the topic.

The point is that the free play of scientific inquiry is assured of its realist import in (and only in) *holist* terms, regardless of its punctuated convictions along a historical career. As a consequence, holism can contribute nothing to the principled assessment of distributed realist claims – *along the lines of salience, within the scope of holism,* where for instance the analytic/synthetic distinction functions (if it functions at all) and must function (if there is to be any science at all). This explains the fundamental difference between Quine's holism and Pierre Duhem's. They are utterly different notions; for Quine's "blurs the supposed contrast between the synthetic sentence, with its empirical content, and the analytic sentence, with its null content,"[22] and Duhem's does not. Hence, Quine's holism addresses the conditions under which science itself is conceptually possible – radically, in the pragmatist sense; and Duhem's is concerned rather with alternative theorizing options confronting the working scientist *within* the space of (what Duhem regards as) the holist paradox *of his empirical findings.*[23] In this sense, Quine (though he would shudder at the use of the term) *is* addressing transcendental questions, whereas Duhem is not.

One late passage from Quine repays close reading:

> Another notion that I would take pains to rescue from the abyss of the transcendental is the notion of a matter of fact. A place where the notion proves relevant is in connection with my doctrine of the indeterminacy of translation. I have argued that two conflicting manuals of translation can both do justice *to all dispositions to behavior,* and that, in such a case, there is no fact of the matter of which manual is right. The intended notion of matter of fact is not transcendental or yet epistemological, not even a question of evidence; it is ontological, a question of reality, and to be taken naturalistically within our scientific theory of the world. Thus suppose, to make things vivid, that we are settling still for a physics of elementary particles and recognizing a dozen or so basic states and relations in which they may stand. Then when I say there is no fact of the matter, as regards, say, the two rival manuals of translation, what I mean is that both manuals are compatible *with all the same distributions of states and relations over elementary particles.* In a word, they are *physically equivalent.*[24]

Here, clearly, in the course of aligning himself with Duhem, Quine *breaks* with his own rejection of the analytic/synthetic contrast: for, to make the indeterminacy thesis work, he must posit behavioral dispositions neutral to his own argument (he must posit empirically detectable behavioral dispositions that, as such, are not open to question in terms of the analytic/synthetic dogma). That *is* Duhem's way with theoretical hypotheses in the face of empirical findings. But, *if* the dogma is rejected at the level at which the very possibility of science is being considered, then even the Duhemian puzzle *does not yet arise;* it is not even admissible. Quine holds that "there is no fact of the matter" ontologically, but he obviously wishes to hold (inconsistently or at least without clear defense) *that there are facts* in the empirical sciences. There cannot be such, however, without some distributive strategy regarding what is salient *within* a Quinean holism; and that is a matter Quine never discusses (though Duhem does). The same difficulty arises in Quine's inviting us ("to make matters vivid") to consider what follows "if" we commit ourselves to the achievements of basic physics. A Duhemian resolution of competing distributed claims within a first-order science does not need to appeal (and finds no use in appealing) to a second-order rejection of a principled analytic/synthetic distinction; and a Quinean insistence on such a distinction is irrelevant (and even inimical) to selecting, within a first-order science, salient, genuinely factual, holistic sketches of sectors of the world with respect to which differently distributed claims may yet empirically compete. To dub both of these approaches "holistic" is to risk, by equivocation, losing the distinction between first- and second-order inquiry within the very continuity of science and philosophy that the introduction of the term was originally meant to fix.

Here, Bachelard's instruction is decisive. For Bachelard emphasizes that the provisional standing of conceptual schemes, theories, geometries, methods of inquiry, ontologies and the like "depend[s] on empirical results." The indeterminacies of *Duhem's* would-be conventionalism are, therefore, of minor significance (even if admitted), once we see that such indeterminacies are themselves restricted by the saliencies of the empirical. But, then, the saliencies of the empirical do provide a ground (escaping the charge of privilege) for the *operative* contrast between the analytic and the synthetic – without in the least disallowing Quine's entirely *formal* analysis of the dogma. (By "formal," here, is meant an inquiry that does not first concede the empirically operative distinction – the distinction in actual inquiry and scientific practice – that any purely syntactic or logical or mathematical systematization would, responsibly, have tried to *fit*, even if in doing so it would inevitably have altered that practice in some way. Insistence on this initial relevance constraint regarding purely "formal" studies provides the precise sense in which – surprisingly, given his view of the alleged dogma – Quine champions, as in his Millian reading of logic, even logic and mathematics as "empirical" disciplines. The original of this conviction, whatever its pretensions, is undoubtedly the one Aristotle expresses.) In this sense, contrary to what Quine never fails to insist on, "there *is* a fact of the matter" – or at least there is the possibility, if we concede (as well we must, if Duhem is close to the truth) that the "fact of the matter" may have to yield somewhat in the direction of a moderate relativism even within the empirical sciences. A Duhemian-like relativism, whether conventionalist or not, is *not* the equivalent of Quine's ontological relativism: the one is internal to the empirical sciences, the other is "external" (in the projected, transcendental sense) regarding the conceptual conditions of science itself. Alternatively, where Quine's ontological relativism *is* internal (as well it may be said to be), Quine is simply mistaken in denying *sans phrase* that "there is a fact of the matter." There may well be no fact-of-the-matter that is so determinately crisp that all Duhemian-like alternative readings of a world-be factual saliency can be ruled on exhaustively and once and for all; but that is not to say that there is no fact-of-the-matter. Facts-of-the-matter *may*, at least in certain sectors of science, have Duhemian contours. Quine cannot really deny this, if, as is clear, he means to champion "ontologically" the scientific accessibility of reality and the continuity of science and philosophy. But to act thus is certainly not to endorse the charge of the analytic/synthetic dogma *at that point* – nor does it require abandoning the charge either. It simply restricts its relevance to exposing the impossibility of capturing, in principled second-order terms, the first-order distinctions of an operative science that (on second-order reflections) are thought to be codifiable as analytic and synthetic. What we may draw, then, from Bachelard's prescient remark includes at least the following:

1   we cannot detach, as Quine seems inclined to, the theoretical relevance and power of the formal analysis of concepts (as of the analytic/synthetic distinction) from some range of salient experience or empirical findings within which it has its function (however provisional its priority may be admitted to be);

2   the revision of science, of our conception of the world and of our cognitive achievement with respect to it, is essentially continuous, formed and reformed in the light of the changing salient experience of investigative communities;

3   discontinuities, therefore, of method (in Bachelard's sense), research programs (in Lakatos's), paradigms (in Kuhn's), hypotheses (in Duhem's), ontologies (in Quine's) are themselves empirically detected and formulated within what is salient for such discerning societies;

4   since science is itself distinctly constructive in whatever sense it may be assigned a realist import, the scientific imagination must always overcome the restricting habits of mind encouraged by

adherence only to the salient, only to the immediate "givens" of experience and *praxis*;[25] and

5   the processes of such revision are ceaseless, never totalized, incapable of final closure, within the conditions of historical existence.[26]

These five findings effectively represent the philosophically viable middle ground between older theories of cognitive privilege and the radical, more recent attractions of the spoiler mentality of (much) deconstruction. The strategic advantage of putting matters this way is simply that it permits us to see at once that the characterization of the self cannot fail to be congruent with the characterization of what, through our science and philosophy, we take to be the real world; *and*, that the characterization of the latter cannot be segregated from our view of the role of the former.

This is the reason why Quine's well-known repudiation of propositional attitudes and intentionality is merely cavalier, however disciplined and intriguing. For, on the argument given, Quine *cannot* report what may be said "in the strictest scientific spirit" without acknowledging the symbiosis of self and world.[27] The repudiation of intentionality is advanced only in the pejorative "formal" sense just introduced; hence, Quine *never* discusses the matter in the context of the actual work of the empirical sciences — a fact of some importance given that, on Quine's view, "epistemology merges with psychology, as well as with linguistics."[28] Worse, these "formal" pronouncements lack altogether a reasoned defense: we are never told why we should presume that they have the telling force they have, why the world should be taken to be such that such findings could collect essential constraints governing any and every realist science. Quine is more correctly treated as a prime target for deconstruction than the proper darling of recent deconstructionists. Alternatively put, Quine's pragmatism, however holistic, is not at all thoroughgoing, for there is a profound and fixed presumption in Quine that the world caters for a totalized extensionalism. Curiously, it is precisely that extensionalism that spawns Quine's ontological relativism, though the relativism could be and could have been defended in terms similar to Duhem's; curiously also, recent deconstructive approval of Quine's relativism has pretty well ignored its conceptual linkage with a doctrine that, "canonically," could only have invited the charge of logocentrism.[29]

If we reverse these objections, then deconstruction itself will be seen to commit either of two fatal blunders or to reduce to little more than an inflated version of the sensible and accurate judgments already given by Bachelard. For, if the distinction, introduced earlier, between an *objectivism of the salient and an objectivism of the privileged* be admitted, then the charges of the deconstructive critics are easily turned: all that needs to be done is to bracket truth-claims within the confines of the salient. Either, then, the deconstructionists conflate the two sorts of objectivisms (what may be called Nietzscheanism[30]), holding that *any* pretension toward science, factual inquiry, objective truth is fatally poisoned by assumptions of correspondence, transparency, cognitive privilege or the like; or else they suppose that a demarcation can in principle be drawn between legitimate science (within the pale of the salient) and "philosophical" excesses that could not possibly be similarly construed. Quine's own reasonable conjecture, generalizing from the continuity of epistemology and psychology, to the effect that philosophy and science are continuous and inseparable,[31] should have decisively marked the impossiblity of a preemptive defense of his own otherwise attractive extensionalist strategy – should have marked it, that is, in a way analogous to the logocentric charge. (Of course, Quine's continuity thesis is entailed in his attack on the analytic/synthetic distinction.)

Nevertheless, in the most recent views offered by Richard Rorty – perhaps not unfairly labeled the most prominent American philosopher-deconstructionist – Nietzsche and Davidson (effectively, Nietzsche and Quine) are linked together in an oddly Darwinian reading of the second sort of deconstructive strategy. Rorty remarks,

> I can develop the contrast between the idea that the history of culture has a telos – the discovery of truth, or the emancipation of humanity – and the Nietzschean and Davidsonian picture by noting that the latter picture is compatible with a bleakly mechanical description of the relation between human beings and the rest of the universe. ... Positivist history of culture ... sees language as gradually shaping itself around the contours of the physical world. Romantic history of culture sees language as gradually bringing Spirit to self-consciousness. Nietzschean history of culture, and Davidsonian philosophy of language, see language as we now see evolution, as new forms of life constantly killing off old forms – not to accomplish a higher purpose, but blindly.[32]

In effect, on Rorty's view, "Our language and our culture are as much a contingency, as much a result of a thousand small mutations finding niches (and a million others finding no niche), as are the orchids and the anthropods."[33] But, *if* science has a reasonably systematic career *within* the range of the salient and in a way (let us concede) disconnected from philosophy, then science need not succumb to the (logocentric) presumptions of either prositivism or romanticism; the two objectivisms can and should be sorted; *and* the "contingency" of science cannot possibly be the same as, or have the same import as do, evolutionary contingencies obtaining in nature prior *to the emergence of self*. The purposiveness of science simply is *not* the equivalent of an inherent teleology in physical and biological nature (which must rightly be opposed).

Here, it is unclear whether Rorty is pursuing the second (deconstructive) strategy – against philosophy, say – or falling back to the first, in the sense in which normative, critical, self-corrective, justificatory, legitimating efforts with regard to the ongoing truth-claims of science (in effect, philosophy) are somehow taken as inevitably repudiating, violating, or attempting to surpass the natural contingency of a scientific culture. When, therefore, Rorty says, "the *world* does not provide us with any criterion of choice between alternative metaphors, ... we can only compare languages or metaphors with one another, not with something beyond languages called 'fact'" (his gloss on Davidson's theory of metaphor), he clearly falls back to the first strategy at the same time as he fails to vindicate the second.[34] For an objectivism of the salient preserves the function of science without the presumption of privilege (or at any rate, that is its claim), and the idiom of "truth" and "fact" (therefore, the idiom of second-order "legitimation") has a perfectly coherent function within that space.

Rorty, however, may well have succeeded in pinpointing (contrary to his own purpose) the fatal irrelevance of *his* kind of deconstruction. For, whatever else may be said about the viability of science and philosophy, the discontinuities of systematized science – which (applauded by Rorty) have been so popularized by Kuhn – have nothing *pertinently to do with the blind contingencies of biological evolution* (the excessive theme read from Nietzsche's own excessive metaphors). This is just what Bachelard had implicitly foreseen. This, therefore, is just the sense in which the deconstructive program (*this* sort of deconstructive program) is no more than an inflated and misleading advertisement of what we have already collected (from Bachelard's statement) as the themes of an objectivism of the salient. To hold that there is *no* purposiveness in nature is to hold either that human beings are not natural creatures or not natural insofar as they are purposive or that their purposiveness is a delusion. To admit the ineliminability of self is to admit purposiveness in nature (though not of nature *en bloc*): hence, within at least the scope of the salient, it is to admit the purposive self-discipline of science and philosophy.

III

One final consideration will serve to demonstrate the remarkable power of the seemingly modest thesis of the ineliminability of self. If we admit the emergence of self and ascribe to it no more than an objectivism of the salient, then the single decisive feature that may be ascribed to any science lies with its *inherently restricted scope* and inescapably *constructive nature*. Again, these are only the dual themes we have conveniently elicited from Bachelard's review of the history of modern science. We may say that this dual feature marks the minimal *intentionality* of science – or self – or science as the work of a society of selves. A God's-eye science would obviate such intentionality.

Now the *scope* of a human science may be admitted to be more restrictive than whatever is regarded as the full compass of the actual world, but it may be thought to be of a suitably congruent sort. That, one supposes, is both what cognitive privilege, correspondentism, transparency, logocentrism and the like affirm and what a throughgoing extensionalism affirms – and what, therefore, signifies the logocentric presumption of the latter. But, if we are confined to the salient (in the sense intended), then the "world" or "worlds" explored within the scope of the salient cannot possibly be the same as the *world* (mentioned in Rorty's remark, cited above) that the extensionalists and the partisans of cognitive transparency posit as the object of their inquiries. The deconstructionists (certainly Nietzsche) hold that *any* truth-claim is a "distortion" of the way the world is (whatever that may impenetrably be). The advocates of salience would cast the matter rather differently, in terms of the constructive nature of science, also of the symbiosis of self and world – for they (but not the deconstructionists) are committed to accommodating the realism of science.

This supplies a reason for our earlier worry about the more than incipient drift toward an extreme idealism in Husserl's late philosophy. Its counterpart appears, surprisingly, in what we are here calling pragmatism, in recent views expressed by Nelson Goodman (who, with whatever sympathies for William James's empiricism, would repudiate the label). Goodman's account invites a small detour.

Goodman is a "constructionalist" on his own account, which is at least to speak (with James) not "in terms of multiple possible alternatives to a single actual world but of multiple actual worlds" – in accord with the dictum, "We can have words without a world but no world without words or other symbols."[35] Goodman fails, however, to explain how to individuate alternative worlds, how to distinguish between incompatible actual worlds and apparent incompatibilities within any supposed world, how we shift or how we may justify shifting from one world to another, how we know we are shifting, how the undertakings of science are affected by the theory of plural worlds, what the relationship is between "our" Earth and the plural "earths" of these plural worlds, what the sense is in which any putative such world is formulably or recognizably a "totality," and how we may distinguish between "true" and "false" worlds (or, "true versions" of "actual worlds" as opposed to "false versions" of "nonworlds").[36] Also, Goodman never explains how the various "worlds" rightly constrain our words.

Goodman's uncompromising thesis is that, "if there is any actual world, there are many. For there are conflicting true versions and they cannot be true in the same world."[37] But Goodman fails to explain how "true" is to be used within *any* "world" or across "worlds." He clearly opposes correspondence and coherence theories – correctly, as a constructionalist. Truth, then, is demoted but not eliminated in his account, as by featuring "validity of inductive inference"; but inductive validity itself depends on "right categorization" (Goodman is thinking here of his familiar example of "grue"/"bleen"[38] – which is not restricted to questions of truth (but accommodates them). And, yet, right categorization itself depends on "entrenchment": "Rightness of categorization, in my view," he says, "derives from rather than underlies entrenchment."[39] But Goodman has no account of entrenchment fitted to plural worlds.[40] If, then, entrenchment is itself an artifact of "making" worlds – by which we make and remake ourselves – *how* can we ever provide a ground for a realist science (however attenuated) that confirms at least "right versions of actual worlds"? Certainly, one can see the implicit danger, in Goodman's ingenious vision, of a florid and nonconverging idealism completely out of conceptual control. Perhaps it is not out of control, but there is as yet no way to make that clear. *If* "entrenchment" approaches what we have been calling "salience," then one world will do and the diversity of incompatible "true versions" can be readily demoted to relativistic accounts of that one world. Furthermore, there is no explanation, on Goodman's part, of the reality of the imaginative selves that "make" the various worlds they do. The parallel between Goodman and Husserl is a little chilling. Whether he wishes to encourage the impression or not, Goodman's doctrine of plural actual worlds drifts in the direction of an extreme idealism. It does so simply because Goodman lacks a sufficiently robust account of entrenchment. This is most surprising, in view of Goodman's own sense of the puzzle of induction. But the fact remains that, *if* that puzzle could be solved, it would require a suitable context for entrenchment;

and that context could hardly be the proliferating worlds Goodman now espouses. In a sense, therefore, Goodman makes it impossible to discipline the symbiosis of self and world for the purposes of science – by dint of his "irrealist" (or idealist) proliferation of worlds. There must be a leaner constraint on all those (logocentric) strategies that effectively preclude or eliminate the constructive role of the self.

We may, however, look at the symbiosis of self and world more promisingly – with a view to explicating what we have provocatively termed the intentionality of science. (Recall, here, that we are concerned with the inherently restricted scope and constructive nature of any science – what we have traced, through Bachelard, to the symbiosis of self and world. This, after all, is the essential connective theme between the ineliminability of self and any theory of science.) What we need at the very least is a frontal demonstration of just how unconvincing it is to eliminate the self's role just where the essential human features of a science are collected – deliberately precluding that role from its constructive function or deliberately restricting its function within its full constructive range. There is at least one specimen theory that will serve our purpose.

Consider (finally, then) Fred Dretske's views. Dretske has attempted what may be the most ramified account to date linking cognition to information theory – so as to provide for intentionality *in* physical systems and to explain science or cognition as a more specialized, higher-order intentionality than that which obtains at a mechanical level. "The distinctive character of our cognitive states," Dretske holds, "lies, not in their intentionality (for even the humble thermometer occupies intentional states), but in their *degree* of intentionality."[41] Hence, Dretske clearly means to supply a fundamental precondition of science that completely escapes the symbiosis here proposed. Dretske believes, incidentally, that cognitive states need not always be propositional in character, but he does not explain whether he means by that that they may have nonpropositional objects in addition to propositional objects or whether they may, independently, have nonpropositional objects only. The latter view is peculiarly difficult to defend[42] (though it has an entirely derivative use in neurobiological contexts).

In any case, following the general lines of standard information theory,[43] Dretske adds a congruent supplementary notion of semantic content – *not* to be construed in psychological or epistemological terms.[44] Here, precisely, is his most challenging and controversial move. For Dretske means to proceed "bottom–up" – *from* information *to* meaning *to* cognition; whereas the thesis we have been developing, the ineliminability of self and the symbiosis of self and world, requires a "top–down" strategy of abstracting or factoring information (if we choose to introduce an information-theoretic idiom at all – that is, one that, like Dretske's, treats informational content *propositionally*) *within* the complex framework of functioning (molar) selves. On the view we are favoring, intentionality is assignable only to selves (or their adequate surrogates) and to the space (and the artifacts of that space) they create – in effect, their culture – that grooms their emergence through successive generations: on that view, any intentionality that ascribes information *sub*-psychologically or *sub*-epistemically (as in Dretske's account) either suppresses a conceptual dependence on the role of selves or else is open to a charge of incoherence or irrelevance. So Dretske's proposal is particularly intriguing. It would neutralize, without eliminating, the complicating role of cognizing selves. In this sense, it affords a possible rationale for an extensionalism like Quine's; for Dretske explicitly posits a world with a determinate structure (including an informational structure) that, latterly, our science can be expected to master moderately well. The world he posits is noticeably hospitable to a thorough extensionalism – which Quine also favors. But Quine had tried to extract a reliable extensionalism (logocentrically) *from* an admitted symbiosis of self and world; and Dretske (inadvertently) concedes the impossibility of doing so, by dint of *his* first grounding such an extensionalism in a world not thus encumbered. It is, in this respect, a minor difference between Dretske and Quine that Quine exaggerates the holist theme and Dretske exaggerates the prospects of a strongly objectivist science.

First, then, Dretske introduces an "information condition":

(A)   The signal carries as much information about s as would be generated by s's being F.

Then, he adds a further condition:

If a signal carries the information that s is F, it must be the case that:

(B)   s is F. [45]

Given these provisions, the opening lines of Dretske's account are particularly instructive: "In the beginning there was information. The word came later. The transition was achieved by the development of organisms with the capacity for selectively exploiting this information in order to survive and perpetuate their kind."[46] Clearly, on this view, information *is* in the physical world, quite apart from and prior to human intervention, *and* information is propositional in character:

"Meaning, and the constellation of mental attitudes that exhibit it, are manufactured products. The raw material is information."[47] Dretske says nothing, however, about conditions affecting the human discrimination of the intended match or about the import of those conditions on his theory. In fact, he goes on to characterize "informational content" in the following very strong way:

A signal r carries the information that s is F = The conditional probability of s's being F, given r (and k), is 1 (but, given k alone, less than 1).[48]

From this it is clear that information about a state of affairs (s's being F) is perfectly matched with that state of affairs, has a conditional probability of 1 – where the state of affairs and the information about it are apparently distinct. r is a physical signal, and k may be construed as a measure of the "knowledge" the receiver has (if any) about pertinent possibilities that exist at the source. So seen, Dretske's notion of objective information (information in the "external" world) is little more than a mythic or heuristic device for converting discourse about the physical order of things into discourse about an informational order. It introduces, roughly speaking, a kind of generalized transducer-operation – for which, apparently, there is no independent evidence or even a need for evidence, except that it rather cleverly ensures the extensionalism wanted. In this sense, Quine's elimination of the intentional *within* his holism answers perfectly to Dretske's neutralizing the intentional in an informational way *prior to* the further intensionalizing complications of human inquiry.

Dretske's purpose is a double one: first, to construe a signal's informational content *de re* rather than *de dicto*; and, secondly, to facilitate the thesis (the one he means to favor) that "a signal's *de re* information content suffices for the analysis of our *de re* beliefs and knowledge."[49] Both these considerations confirm that Dretske does not construe the assigned information *or* the *de re* cognitive states of "receivers" as subject, in principle, to the constructive role that we have found ourselves obliged to acknowledge as a consequence of (conceding) the symbiosis of self and world. (This is the upshot of Dretske's thesis that information need not be languaged or linguistic.) As Dretske says, "it is assumed ... that *de re* beliefs and knowledge are more fundamental than their *de dicto* counterparts."[50]

If, however, information cannot be said to obtain unless it is assignable in some determinate way (say, very roughly, in accord with Michael Dummett's constructivist view[51]), and *if*, in being thus assignable, it cannot but be encumbered (*in addition*) by the symbiosis of self and

world, then Dretske's information-theoretic account is altogether too
sanguine, or is simply thus defined but never discernibly found to be
such *and never operationally* thus *employed*. Furthermore, Dretske is
quite explicit that, although knowledge requires causality in "the sense
in which it requires information," informational relationships are not
and do not as such require causal relationships, though they usually
depend "on underlying causal processes."[52] Dretske's view, in sum, is
that "information" is quite distinct from "meaning," is not an "artifact"
at all – not intentional in the express sense of being "a way of describing
the significance *for some agent* of intrinsically meaningless events" – but
is "an objective commodity, something whose generation, transmission,
and reception do not require or in any way presuppose interpretive
processes."[53] (A Peircean, for example, would have insisted that
information, like signs, is inherently triadic – and therefore requires
[intentional] reference involving selves or suitable surrogates.)

For cases in which "there is a positive amount of information [cases
of initial perception, say, that do not depend on K's antecedent knowledge
or belief] associated with s's being F," Dretske holds,

> K knows that s is F = K's belief that s is F is caused (or causally
> sustained) by the information that s is F.[54]

But this *is*, precisely, an information-theoretic account of cognitive
privilege, even, or particularly, if given Dretske's adjustment to the
effect that K's belief may only have been "causally sustained" but not
actually caused by the information in question – that is, if it were
originally caused by some episode of human discourse or communication.
For, for K to know that s is F "require[s]" K's "receipt of the appropriate
piece of information":[55] there must have obtained the requisite information and *it* must have been received; and that information, on
Dretske's view of "informational content," conveys (nonlinguistically)
"that s is F" – with a probability of 1 of s's actually being F.

Dretske had held at the beginning of his account, it will be recalled,
that the "artifactual" view of information "rests on a confusion, the
confusion of *information* with *meaning*."[56] But what Dretske could
only have meant is that a "natural," non-anthropocentric account of
information was possible. Now, we find that it is possible on three
conditions that themselves seem clearly linked to the problem of the
symbiosis of self and world – conditions that threaten its very viability:

1 that information is propositional (though not linguistic);
2 that knowledge requires the causal reception of information in a

form tantamount to that favored in claims of cognitive privilege and transparency; and

3 that the dialectical persuasiveness of the theory itself depends on "its usefulness in organising and illuminating the material to which it is applied."[57]

Condition 3 might well be taken to concede implicitly that, however "fundamental" knowledge *de re* may be supposed to be, the required assessment cannot in principle (one may perhaps assert this along Quinean lines) be pursued except by first nesting *de re* claims inextricably in *de dicto* claims: a *fortiori*, in *de dicto* knowledge – which would reverse Dretske's own priorities. For, if *de re* claims cannot be initially fixed except within *de dicto* claims, then the informational notion Dretske himself proposes cannot but be an abstraction projected from within *de dicto* contexts. Effectively, that would mean that the higher-order intentionalities of *de dicto* cognitive states would inevitably infect *ab initio* whatever *we* propose to treat as *de re* knowledge. (Dretske's view about *de re* knowledge may now be seen to be rather sympathetic to an empiricist-based realism.) *De re* knowledge is more "fundamental" to a realist science than *de dicto* knowledge, it is true; but, *on any theory that eschews transparency and privilege, de re* knowledge is itself an artifact or a reasonable abstraction formed *within* some holist or pragmatist account of a realism of *de dicto* beliefs. In short, it is not an independently operable notion just in the sense in which it is fundamental.

These distinctions set the stage for a fair appraisal of two competing conceptions of intentionality: Dretske's, and the one we are favoring. Against Brentano's well-known thesis, Dretske claims that "intentionality, rather than being a 'mark of the mental' is a pervasive feature of *all* reality – mental and physical": it is ubiquitious simply because information is ubiquitous; and information, on Dretske's view (as we have seen), "depends on the set of conditional probabilities relating events at R [receiver] and S [source]." In accord with his adopting Claude Shannon's mathematical theory of communication, Dretske holds that "Every conditional probability [relating S and R] must be either 0 or 1 – strict nomic dependence between the events occurring at S and R."[58]

So information (which causally informs *de re* knowledge) itself depends on nomic universals, whether statistical or deterministic. Where, then, does the intentionality obtain? The answer is quite straightforward:

Information has an intentional structure that it derives from the *nomic* relationships on which it depends. Since a nomic relation between properties (magnitudes) F and G is an intentional relationship, information, understood as the measure of this mutual dependency, inherits this structure. If F is lawfully related to G, and "G" is extensionally equivalent to "H", F is not necessarily related in a lawful way to H. If it is a natural law that things having the property F have the property G, it does not follow that there is a law relating the property F to the property H just because, as a matter of fact, everything that is G is also H (and vice versa).[59]

The possibility of cosmic accident confirms the intentionality of information. The trouble is that – quite apart from the defensibility of real nomic universals, apart from the defensibility of holding that nomic universals are not linguistic artifacts, apart from the defensibility of holding that actual nomic universals can be satisfactorily detected in specifying information and *de re* knowledge – these considerations are nowhere reconciled with the deep methodological questions posed by the very enterprise of science (for instance, regarding the import of probabilities assigned within the range between 0 and 1).[60]

Cognitive, as opposed to informational, capacity, then, constitutes at least minimally a *restriction* of some sort *on* informational capacity. (Dretske makes no claims about the adequacy of his account for all forms of human knowledge.) This of course is just what a bottom-up account would require and propose. "What is known," Dretske explains, "when something is known differs, not only from extensionally equivalent pieces of information, but from nomically equivalent pieces of information." This is the reason why, for instance, the galvanometer, though it does convey the information (possesses the intentional state to the effect) that, say, there is a current-flow between points A and B, does not *know* that there is that current-flow: it is "not *intentional enough*"; its intentional states "carry *too much* information, have *too much* content, to qualify as genuine cognitive states."[61] Informational systems cannot discriminate between nomically equivalent pieces of information – for instance, that there is a flow of current between points A and B and that there is a voltage difference between points A and B. A genuinely cognitive system "has ... a representational or coding system that is sufficiently rich to distinguish between something's being F and its being G [think of the galvanometer case] where nothing can be F without being G."[62] In effect, it must have a system that is "at least as rich in its representational powers as the language we use to express what is known." Hence, to construe cognitive capacity in terms of "*sentences* used [by human agents] to describe such phenomena" effectively fixes (for Dretske) the notion of *intensionality*: that is, intensionality is just intentionality expressed as a property of sentences.[63]

At the fundamental level of *de re* knowledge, therefore, intensional complications that depend, say, on the empirical underdetermination of theory *and* on the inseparability of the acquisition of perceptual knowledge from that condition *and* on the further inseparability of that inseparability from the historical contingencies of investigative communities – in sum, complications that depend on the symbiosis of self and world – are simply set aside by Dretske. The scheme is admittedly most ingenious. But it is more instructive in indicating what would have to obtain *if* the puzzles of intentionality and intensionality *could be resolved* – in a way that would either permit the elimination of selves or permit their behaving benignly enough as physical systems, at least through the "fundamental" range of initial perceptual knowledge on which a theory of science might ultimately rest – than it is in providing any reason for believing that things do thus obtain. On the argument already advanced, even if something like Dretske's view were empirically viable, its actual defense would have to be processed through the complications generated by such imaginative and investigative selves as Bachelard for one has described; and those complications cannot (as we have argued) be construed merely as noise affecting the essential cognitive process. They would be constitutive of that process itself. This, in a word, is the reason meaning and cognition cannot be convincingly construed (bottom–up) as *restrictions* on antecedent information: they concern the very constructive role of a community of inquiring selves with respect to which information is first specified.

Our argument, therefore, is not as rambling as it may appear. But even its ramble is meant to fix a fair impression of the extraordinary energy with which bottom–up visions of science press to eliminate or to neutralize the disturbing intrusions of actual, of cognizing, selves. The symbiosis of self and world bears in a crucial way on the hopes and fears of the unity of science. Dretske and Quine establish the polar extremes of analytic philosophy within which a thoroughgoing extensionalism may hope to be recovered. That objective, however, is in a sense no more than a certain appetizing benefit to be gained by privileging certain approaches to the symbiosis of self and world within the larger current of Western philosophy. We have pursued this improbable theme through certain particularly strategic (and ingenious) maneuvers meant to dismiss altogether the constructive role of the self within the work of the empirical sciences – or, alternatively, to make it omnipotent there or to retreat to an aporetic skepticism in the presence of its labors. But the truth is that it is ineliminable in a balanced view of science and that its admission both precludes certain fashionable options (a logocentric extensionalism but not a pragmatized one) and warns us about its own tendency to encourage conceptual extravagance (idealism and deconstruction). To be thus forewarned, however, is to give up nothing of the admirable rigor of science itself.

Notes

1   As posed in early empiricism (Locke) and in early rationalism (Descartes)
    – resolved (by Locke) only by an arbitrarily favorable representationalism
    and (by Descartes) by a nondeceiving God.

2   This seems to be the most recent, strenuous form of Richard Rorty's
    deconstructive turn. See Ricard Rorty, "The Contingency of Language,"
    London Review of Books, April 17, 1986, pp. 3–6.

3   See Aristotle, *Categories*, chs 3–4 particularly. See also Emile Benveniste,
    "Categories of Thought and Language," *Problems of General Linguistics*,
    tr. Mary Elizabeth Meek (Coral Gables, Fla: University of Miami Press,
    1971); and Jacques Derrida, "The Supplement of Copula: Philosophy before
    Linguistics," *Margins of Philosophy*, tr. Alan Bass (Chicago: University of
    Chicago Press, 1982).

4   See W. V. Quine, *Word and Object* (Cambridge, Mass.: MIT Press, 1960),
    sections 15–16. It is fair to say that, in his *The Roots of Reference* (La Salle,
    Ill.: Open Court, 1973), particularly pt I, Quine's resistance to a full
    naturalism accommodating biology is at best equivocal.

5   See for example Edmund Husserl, *Cartesian Meditations*, tr. Dorian Cairns
    (The Hague: Martinus Nijhoff, 1960), the famous Fifth Meditation in
    particular.

6   See for instance, Edmund Husserl, *Logical Investigations*, tr. J. N. Findlay
    (London: Routledge and Kegan Paul, 1970), p. 596. I have benefited
    enormously from the account in Izchak Miller, *Husserl, Perception, and
    Temporal Awareness* (Cambridge, Mass.: MIT Press, 1984), ch. 1.

7   This marks the excessive zeal of certain contemporary Nietzscheans, notably
    Jacques Derrida: see particularly his *Of Grammatology*, tr. Gayatri Spivak
    Chakravorty (Baltimore: Johns Hopkins University Press, 1976), and "The
    Ends of Man," *Margins of Philosophy*; also, more remotely, perhaps
    through a Derridean or Heideggerian reading of John Dewey and Ludwig
    Wittgenstein, Richard Rorty: see particularly his *Philosophy and the Mirror
    of Nature* (Princeton, NJ: Princeton University Press, 1979).

8   I find a somewhat companionable adjustment, written however within an
    entirely different idiom, in Gianni Vattimo's notion of "weak thought" (*il
    pensiero debole*), which appears to recover (the point may be contested)
    metaphysics *within* the very stricture of Heidegger's accusation against
    metaphysics: "the unviability of what metaphysics has always ascribed to
    being, namely, its stability in presence, its eternity, its 'thingness' or *ousia*
    [which] Heidegger exposes as a 'confusion', as a 'forgetfulness', because it

derives from the act of modeling being on beings, as if being were merely the most general characteristic of that which is given in presence" – "Dialectics, Difference, and Weak Thought", *Graduate Faculty Philosophy Journal*, X (1984), 156. See also Gianni Vattimo, *La Fine della Modernità* (Milan: Garzanti, 1985); and Reiner Schürmann, "Deconstruction is not Enough: On Gianni Vattimo's Call for Weak Thinking," *Graduate Faculty Philosophy Journal*, X (1984).

9 Rorty, "The Contingency of Language," *London Review of Books*, April 17, 1986, p. 3.

10 Ibid., p. 4.

11 Quine, *The Roots of Reference*, section 7.

12 Martin Heidegger, *Being and Time*, tr. John Macquarrie and Edward Robinson from the German edn (New York: Harper and Row, 1962), pp. 12–13 (in the German edn); pp. 32–3 (in the English).

13 See Edmund Husserl, *Phenomenology and the Crisis of Philosophy*, tr. Quintin Lauer (New York: Harper and Row, 1965).

14 See for instance Michel Foucault, *Discipline and Punish*, tr. Alan Sheridan (New York: Random House, 1977).

15 See Heidegger, *Being and Time*, pp. 24–5, 64–5 (in the German edn); pp. 45–7, 92–4 (in the English edition); and section 76. For a Heideggerian reading of Kuhn, see for example Joseph Rouse, "Kuhn, Heidegger, and Scientific Realism," *Man and World*, XIV (1981).

16 Gaston Bachelard, *The New Scientific Spirit*, tr. Arthur Goldhammer (Boston, Mass.: Beacon Press, 1984), p. 136 (italics added).

17 Ibid., p. 171. See also Ludwik Fleck, *Genesis and Development of a Scientific Fact*, tr. Fred Bradley and Thaddeus J. Trenn, ed. Thaddeus J. Trenn and Robert K. Merton (Chicago: University of Chicago Press, 1979) – a text contemporary to Bachelard's; and Ian Hacking, *Representing and Intervening* (Cambridge: Cambridge University Press, 1983).

18 See ch. 12, below.

19 I have had the benefit, here, of the critical comments of David Rosenthal, who served as respondent to a version of this chapter, first presented at the Boston Colloquium for the Philosophy of Science, fall 1986. Rosenthal helped me to see (though that was not his own intention) that standard Anglo-American analytic accounts of philosophy, more or less hospitable to Quine's view, would not be easily persuaded that a thesis such as the indeterminacy thesis, or a determined extensionalism, is itself as thoroughly hostage to the symbiosis of self and world as I am claiming. This, I dare say, is the proper force intended in holding that Quine's thesis (not unlike Rorty's, though for very different reasons, and despite its incomparably richer significance) is ultimately open to the charge of logocentrism. In fact, the validity of the charge can be effectively traced through a very large swathe of current analytic philosophy. We shall, in subsequent chapters, pursue the issue in pursuing more frontal questions.

20 See W. V. Quine, "Five Milestones of Empiricism," *Theories and Things*

(Cambridge, Mass.: Harvard University Press, 1981).

21 See W. V. Quine, "Epistemology Naturalized," *Ontological Relativity and Other Essays* (New York: Columbia University Press, 1969), and *The Roots of Reference* (La Salle, Ill.: Open Court, 1973).

22 Ibid., p. 71.

23 The most convenient collection of views on the issue is provided in Sandra G. Harding (ed.), *Can Theories be Refuted? Essays on the Duhem — Quine Thesis* (Dordrecht: D. Reidel, 1976).

24 Cf. Quine, "Things and their Place in Theories,' *Theories and Things*, p. 23 (italics added).

25 Bachelard, *The New Scientific Spirit*, p. 40: "Psychologically, the modern physicist is aware that the rational habits acquired from immediate knowledge and practical activity are crippling impediments of mind that must be overcome in order to regain the unfettered movement of discovery."

26 Ibid., pp. 135–7. Here, Bachelard summarizes the views of George Urbain: "every method eventually loses its initial fecundity"; "Science concepts themselves may lose their universality."

27 See particulary Quine, *Word and Object*, section 45.

28 Quine, "Epistemology Naturalized," *Ontological Relativity and Other Essays*, pp. 89–90.

29 *If* Rorty is an American deconstructionist (as it seems he is–sometimes Heideggerian, sometimes Derridean), then he has failed to grasp the logocentric themes in Quine's philosophy as well as in Donald Davidson's (which pretty well follows Quine's in this regard); see *Philosophy and the Mirror of Nature*. See also Christopher Norris, *Contest of Faculties: Philosophy and Theory after Deconstruction* (London: Methuen, 1986), chs 2, 8.

30 To assign Nietzsche his own disciples may be somewhat barbarian, but there is a clear precedent in the pertinent literature. In any case, there is also some textual basis for the reading. See particularly Friedrich Nietzsche, "On Truth and Lies in a Nonmoral Sense," in Daniel Breazeale (ed. and tr.), *Philosophy and Truth; Selections from Nietzsche's Notebooks of the Early 1870's* (Atlantic Highlands, N.J.: Humanities Press, 1979).

31 See Quine, *Word and Object*, section 56.

32 Rorty, "The Contingency of Language," *London Review of Books*, April 17, 1986, p. 6.

33 Ibid.

34 Ibid. (italics added). See Donald Davidson, "What Metaphors Mean," *Inquiries into Truth and Interpretation* (Oxford: Clarendon Press, 1984).

35 Nelson Goodman, *Ways of Worldmaking* (Indianapolis: Hackett, 1978), pp. 1, 2, 6.

36 These puzzles are all reasonably posed by Goodman's account. He himself appears to take them to be such; see for instance Nelson Goodman, *Of Mind and Other Matters* (Cambridge, Mass.: Harvard University Press, 1984), pp. 30–2. See also Hilary Putnam, "Reflections on Goodman's *Ways*

of Worldmaking," *Philosophical Papers*, vol. 3 (Cambridge: Cambridge University Press, 1983).

37 Goodman, *Of Mind and Other Matters*, p. 31.
38 See Nelson Goodman, "The New Riddle of Induction," *Fact, Fiction and Forecast*, 2nd edn (Indianapolis: Hackett, 1965).
39 Goodman, *Of Mind and Other Matters*, p. 38.
40 One can see a somewhat distant similarity between Goodman's present view and William James's empiricism, in the account of James in Elizabeth Flower and Murray G. Murphey, *A History of Philosophy in America*, vol. II (New York: G. P. Putnam's Sons, 1977), ch. 11. Elizabeth Flower drew my attention to the connection, though I confess it still seems slim (in spite of the reference to multiple worlds) as far as Goodman's specific projects are concerned.
41 Fred I. Dretske, "The Intentionality of Cognitive States," in Peter A. French et al. (eds), *Midwest Studies in Philosophy*, vol. 5 (Minneapolis: University of Minnesota Press, 1980), p. 282.
42 See further Fred I. Dretske, *Seeing and Knowing* (Chicago: University of Chicago Press, 1969); and Roderick M. Chisholm, *Perceiving: A Philosophical Study* (Ithaca, NY: Cornell University Press, 1957), ch. 10. Cf. however, Patricia Smith Churchland, *Neurophilosophy* (Cambridge, Mass.: MIT Press, 1986), ch. 9.
43 See Claude Shannon, *The Mathematical Theory of Communication* (Urbana: University of Illinois Press, 1948).
44 See Robert Cummins, *The Nature of Psychological Explanation* (Cambridge, Mass.: MIT Press, 1984), pp. 65–74, for a very clear analysis of Dretske's position.
45 Fred I. Dretske, *Knowledge and the Flow of Information* (Cambridge, Mass.: MIT Press, 1981), pp. 63–4.
46 Ibid., p. vii.
47 Ibid.
48 Ibid., p. 65.
49 Ibid.,pp. 66–8.
50 Ibid.,p. 68; cf. particularly p. 66.
51 See Michael Dummett, "Truth," *Truth and Other Enigmas* (Cambridge, Mass.: Harvard University Press, 1978), especially pp. 18–19. *Constructivism*, it may be noted, is opposed to Goodman's constructionalism; cf. Dummett's "The Structure of Appearance," "Nominalism," "Constructionalism," in the same volume.
52 Dretske, *Knowledge and the Flow of Information*, pp. 39, 26.
53 Ibid., p. vii.
54 Ibid., p. 86; cf. pp. 86–9. *S* designates some individual and "... is *F*" an open sentence; and the content of *r* "is fixed by perceptual (noncognitive) factors" – Dretske's exemplar of *de re* content.
55 Ibid., p. 104; cf. p. 105.
56 Ibid., p. vii.

57 Ibid., p. 106.
58 Dretske, "The Intentionality of Cognitive States," in French et al., *Midwest Studies in Philosophy*, vol. 5, p. 284 (italics added).
59 Ibid., pp. 285–6.
60 See Open Peer Commentary to Fred I. Dretske, "Precis of *Knowledge and the Flow of Information*," *The Behavioral and Brain Sciences*, VI (1983), particularly the comments by C. Ginet, P. M. and P. S. Churchland, I. Levi, R. Cummins, and Dretske's response.
61 Dretske, "The Intentionality of Cognitive States," in French et al., *Midwest Studies in Philosophy*, vol. 5, pp. 288, 289.
62 Ibid., p. 291.
63 Dretske, *Knowledge and the Flow of Information*, p. 75.

5

Top–Down and Bottom–Up
Strategies

There is a near-paradox that is commonplace in the anecdotal psychology of humans: on the one hand, everyone, it is supposed, continuously lives with an inconsistent set of beliefs; on the other, the apt members of a society must, generally, share true beliefs relative to and about their central practices. There is, correspondingly, at the metaphysical level, at the level of theorizing about psychological theorizing, a distantly similar near-paradox: on the one hand, it is widely affirmed that our "folk" psychological theories – that is, theories routinely, perhaps immemorially, embedded in the central grooming and survival practices of a viable society – must be substantially correct about molar psychological functioning, so that a scientific psychology will be largely directed (with relatively minor revision) to systematizing as rigorously as possible our folk theories concerning our (folk-) theoretically informed reflection on our own behavior and mentation; on the other hand, the march of scientific psychology – perhaps chiefly along the lines of neuropsychology and cognitive science – inclines one to believe (by analogy with the history of "folk" physics and the like) that folk psychology cannot fail to be seriously mistaken and misleading and cannot fail to require radical revision, perhaps even the elimination of much that we take for granted. Add to this second story the largely convincing truism that the human is paradigmatic for psychology: we find ourselves confronted by the need to organize the space of psychological theorizing, so that we neither lose the initial bearings that an intuitive grasp of human functioning affords nor deprive ourselves of the speculative riches of viewing psychological functioning as somewhat alien, not well understood, possibly explicable in terms of sub-processes that are themselves not normally accessible to the folk-theoretic approach.

There is justice on both sides, and there is considerable misunderstanding about the asymmetry of these two perspectives. Correcting our sense of their relationship is certainly not substantively dependent on any particular empirical thesis. Also, it pays to remember the important truth, sober and hilarious by turns, that theory is drastically underdetermined by data – in psychology as well as in physics. The swings of psychological theory, forever proliferating along divergent lines, have somehow managed to save themselves without producing utter chaos, and without ensuring any uniquely progressive scruple.

It is quite unclear, at the present time, what the serious prospects of a scientific psychology are – both in the sense of whether really fine-grained, strongly quantified explanations rivalling in precision the explanations of physics can reasonably be expected (or how much patience we can afford in waiting) and in the sense of what effect such maverick sciences as psychology and sociology may have on our convictions regarding the methodological unity and conceptual uniformity of all the disciplines we elect to call sciences. Psychology may not be a science – on empirical evidence serving the canonical picture; or the evidence there is may well support the natural diversity of the inquiries we are unwilling to do without. Still, the intellectual cleavage suggested is not so much a sign of the inherent disorder of psychology as it is of a fresh and comprehensive charting of the global options that confront us. The geography of metapsychology favors no single Mercator projection. The single most important truth about any would-be mapping is, however, that, if human psychology is the principal text (and if in this sense animal psychology is anthropomorphized *both* by directly adjusted modeling and by the use of a human methodology subterraneanly linked with such modeling[1]), then, contrary to one of the deepest themes of scientific discipline, the studies and the students are the same; we examine ourselves, reflexively, perhaps too intimately (and yet inextricably) involved in the transaction, interested parties on both sides of the rheostat and the survey.

This imposes a profound asymmetry on all the options of psychological procedure, quite apart from any quick rejection of the scientific standing of the discipline. There is, first of all, an essential *equilibration* that psychological studies cannot fail to accommodate: the fitting of our theoretical pictures to what we take to be the least easily dislodged intuitions about our inter- and intrapsychological processes and the careful, even if grudging, changes made under the pressure of somewhat alienized studies, within otherwise privileged self-images. Quite apart from the assessment of their respective work, this is surely the obvious function of Freud's great clinical speculations and of the smaller-scale but perhaps for that very reason even more arresting conclusion of

Stanley Milgrim's now-classic experiments.[2] But it is not merely equilibration that we must concede. All the sciences require that much – except among those innocents who still genuinely believe in the cognitive transparency of nature *and* ourselves, and who make no concessions regarding the theoretical freighting of observation and experiment. No, the equilibration is subject to the influence of what we may call *salience* – not necessarily a pernicious influence, in any case not an eliminable influence (once we concede psychology to be a *reflexive* discipline). There is a limit to how far we can alienize ourselves and our studies of ourselves *and* to how far, to the extent we actually succeed, we can thereupon claim to have examined the essential beast. Is there a human ecology that collects the salient? And is the laboratory mentality of an observational psychology the unhappy source of tachistoscopic blips?[3] Part of the answer is a function of the dialectic of competing intuitions at (may we say?) the molar level – between well-entrenched folk conceptions and such relatively alienized but folk-susceptible adjustments as the Freudian and the Marxist – and part is a function of the more radically alienizing effect of reinterpreting the molar in terms of interlevel theoretic speculations centered in such sub-processes as the neuronal and the informational.

It is precisely because of this dual complication that current psychology is obliged to address the puzzle of choice between so-called *top-down* and *bottom-up* conceptual strategies. These are terms of art fashionably drawn from the lingo of artificial intelligence and cognitive simulation; but they have an apt association as well with the deeper currents of the general unity of science movement (from its antecedents in the visions of Ernst Mach and Hermann von Helmholtz to its full flowering in Rudolf Carnap's Vienna Circle[4] and beyond); and they have a looser association with such inviting inquiries as are favored in Freud's original *Scientific Project*[5] or, more recently and more argumentatively, in Michael Gazzaniga's cognitive neuroscience,[6] Noam Chomsky's linguistic nativism,[7] Jerry Fodor's conceptual nativism,[8] Zenon Pylyshyn's computational model of cognition,[9] Daniel Dennett's homuncularism,[10] and the even more radically eliminative strategies of Paul Feyerabend,[11] Stephen Stich,[12] and Paul and Patricia Churchland.[13] There may be partisans among us, of one but not another of these separable fiefdoms; but the importance of such honorable prejudices is greatly overshadowed by the primary need to understand the conceptual choice between the top-down and the bottom-up. (We shall, incidentally, examine all of these views in due course.) The distinction is much muddled, frankly, in the philosophical literature, where it has been rather ardently pressed –

largely because of a haste and preference too sanguine, on either side, for the existing evidence.

The important point about the choice is that, in a sense, everything in psychology depends on it, that it is asymmetrically skewed in principle, that it is in another respect contingently (some would insist, unfairly but ineluctably) skewed at the present time, and that it can hardly fail to be skewed in variable and novel ways as a consequence of diachronic shifts in the actual work of professional psychologists and other specialists. For one thing, the top-down and the bottom-up are not formally symmetrical or the logical reverse of one another; and, for another, they are (certainly for the foreseeable future, arguably in principle) conceptually interconnected. There are in fact two distinct considerations affecting the choice. One is the admission of the salient, already remarked; the other is the actual achievement and realistically assessed programmatic promise of alternative lines of work. Both support a reasonable and reasonably generous sense of the empirical seriousness of the opposing strategies.

Admission of the *salient*, in psychology, signifies admitting the entire body of what (prejudicially) has come to be called folk psychology[14] in more than a merely initial and provisional sense. The admission *is* initial and provisional, but the displacement of the more pervasive theory-linked conceptions of our intuitive self-description and explanation – for example, our more-or-less holistic model of rationality linking belief, desire, reasoning, and action, or our intentional characterization and linguistic modelling of these phenomena, or our conception of choice based on these notions – *cannot*, for any now empirically promising reasons, be downgraded merely to one contingent possibility among many, to the status of a sort of temporary *primum inter pares*. If such notions are fairly regarded as the core of "folk" psychology, then the scales are clearly tipped in their favor: not because they are inviolate (they *are* provisional) but because they fix, in the only sense in which any early science is empirically responsible, the salient phenomena that require description and explanation and because they do afford a recognizably disciplined and relatively successful explanation of those phenomena – *to the extent that we have a description and explanation of them at all*. (It would not be unreasonable – or amiss – to suggest that the Wittgenstein of the *Investigations* is the single most important philosophical champion of the folk-psychological orientation, even of its principled stubbornness in the face of scientific advances that it can and normally must accommodate.[15])

The fact is we lack very nearly completely a description of any of the details of mental life and psychological process that is genuinely *independent* of the governing conceptions of our folk psychology. The idea that we *could* scrap that kind of theory arid move directly to a "scientific" description and explanation of what is only "figuratively" hinted at by the use of that idiom is surely just what has been shown to be hopelessly premature as well as logically inadequate in what may be called B. F. Skinner's behaviorist peripheralism (that is, Skinner's thesis that behavior may be studied without reference to central states, *a fortiori* without reference to intentionality).[16] Furthermore, whether central-state materialism can be largely freed of folk-psychological notions (for example, as heroically championed by the Australian philosopher D. M. Armstrong,[17] or perhaps more promisingly in an empirically detailed way by Gazzaniga and his associates) frankly remains to be seen. But that very admission is a concession to the saliences of our folk psychology. On that score, the only reasonable concession that may be drawn from the asymmetry in place is of course that, under the pressure of a developing science, whole portions (for all we know) of our folk theories and folk descriptions may have to be *modified* (which hardly affects the essential asymmetry: witness Freud) or may be *eliminated* (with what consequences on the future of that asymmetry we cannot yet say, simply because we have hardly succeeded beyond heuristic first sketches). The prospect is there, however; it cannot be gainsaid, and there is no call for professional sneers. But for the foreseeable future – let us for the moment leave the question of principle aside – the general effect of empirical salience is to favor lopsidedly the folk-theoretic and the entire range of the experiential or qualitative that has been so completely integrated with the sort of models already mentioned. To insist on the point may be terribly misleading, however, because it does not yet say what the nature of needed revisions is likely to be and it hardly dispels impossibly dualistic notions of the elements of our folk psychology that are likely to remain.

It must be admitted, then, that the scientific programs of what we are calling top-down and bottom-up disciplines are capable of a measure of reconciliation, are also capable of being pursued in radically opposed and irreconcilable ways, are asymmetrically favored and disfavored *by the effect of salience itself*, but are not (unless trivially construed) committed to the very same explanatory project. What hangs on the issue is the principled unity or bifurcation of the sciences and the fate of certain tenets hospitable to the unity model even when it itself is abandoned.[18] In pursuing the matter, however, we must be clear

1   that the unity program may attempt to reconcile the bottom-up and top-down as symmetrical alternatives;

2 that the unity program and the bottom-up strategy strongly converge only where a radical reduction or elimination of molar psychological phenomena is attempted; and

3 that the bottom-up strategy may not actually be committed to any version of the unity program, but only to certain tenets normally shared by versions of that program.

A further small aside to the reader is advisable here. The top-down/bottom-up issue is a decisive one. But it is much too complicated and much too ramified to be resolved in the conversational manner here adopted. We shall pursue it in a frontal way in two installments: here, largely in the informal manner already favored, except where a natural opportunity presents itself to examine one or two important positions bearing on its fate; and then again, immediately after, essentially by way of a detailed canvass of a variety of the most influential recent views of the matter, largely skewed toward promoting the bottom-up strategy, even where there is apparent sympathy for the top-down.

I

The essential contrast between the top-down and the bottom-up is simplicity itself: top-down preferentially installs folk-theoretic models as the metachoice of psychological description and explanation – not, let it be said, particular or unalterable models or models restricted in any chargeably narrow way to the level of discourse at which the functioning agents of folk psychology allegedly function, the so-called molar level, the level of full-blown persons or selves (just to mention the key terms of that picturesque jargon). *All* other psychologically pertinent states and processes and events located at more "fundamental" levels of physical or biological (or, for that matter, electronic or informational) levels, at which the would-be molar agents are either not discernibly at work or, by adverse hypothesis, not even engaged, are *in principle assigned* pertinent functions as the *sub*-functions *of* molar funcitoning. The result is that top-down psychologies capture all fine-grained analysis *factorially* – for example, cerebellar facilitation of sensorimotor coordination triggered (in some as yet quite unexplained way) by cerebral input that is itself folk-theoretically interpreted (read here: I decide to have another bite of buttered peas and my hand smoothly conveys a laden knife to my mouth).

The point is that, on the folk-theoretic view, the smooth, automatized micro-functioning of the cerebellum has no psychological relevance in principle except *on* a theory's assignment of, say, extended "intentional" import from "above" (where it is more natively and more paradigmatically first introduced) *to* a sheaf of such micro-processes; *and* that that assignment identifies the kind of conceptual plasticity of all folk-theoretic strategies. Since the meta-choice is essentially formal, no empirical possibilities stemming from close studies of the brain (or of genetic or hormonal processes, for that matter) could possibly be precluded from contributing to our understanding of psychological phenomena merely by favoring the option itself. It would of course distort the picture *if* all top-down strategies were dead wrong, but that's a risk that goes with the territory. In any case, so construed, *no* contending models of cerebellar facilitation, for example the tensor-network theories of Andras Pellionisz and Rodolfo Llinás[19] or more radical first steps toward an alphabet of cellular plasticity, for example the studies of Eric Kandel and his associates,[20] are likely to bear directly *on the primary choice* bewteen top-down and bottom-up strategies, though, from time to time, they certainly could and would be likely to require adjustment and fine-tuning of particular top-down models, and could of course contribute to the eventual overthrow of the top-down strategy itself.

The point is that the conceptual role of such studies within the larger picture of things ought not to be confused with the precision and accuracy of those studies themselves: adjustment with respect to the meta-choice has to do with coherence, explanatory promise, salience and the like; but the folk model is already self-consciously receptive to whatever we can imagine may flow, nor or later, from such detailed studies. It is on the face of it a logical blunder to suppose that folk psychology can be subverted by the mere accumulation of powerful studies of the sort in question. Not that the folk model cannot be subverted; only that that would require a drastic change of a *kind* of theory (not merely a drastically different theory of the same kind), and that we are nowhere near understanding what such a change would in detail commit us to.

It may seem a mere niggling *caveat* – but it is nothing of the sort – to insist that *if* the accumulating achievement of sub-molar-level processing, whether neurological or informational, is to effect a reversal of the folk model, *it must do so by radically reinterpreting the phenomena it explores at given micro- or sub-levels.* The subversion of theory, here – of change at the level of a meta-choice affecting promising kinds of theories – *cannot subscribe to the salient factorial idiom*: that is, the idiom on which forms of functioning discerned at more and more fundamental levels are *first* and continuously identified *in psychologically pertinent terms as sub-functions of whatever functions in a molar way* (for instance, as a rational or purposive or volitional agent). The shift entailed is certainly theoretically possible, but it must be scientifically motivated and suitably explicit. Such a shift cannot possibly be so motivated merely by ignoring the salient data of our psychological space. It must reinterpret all such data conformably. To put the point in a way that catches up an earlier discussion,[21] we cannot study psychological phenomena detachedly – as a merely observable domain neutral to our observational powers: not because we cannot observe ourselves but because scientific inquiry is, ineliminably, a *psychologically* implicated, effective factor within any such observed phenomena. We cannot pretend to bracket this complication "for a later time," to make "finer" adjustments in our first efforts at objectivity. The very objectivity *of* our psychological sciences *is* a function of the unique (usually ignored or marvelously defused) circumstance that we observe ourselves and observe ourselves observing ourselves. There is nothing, on the canonical view of the physical sciences, that corresponds to this circumstance. (Perhaps, for the sake of simple continuity and the continuity of an initially simple account, we may accept a claim check here in place of the philosophical goods that are needed. We shall see, a bit later on, that confusion about the essential disjunction between the two approaches fatally disables Daniel Dennett's extremely popular and influential elimination of personal- or folk-level discourse. Dennett's mode of arguing may then serve as the example for such accounts as those of Stich and the Churchlands.)

The fact is that the top-down/bottom-up contrast is easily miscon-strued. Both strategies are surely concerned with theories of a hierarchy (or at least a network) of levels of analysis ranging from what is salient at the molar level to what, on a theory, may be taken as the fundamental level of relevant analysis (perhaps the neuronal); and both are meant to be fully compatible with the data of biological and psychological evolution. This may come as a surprise. For it is sometimes supposed that, *if* the human has evolved continuously from lower forms, then bottom-up strategies *must*, exclusively, constitute the right meta-choice: the top-down must be discontinuous in an evolutionary sense, must therefore smuggle in the vitalistic, the dualistic, or the discontinuously *sui generis*. But to draw that conclusion is an enormous and elementary *non sequitur*. It is entirely possible that what is salient at the highest psychological and cultural level – what, in effect, *is* "*sui generis*," *within the evolutionary range*, may be reflexively analyzable *only in terms of that emergent level*, though without at all precluding, ignoring, dismissing, or failing to accommodate the biological continuity *in* which, functionally, human psychology is fully manifest. If that were true, then psychological explanation – *a fortiori*, sociological, historical, cultural explanation –

would have to be top-down, discontinuous (if you please) in terms of *explanatory* adequacy but not discontinuous in terms of biological evolution itself. Roughly speaking, it is the denial of *that* possibility that constitutes the bottom-up vision, *not* the incorporation of the ingredients of bottom-up explanations in physics, biochemistry, neurophysiology, or information theory *in the top-down explanations of psychology and the social sciences.* If we admit a real-time asymmetry favoring what we have called psychological salience, the only prospect for bottom-up theories is to *replace* the top-down entirely – or to do so so effectively in clearly essential areas of dispute, that the outcome cannot fail to be clearly seen. On the evidence, no one seriously supposes that that is now possible. So the opposing bets remain as before.

Bottom–up strategies are *not* opposed to the top-down in any way bearing on the division of the relevance of empirical studies of *any* kind. They are certainly not opposed in terms of the division of interest between the neurophysiological and the informational or between either of those sorts of disciplines and disciplines of the folk-theoretic sort. They are certainly not opposed in what is sometimes called the "data-driven" sense and the sense in which there are familiar forms of resisting data. They are opposed purely and simply in terms of governing visions of what is likely to be the master organization of the explanatory powers of psychology together with the neighborly contributions of biology and physics and the social sciencs to its own central puzzles. That's all. Bottom–up strategies are committed to some version of the unity-of-science model of explanatory programs or at least to some sub-set of its characteristic tenets,[22] and top-down strategists are betting that those alternatives won't work.

The contest is an honorable and important one, not one to be defrauded by charges of scientific illiteracy or abuse of empirical data: there are grounds aplenty for civil suits on either side. The top-down is frankly committed to the bifurcation of the sciences; the bottom-up is opposed. The top-down treats the relevant use of materials from the physical and life and information sciences *factorially*; the bottom-up treats the ultimate reduction of the psychological *compositionally*. The top-down theorist denies that there is likely to be found a finite, fundamental alphabet of empirically discernible processes and structures of any sort that, by hierarchical organization of a confirmably nomic kind – even with a fair measure of tolerance for idealizing and improvisational play – will succeed in integrating the phenomena of psychology (and the human sciences) with the best of any thus-characterized theories in the physical and life sciences: the top-down theorist denies this *on principled grounds*, not merely as a likely contingency. The bottom-up theorist rejects that meta-choice and pursues the reductive, the essentially compositional project. And neither can quite demonstrate that the other could not possibly be right.

Nevertheless, the opposition between top-down and bottom-up theorists is not (or at least need not be – and is best not construed as) restricted to the mere contingency that a directly bottom-up strategy, pursued by way of analogy with work in chemistry and biology, will not avail the bottom-up-*minded* psychologist. For example, Dennett argues against a bottom-up procedure but he does so for ultimately *bottom-up* reasons. Hence, on Dennett's view, top-down and bottom-up are conceptually symmetrical alternatives differing only in their "engineering" payoff at this or that particular moment in the history of the science.[23] On the argument being developed, Dennett could not be more mistaken; and, on the evidence, he does not actually show that his own conjecture is at all a reasonable one.

Furthermore, the very characterization of what it is to be a science is at stake. It is no use, therefore, declaring first what a science is and then confirming (the then-trivial consequence) that top-down strategies fail to meet the test. The truth is that *what a science is is itself a profoundly disputed empirical question*, as the quarrels between the positivists and such critics of them as Popper and Kuhn and Feyerabend, or between conventional scientific realists and anti-realists, have made so very clear – quite apart from the validity of their own affirmative views.[24] Once one appreciates the nature and scope of the dispute, the nerve of folk-theoretic psychology may be effectively exposed. The point is that the folk-theoretic concepts of empirical psychology *are not systematically expressible in terms of ultimate theories of the bottom-up kind*. Bottom-up and top-down *metatheories* are not logically reconcilable or compatible as theories of the real structure of the world relative to our explanatory undertakings; but the *theories and the theory-freighted data* of any particular empirical science *are* (as far as one can tell) entirely reconcilable with whatever – folk-theoretically or otherwise – may be gleaned from empirical psychology. In a word, there is absolutely no evidence to show that any first-order scientific inquiry cannot be *interpreted* in a way to fit either top-down or bottom-up readings: the presumptively compositional role of a neuronal alphabet, for instance, can *always* be suitably reread within a top-down factoring of a given range of molar phenomena; the only question is whether, at that moment of inquiry, there are local or global reasons of weight for pressing in one direction rather than the other. It is, similarly, always possible to read the data bottom-up; but, conceding the asymmetrical advantage (to top-down views) of a good part of the salient phenomena, bottom-up strategies regularly plead the prospects of a future breakthrough.

The essential clue is this. On the top-down view, what the empirically best descriptive categories of psychology will prove to be is an open question all right; but the bet is that all revisions will remain centrally committed to folk-theoretic or molar concepts. *If* that proves true, then the fine-grained contributions of every pertinent science addressed to admittedly more fundamental physical, biological, or informational "levels" of analysis will be construed as describing or explicating *sub-functions of some molar functioning*. It is for example quite possible – indeed, quite likely – that our gross-grained views of memory and learning (folk-theoretic views) will have to undergo radical revision as a result of new studies of paradoxical or hitherto utterly unfamiliar phenomena concerning reasonable applications of such descriptive terms as "memory" and "learning" – for instance, as in speaking of learning at the cellular level or as in attempting to understand the strange cognitive capacities of profound amnesiacs.[25] It may also be that the required vocabulary will not be exclusively unified along the lines of either top-down or bottom-up preferences. But the single most important consideration is this: that, on the top-down view, sub-functional processes and structures are, *for logical reasons, relationally* identified in terms of and only in terms of (however much we relax or lose sight of the distinction) molar, holistic, or integral processes and structures. The sub-systems *are* sub-systems only on the sufferance of being sub-systems *of* some pertinently superior system: they are subaltern relative to a more inclusive system in which they are only (identified as) functioning parts. It is in this sense that top-down theories are factorial. Bottom–up theories are compositional, therefore, in treating putatively more funda-mental "elements" *as* context-free, alphabetic, relatively independent and fixed elements – from the law-like hierarchical organization of which (along whatever empirically promising lines may develop) the complex phenomena of a folk-theoretic psychology will ultimately be shown to be properly generated (and to which it will ultimately be reduced, or reduced with whatever eliminative discard will prove necessary).

There is no question that these opposed visions subscribe to opposed theories of empirical laws and rigorous explanation in science; but there is little reason to think that current disputes about such matters are merely reflections of the top-down/bottom-up fracas and would subside if one or the other party were to prevail through the entire scientific community. We simply don't know what the best account is of causality and nomologicality in the psychological and human sciences; and we are hardly more certain in the physical and life sciences. For example, the logical and methodological status of universal covering laws is really a scandalous mystery in the whole of the empirical sciences, regardless of what the partisans on any side may choose to say. Also, contrary to canonical practice, it is quite feasible to admit the empirical regularity of causal processes without invoking lawfulness at all – or, at least, without invoking it everywhere. A fundamentally factorial approach such as the top-down, fitted to the peculiar contingencies of human psychology, still construed in terms of a realist science, may well have to sacrifice (or revise in a drastic way) the supposed logical connection between causality and nomologicality. But that, frankly, would be a strong piece of evidence in favor of bifurcation of the sciences, if it were suitably motivated empirically. It may be fairly claimed, therefore, that there is no canonical picture of science or of the sciences that may be neutrally invoked to decide between top-down and bottom-up conceptions. There is a dialectical connection between such speculations. Not only may top-down and bottom-up theories interpret the admitted psychological data with a fluency fairly matched between them; but in doing so they lend support to and gain support from eccentrically linked accounts of what counts as a science in the first place. (Needless to say, we shall return to the question of causes and laws.)

No psychology at the present time can be responsibly dualistic in a Cartesian sense – notoriously, in the manner of John Eccles;[26] every pertinent theory is bound to posit *incarnate* processes[27] if it means to avoid reductionisms of the materialist sort – of either the identity or eliminationist variety. But that means only that psychology is committed at the very least to the view that its phenomena are indissolubly complex *if* they exhibit features not expressible or reducible in the idiom of physics or biochemistry: for instance, if they are thought to exhibit irreducibly informational or representational or mental or functional or cognitive or cultural features. But the thesis of incarnate processes, of emergently and indissolubly complex (not compound) processes, *is* the only option top-down strategies can convincingly embrace, if they are to remain realist and to admit causal interaction; it does in fact favor the top-down over the bottom-up; and it can accommodate, by a factorial reading of would-be bottom-up research, any otherwise empirically supportable finding of such research. In fact, in terms of what we have called salience, it is reasonably clear that the top-down is effectively the provisional orientation of *every* empirical researcher despite idealizing loyalties (and dreams) of a subversively bottom-up kind. *There simply is no genuinely bottom-up psychology at the present time*: there are only top-down psychologies occasionally offered as potentially reformulizable in bottom-up terms if and when an effective bottom-up theory surfaces.

We know what such a theory would be like, at least in formal terms; but we have no idea at all how to get from here to there – empirically.

So seen, the dispute between neurobiologists and cognitive scientists (those for instance committed to the computer simulation or – better – the real analysis of intelligence) is no match at all. *There cannot be a natural kind* – "informavores" (to use George Miller's trim expression)[28] – if information or cognition is incarnate, if it is explicable only in terms of the neurobiology of distinct species (themselves distinct natural kinds). Cognitive science could not then be a distinct and autonomous science ranging nomically over information-processing in the abstract, unless *by a lucky hit* there were a bottom-up neurobiological science of incarnate processing ranging over all pertinent species. But the cognitive scientist, Zenon Pylyshyn for instance, could hardly be expected to know that *a priori*; it would still not really do the trick; and the quarrel between such theorists would have absolutely nothing to do with the principled meta-choice between bottom-up and top-down strategies. All the neurobiological record signifies (and it does signify that) is that top-down theorists had better not be simple Cartesians or theorists fixedly committed to their own idiosyncratic folk psychologies *or* had better not be "pure" cognitive scientists.[29]

Another way of putting the point is to concede, with Robert Cummins for instance, that a *bona fide* science must accommodate not only explanations in accord with "causal subsumption" (covering-law explanations) but also explanations in accord with the (noncausal) "analysis" of the properties and systems within which such properties play an essential role.[30] To admit this is *not* (*contra* Cummins) to concede that the causal and analytic aspects of a science are genuinely separable in addition to being conceptually distinct, or that one could proceed analytically independently of or prior to considering the real causal constraints imposed by the objects of any actual sector of inquiry. Cognitive scientists sometimes pretend that they do so proceed (even if not as Cartesian dualists). But it is easy to show that, though we may shift from a functional analysis of the causal behavior of a physical device to an independent "interpretive" analysis of a machine program that that device may (thereafter) be made to instantiate, it may not be (and normally is not) possible to do *that* with linguistic behavior. There is, for instance, *no* clear way to segregate the analysis of the acoustical or syntactic properties of natural language from an analysis of the molar experience of actual linguistic exchange. The acoustical array in which the semantic properties of a given language may be incarnated *cannot* be specified independently of actual linguistic behavior and linguistic history. That the analysis of invented machines may proceed bottom-up – because the relevant dimensions of analysis *are* independent of one another, are guaranteed to behave extensionally, are made to form a closed system – affords no reason for thinking that the analysis of human language or of human culture must yield conformably. This is the utter *non sequitur* that Cummins shares with Dennett (that "justifies" rejecting the "folk" psychologies of Chomsky and Fodor).

Bottom–up theorists such as Dennett and the Churchlands tend to ignore the bearing on the *theory* of science of the reflexive nature of inquiry; top-down theorists such as Fodor and Chomsky hold too simple a view of the realism of the psychological theories they happen to favor. Bottom–up theorists emphasize the fictive possibilities of theory – but are excessively sanguine about these; top-down theorists are often too dogmatic about *particular theories* that do instantiate their more reasonable general strategy – but their theories are hardly the exclusive options they suppose them to be. Bottom–up theorists slight the realist constraints on the theory of science that the agency of human inquiry imposes – which must affect the realism of particular theories; top-down theorists naively cling to essentialist versions of that realism – for which reason they fail to accommodate the underdetermination of theories, of psychological theories in particular. Bottom–up theorists are arbitrarily convinced that they can eliminate the human (the "folk") element from science because they are convinced they can eliminate the human from physical nature; top-down theorists are equally arbitrarily convinced that their nativism is inviolate between language and inquiry ineluctably entail and presuppose the human. Bottom–up theorists honor the vestiges of the unity canon at the same time as they mean to eclipse the realism of the human as radically as possible; top-down theorists extend the same canon to the human already suitably tailored to receive it. The result is a comedy of philosophical manners. (Of course, there are options for bottom-up and top-down maneuvers not captured by those of the eliminationists and the nativists.)

Alternatively put, *no* gains in the physical, biological and neurobiological disciplines that proceed more or less linearly along the lines of present research (even with a considerable tolerance for novel hypotheses) can be expected to undermine the top-down vision in favor of the bottom-up. There is of course always the question of how much in the way of enlarging and organizing such sciences would incline us to abandon a top-down strategy in favor of a bottom-up. But that question is properly not really on target at all. First of all, on the argument given, there is no pertinent research in those disciplines, however inspired by bottom-up thinking, that cannot be plausibly reinterpreted in top-down terms *if core folk-theoretic concepts remain salient in psychology*. If such notions as those of person, self, agent, choice, commitment, reasoning, invention, speech, knowledge, belief and the like are characteristically centered on the higher, culturally significant achievements of humans, then no accumulation of new data or new theories will affect the meta-theoretic choice unless and until those very notions are convincingly reduced in empirical terms (by identity, replacement, or elimination) to such terms of the disciplines mentioned or others like them that would be noticeably congruent with and favorably disposed toward strictly bottom-up strategies. But there are really no such prospects at the moment and none that we have any reason to think will soon loom on some future–present horizon.

Secondly, it is very likely true that the progress of psychology is in good measure a function of the dialectical contest between these two approaches. There is certainly no reason to think that promising research is stifled either because there is a contest between the two or because researchers are actually inspired by the one approach or the other. One must realize that the expansion of a modern discipline and its professionalization incline us to honor remarkably circumscribed, small-gauge research – where, admittedly, the stability of the phenomena in question encourages an attitude sympathetic to the bottom-up strategy. For example, one may attempt to study the electrochemical properties of the large axon of the squid, because of its size and regular behavior; but success there neither obviates a need to attend to the full setting in which the living creature itself lives (J. J. Gibson's ecological orientation, say), nor augurs in any way what may be said about the major nervous functions of man in spite of the apparent uniformity of the explanatory principles underlying neuronal functioning throughout the animal world.

Finally, a third consideration arises. Top–down analyses are not in the least inimical to strong law-like regularities of the sort usually proposed for sub-molar functioning of an increasingly fine-grained sort. (But it needs also to be said that there are reasonable questions to be raised about what should count as a natural law in a molar biology or psychology or sociology – or even whether causal regularities need be law-like at all or need be invariably expressible in law-like ways. The matter is certainly not settled – certainly hardly even clear.) On the contrary, the admission that psychological processes are incarnate encourages us to suppose that whatever, identified at given physical or biological levels as the sub-functioning parts of whatever incarnates molar functioning at the full psychological level, may continue to be usefully analyzed by any strategy favored by bottom-up theories, without in the least disturbing an ultimate preference for the top-down (at least on the conditions already given). Hence, if the psychological is incarnate, then it is a foregone conclusion that progress in the development of a unified neurobiological theory cannot fail to enhance the effective revision of a folk-theoretic psychology – unless, in the process of doing that, it suitably reduces (in bottom-up terms) the salient concepts of most folk psychologies. Otherwise, it is simply a mistake to suppose that the mere growth of neurobiology yields any advantage (for that reason alone) to the impending victory of bottom-up strategies. Nothing could be farther from the truth.

II

If we ask ourselves at long last for some of the reasons why the top-down vision persists so vigorously – even thrives and even appears to have the advantage – in a world of science in which bottom-up speculation prevails outside psychology and the rest of the human and social sciences, we may attempt a partial tally that should be of some service at least. Certainly, given that the very achievement of any science, as itself a natural phenomenon, needs to be explained – is presumably a large datum within the scope of some appropriate science – it is hard to see that there are any prospects at all for recovering the account required in bottom-up terms. Remember that the bottom-up must either eliminate the executive molar level of the top-down strategy or else demonstrate that whatever obtains at that level is not merely altered by fine-grained work but actually satisfactorily analyzable in terms of a genuinely bottom-up strategy.

There are of course many features that make it difficult to eliminate or reduce folk-theoretic notions (granting always that they are clearly subject to considerable revision, on the achievement of provisionally subaltern sciences). The intentionality of high-level cognition, *not* of every facilitating neurophysiological structure in which such cognition is incarnate, appears to require a realist analysis (usually modelled sententially)[31] that is simply not straightforwardly or obviously reconcilable with the usual bottom-up strategies. Such cognition is, for various reasons, not in accord with the strong extensionalism favored in the unity-of-science program. In the first place, models of rationality at the level at which such higher-order cognition and activity obtain are normally construed as holistic, as linking in a strongly relational way (as far as characterizing ascriptions are concerned) intentions, beliefs, desires, and actions;[32] as a result, salient folk-theoretic phenomena cannot be sufficiently segregated in terms of their intentional complexity so as to be straightforwardly studied by means of extensional tactics.

Secondly, since human behavior and mentation are strongly lingual, if not linguistic – in the sense of presupposing linguistic aptitude where they are not specifically linguistic themselves (dancing or recalling images, for instance) – ascriptions in accord with the holistic model linking belief and desire and action are inevitably infected with all the well-known puzzles of intensionality. Hence, thirdly, wherever the psychologically incarnate is admitted to have a causal role (which, in effect, is ubiquitous), familiar concepts of the extensional nature of causality will be frontally faced with an obvious inconsistency.[33] Once these themes are in place, it is hopeless to speak of reconciling top-down and bottom-up strategies. It is not, however, hopeless to speak of reconciling, *on an interpretation*, whatever reasonably rigorous findings may be yielded by first-order disciplines inspired by the one vision or the other. Thus is the contest joined.

One enormous area of importance remains to be mentioned, almost completely neglected by contemporary bottom-up theorists of every stripe: the dimension of cultural fluency among the linguistically apt members of distinct human societies. Quite clearly, it is the phenomena of this entire field that provide the saliences on which the strength of the folk-theoretic approach depends. One may easily say that, in time, some hierarchically organized sequence moving smoothly from neurons to supersystems of sub-systems of neurons will capture the complexities of a cocktail party or an act of political terrorism or the reading of Marcel Proust or the emergence of post-modernism. To put the point in terms of a well-known dispute, Skinner's and Chomsky's reductionisms fail, in rather different ways, to come to terms with the problem of methodological individualism.[34]

Thus, language, history, cultural rules, styles, genres, institutions, traditions, practices, consensus require a richer framework within which psychological description and explanation can flourish than can be got from *any* complexification of the putative structures of *aggregated* systems. But, on folk-theoretic views, it is reasonably argued that the use of such notions as those of person, self, rationality, language and the rest *presuppose and entail a conceptual symbiosis of self and society*. The upshot of this, of course, is that psychology is not and cannot be an autonomous discipline and cannot be satisfactorily analyzed in neurobiological terms even where no satisfactory analysis could preclude the contributions of the neurosciences. If that were admitted, *then bottom-up theories could not but be ultimately mistaken*. Psychology, so seen, would be a distinct but hardly autonomous discipline, situated at the intersection of the neurobiological and the societal or cultural.[35]

There are at least three distinct interlocking themes that may be mentioned here, that confirm the clear power of folk-theoretic models elaborated in terms of the symbiosis of self and society at the full cultural level. First of all, with respect to language and cultural practice and tradition, one must acknowledge a profound *division of labor* that is culturally diverse, historically contingent and variable, hardly uniform species-wide, and therefore not likely to be explicable in genetic or neurobiological terms even where (on the admission of the psychologically incarnate) there must be some suitable account of neurobiological facilitation. If societal or cultural divisions of labor cannot be suitably reduced bottom-up, then it is a foregone conclusion that bottom-up theories will ultimately wither. One cannot, on the division thesis, explain the actual linguistic and lingual performance of any individual agent on physical or biological grounds, simply because no individual can be supposed to have internalized all the societal structures on which such performance depends. The description and explanation required will include structures *that are societally* (not individually) *centered*, where those structures are themselves, however realist our characterization of language and culture, distinctly emergent and *sui generis* with respect to the physical and biological processes in which they are (and must be) incarnate. That is, *if* the division thesis holds *at the cultural level* (here, one may usefully contrast the genetically determined division of labor of the bee hive), then it is impossible to deny a crucial dimension of psychological emergence (in the strongest sense of emergence: one that is incompatible with all forms of the unity-of-science program); and, if we admit that, then bottom-up theories must fail utterly. At the very least, it is an important stumbling block that bottom-up theories characteristically ignore. The real properties of *psychological* phenomena may include a significant range that are ascribable only in terms of *cultural* or collective relations of an intentionally complex sort.

Secondly, if the division thesis obtains (at the cultural level), then the individual members of a viably functioning society must have been effectively groomed in acquiring their natural language and "natural" culture (if we may extend that term of art in an obvious way) to be able to *interpret* satisfactorily the often improvised performances of apt individual agents and to arrive at an *interpretive consensus* regarding the properties and significance of such performances. But, again, there is no plausible way to understand such an aptitude in reductive terms. (Here again one may usefully contrast the genetically determined song template of many species of birds, accounting for neighborhood variability within species and the gradual learning of local songs by resettled birds of the same species.[36])

Perhaps the most famous exploration of this culturally emergent pattern of social grooming – one that the bottom-up theorist has hardly even acknowledged and that, failing reduction, must profoundly affect the meta-choice between top-down and bottom-up strategies – is to be found in the enormously influential speculations of Wittgenstein regarding what Wittgenstein calls "forms of life" and "language games." One cannot of course find in Wittgenstein any attempt to explore the full relationship between the neurobiological and the historically societal, but *we* can see (and explore) the connection. It is a connection that, with some imagination, we may concede has also attracted contemporary Marxist, Hegelian, phenomenological, hermeneutic, and semiotic currents – currents also committed to an ineliminably culturally collective dimension of psychological life. Perhaps the figures of Maurice Merleau-Ponty and Pierre Bourdieu may be allowed to serve as markers for that immense Continental European literature that, with Wittgenstein, adheres to top-down strategies in ways the bottom-up has very nearly refused to consider.[37]

There is at least one extraordinarily subtle consequence of embracing our second theme: on the argument, it is quite possible that the correct or a reasonably admissible description of the psychological behavior and states of individual members of a human society may not *in principle* be correlated with any neurobiological processes detectable in a given individual *in law-like terms confined to any biological range*. The reason is that, although the states and behavior in question must be incarnate, their emergent features (on the folk-theoretic account) are rightly assigned only on the basis of an interpretive consensus (which, by the way, is very close to Wittgenstein's master theme); and that means that their description may not be able to be falsified or nomically collected by reference merely to the *internal* neurobiological processes of the agents in question. There is no natural priority favoring neurobiological descriptions, and there is no known bottom-up reduction or elimination of the folk-theoretic notions we are bound to invoke in culturally complex behavior. To press the point is simply to draw attention to the methodological preferences of the factorial idiom of top-down theories that address culturally emergent phenomena. The upshot is that, on that view, the neurobiological is necessarily subaltern – but not, because of that, altogether without clout in forcing a revision of folk-theoretic notions.

The third theme is in a way the most important. All bottom-up strategies suppose that physical nature is a *closed system* and that human psychology must fall somewhere within that system. In the limit, scientific explanation is, on that thesis, *homonomic*:[38] that is, the vocabulary of description for a given domain may be satisfactorily and completely drawn from the vocabulary suited to the explanation of the phenomena of that domain, viewed under covering laws. On the unity-of-science reading, provisionally homonomic systems will prove to be sub-systems of a completely unified system. But, of course, the admission, on empirical grounds, of the culturally emergent signifies that human psychology (1) cannot as such be homonomic at any physical, biological, or computational level, (2) may not be uniformly law-like though it is causally efficacious, and (3) cannot form a closed system. It cannot (on the argument) form a closed system because the causally efficacious behavior of the members of a human culture include what is improvisationally open to the members of an interpretive community – where what is thus open is not demonstrably generable from the elements of any would-be *system*, whether neurobiological or linguistic or language-like. If the matter is an empirical one (as it seems fair to claim), then once again the bottom-up approach must be the wrong way to go – or, we haven't the faintest idea of how to show that it is the right way to go. Remember that, on the argument, nothing produced by a first-order science inspired in the bottom-up way need be precluded from a suitably enriched top-down psychology: what would happen of course is that all bottom-up presumptions or interpretations of such incorporated items would have to give way to an alternative reading. But this happens everywhere in science.

If the argument sketched holds, then we may anticipate that the contest between the top-down and the bottom-up depends not only on the reducibility or eliminability of the experiential, intentional, and purposive lives of individual selves but also on the reducibility or eliminability of the languaged and cultural dimension of human psy-chology – the symbiosis of the psychological and the cultural and the use of the collective predicates of cultural ascription. On the first line of attack, purely functional or abstract informational models must give way to neurobiological models, whether for top-down or bottom-up reasons; although the salience of folk concepts constitutes what may well prove an insuperable obstacle to any realist version of the bottom-up. Add the second line of attack: one hardly finds any explicit bottom-up strategy that even addresses the pertinent issues. Only the recent development of Noam Chomsky's deeper account of the theory of natural language can claim any pertinent interest here. But it is a theory that itself needs to be defended, and it is a theory (as is also Jerry Fodor's, closely linked with Chomsky's) that the partisan of the bottom-up is inclined to treat as itself unacceptably top-down.[39]

III

It is in fact the fate of the concept of human selves that is being contested by top-down and bottom-up programs. We cannot hope to track all the strategies the bottom-up theorist has invented – but neither need we do so: there is a reasonably perceptible trail that we can map, that all bottom-up riders must travel if they are to arrive home safely. It is a sort of detour of the *sine qua non*, an avoidance of the escarpments of intentionality. Needless to say, the scouts of the top-down vision insist that the road leads nowhere.

For our present purpose, we may safely put aside the full pursuit of the intentional. We need consider only how that essential obstacle is apparently avoided without actually being bypassed. It is a remarkable fact that the principal "solution" of the intentional puzzle, favored in analytic philosophy (whether or not addressed to the top-down/bottom-up issue), is to dismiss it as a delusion or a mere shadow. In any event, it is not to be taken as a hard conceptual tract that we must pass through. For the purpose at hand, Daniel Dennett is probably the most visible champion of the bypass. He has actually compiled (in an as yet unpublished paper) a generous list of those who side with him and against him and even of those who are difficult to classify. Those who oppose him subscribe (on his own summary) to what John Haugeland has dubbed "original intentionality": "The idea is that a semantic engine's tokens only have meaning because we give it to them; their intentionality, like that of smoke signals and writing, is essentially borrowed, hence derivative. ... Genuine understanding, on the other hand, is intentional 'in its own right' and not derivatively from something else."[40] Dennett takes Haugeland to agree with him – against original intentionality – and to oppose the views of Jerry Fodor, Fred Dretske, John Searle, Tyler Burge, and Saul Kripke at least (some of whom suppose that Haugeland sides with them). He also rightly claims that Wilfrid Sellars, Robert Stalnecker, and Paul and Patricia Churchland in effect side with him.[41] Dennett had, quite early in *Content and Consciousness*, already clearly opposed what, in the modern literature since Brentano's work, has (effectively) been seen to be the master theme behind that original intentionality:

> Intentionality is not a mark that divides phenomena from phenomena, but sentences from sentences. ... Intentional objects are not any kind of objects at all. [The tendency to treat them as distinct objects rests on] the dependence of Intentional objects on particular

descriptions [that is, on the thesis that] to change the description
is to change the object. What sort of thing is a different things
under different descriptions? Not any object. Can we not do
without the objects altogether and talk just of descriptions? ...
Intentional sentences are *intensional* (non-extensional) sentences.[42]

This remains admirably clear. Furthermore, we may understand (by the
unpublished paper, 1986) that Dennett pretty well stands by his original
views, which, he also finds, converge substantially with the ramified
theory Ruth Millikan has recently formulated.[43]

Let us step back then for a moment, in order to approach the issue
in a fresh way. In an early, extremely well-known and much-discussed
paper (which he has now pretty well repudiated), Hilary Putnam had
announced that

The various issues and puzzles that make up the traditional
mind–body problem are wholly linguistic and logical in character
... [in fact] it is no longer possible to believe that the mind–body
problem is a genuine theoretical problem, or that a "solution" to
it would shed the slightest light on the world in which we live ...
[it] is nothing but a different realization of the same set of logical
and linguistic issues [as are raised by the "'identity' or 'non-
identity' of logical and structural states in a machine"].[44]

There are problems raised by Putnam's view (as originally formulated)
that are merely local to the mind/body program – for instance, that pain
is somehow characterized in a purely abstract or functional way, and
that mental properties are entirely functional and yet can be assigned a
causal role. Viewed solely in terms of our present concern, it is clear
that Putnam's maneuver could not but be unsatisfactory, in the strict
sense that the mind/body problem is "nothing but" the logical/structural
problem of information-processing machines since (or, if) *that* problem
is (assumed to be) nothing but a manifestation of the generic mind/body
problem. For, for one thing, even if the human mind is deliberately
modeled by reference to a Turing machine, Turing machines and artificial
intelligence are themselves *part and parcel of the actual work and
working of human technology and human intelligence*. For another, on
Haugeland's formulation, the question of "original intentionality" is a
viable and unavoidable one – one that Putnam did not care to address
– even if, on views such as Dennett's, it may (or must) be answered in
accord with the bottom-up strategy.

Putnam had held, further, that "everything is a Probabilistic Automaton under *some* Description,"⁴⁵ which is to say (only) that any finite, informationally qualified *segment* of a system can have its informational properties generated by a Probabilistic Automaton – *not* that informational properties or processes can be reduced to the noninformational properties or processes of *any* physical system in which information is thought to be incarnate, embedded, or realized in some particular way, *or* that the ecologically rich and open-ended *capacities* of humans can as such be modeled by machines. Here, we come directly to the problem of the choice between top-down and bottom-up strategies so dear to the speculations of the theorists of artificial intelligence.

What the aside on Putnam conveniently shows, therefore, is that the reduction or elimination of selves cannot be convincingly mounted solely on the basis of formal (or abstract) functional similarities – there *is* a question of substantive analysis at stake; also, that the formal modelling of mere finite segments of intelligent behavior or performance or processing cannot replace the analysis of intelligence itself – for the very thesis discounts its possessing *any* such relevance. Putnam had doubly trivialized a question of some importance.

Dennett proceeds differently. For one thing, he seriously misstates the intentional thesis – misstates it so badly in fact that the thesis seems hardly more than an idiocy. The "thesis," he says, holds that "to change the description is to change the object." But that is utterly false. The thesis holds, rather, that intentional phenomena are real and actual, but that we cannot provide a satisfactory criterion (or even a fair approximation to one) by means of which we can regularly *tell* whether the same phenomenon is being designated under changing descriptions. Dennett manages to dismiss the reality of the intentional by intruding a bizarre claim that certainly cannot be taken to capture the reasonably well-defined tradition that has steadily explored the puzzle. His implied conclusion, "Can we not do without objects altogether and talk just of descriptions?" is hardly more than an imposture.

There *may* be a sense in which intentionality is rightly construed as a "mark" that divides "sentences from sentences," not "phenomena from phenomena," but Dennett nowhere supplies the argument. Once we allow, with Dennett, that "intentional sentences are *intensional* ... sentences" *and* affirm, against Dennett (in the spirit at least of Haugeland's open question), that intentional phenomena may be actual, the charade is exposed. It needs to be said at once that Wilfrid Sellars had characteristically emphasized that, on his view, the analysis of intentionality was, effectively, the analysis of intentional as opposed to "behavioristic" *discourse*.⁴⁶ But Sellars has also vigorously argued on independent grounds (honorably open to dispute) that the "intentional" order – "the order in which the intellect pictures the world" – "*qua* belonging to the real order ... is the central nervous system, [is] *in propria persona* [not analogically] neurophysiological states." He argued for this thesis on the basis of a quite extreme and well-known form of scientific realism. Dennett develops no such strategy; and Sellars is obliged to argue that persons are merely logical "roles" assigned to what is real, not genuine entities at all.[47]

This contrast – between a *discursive* and a *substantive* characterization of intentionality – may well be the principal contest of our age. On the evidence, though a restrictedly discursive sense is surely eligible, it would be arbitrary to insist that all intentional attributes or phenomena were merely discursive (that is, functionally or heuristically limited to discourse about some domain) rather than the actual attributes or phenomena of some examined domain. In reading Husserl, for instance, the question arises of whether intentionality should be restricted to the "thetic character" of a perceptual "act" (marking it as a perceptual rather than a mimetic act) or whether it applies to the "object side" of the act (affecting the properties of what may be "given" in perception).[48] Certainly it would be difficult to assign intentionality exclusively to the discursive or cognitional dimension (of the symbiosis of self and world, or, within that symbiosis, to the thetic aspect of inquiry) – or, farther afield, to what is said to be eliminable by way of the thinner reductive claims of recent Anglo-American accounts.

There is another well-known claim advanced by Quine that must also have influenced Dennett. Quine argues that "When ... constructions on sentences are limited to quantification and truth functions (that is, logical transformations involving the simplest sentences and complex sentences constructed from them), one law that is easily proved by ... [mathematical] induction is that of extensionality" – that is, "substitutivity of coextensive terms preserving truth."[49] Quine applies this thesis in a fully sanguine way to the entire work of empirical science, reassuring us that "Surviving idioms of an *extraneous* sort – indicator words [e.g. 'this,' 'this water,' and the demonstrative pronouns], intensional abstracts [as in treating classes, attributes, relations, and the like as objects], or *whatever* – can remain buried in larger wholes which behave *for the nonce* as unanalyzed general terms."[50] Nevertheless, whatever formal demonstration may be given regarding extensionality, it hardly follows that the terms of *natural-language* discourse (or of any would-be science) developed in an empirically serious way *can* be reliably shown to yield to such regimentation: the pretty expression "for the nonce" betrays an altogether unexamined assumption. (This is, also, it may be remarked, the weakness of Davidson's attempt to "regiment" natural languages in accord with Tarski's Convention T.[51] The point may be brought to bear on the Davidsonian view of psychology.) In particular, Quine's attempted elimination of the so-called referential opacity of sentences involving "believe" and similar verbs of propositional attitude (that is, verbs that appear to thwart the extensionality thesis) fails simply because the problem requires a *cognitively* motivated solution, *not* a purely formal one that begs the very cognitive question at stake.[52] No one has as yet supplied the answer needed – in terms of natural-language sentences; and Quine's claims about sentences simply ignore the existence of actual intentional phenomena.

Quine's well-known convictions about this matter are curious for another reason: namely, because Quine opposes all forms of foundational-ism and essentialism and insists nevertheless on exposing the analytic/synthetic dogma. Apparently, in spite of that, he is persuaded that our inability to discover "some fundamental set of general terms on the basis of which all traits and states of everything could in principle be formulated" somehow *does not* render in the least doubtful or uncertain the extensionalist project iself.[53] So there is a definite kinship between Dennett and Quine. It is worth remarking that, whatever the difficulties of his own proposed solution of the intentionality of psychological states, Dretske (whom Dennett rightly takes to oppose him) is entirely clear about the distinction between the use of the term "intentional" applied to *phenomena* and applied to *sentences* about phenomena. Speaking of the distinctive features of "the informational content of a signal" (an actual phenomenon), Dretske observes, "Philosophers have a special terminology for describing such phenomena: *intentionality* (or, if we are speaking of the sentences used to describe such phenomena: *intensionality*)."[54]

The upshot is that, if the argument against *real* intentionality is to be won – is to be secured for bottom-up theories – it must be won fairly. There is only one strategy that could possibly count: demonstrate that the intentional idiom designates nothing real or that what is real that it does infelicitously designate (even if at the expense of appearing to designate more than is actually real) may be brought into perspicuous accord with a thoroughly extensional idiom *fitted to what is reasonably taken to be empirically real*. Dennett offers two further strategies bearing on the issue – both of which may be shown to fail; and Davidson offers a most influential alternative solution – which also fails. We may, therefore, round out this preliminary account with a useful summary of these additional maneuvers. For, in a fair sense, Dennett's and Davidson's arguments pretty nearly count as the prime exemplars for the most fashionable bottom-up arguments of the day. We may then turn to make a fresh start on a number of more recent specimen views.

Dennett's first proposal may be put in the form of a self-disabling dilemma. The easiest way to present it is to cite Dennett's own words. Here is his optimistic view of the dismissive fate that, in principle, awaits the psychological sciences: "the personal story [the 'story' of a person's mental states, the 'story' of selves] has a relatively vulnerable and impermanent place in our conceptual scheme, and could in principle be rendered 'obsolete' if some day we ceased to *treat* anything (any mobile body or system or device) as an Intentional system – by reasoning with it, communicating with it, etc."[55] This is surely the objective of Dennett's homuncular program: the elimination of the personal and the mental – primarily because they are intentional phenomena and, as intentional, unreal. Talk of the mental is a mere *façon de parler*; the problem is to save what, beneath the verbiage, is real. But how does Dennett hope to achieve his objective?

He means to proceed *top-down*, by first introducing *sub*-personal "persons" who effectively subdivide the work of molar persons; then by introducing *sub*-homuncular "homunculi" who effectively subdivide the work of sub-personal homunculi:

> One starts, in AI [Dennett says], with a specification of a whole person or cognitive organism – what I call, more neutrally, an intentional system . . . – or some artificial segment of that person's abilities (e.g., chess-playing . . .) and then breaks that largest intentional system into an organization of subsystems, each of which could itself be viewed as an intentional system (with its own specialized beliefs and desires) and hence as formally a homunculus. In fact, homunculus talk is ubiquitous in AI. . . . Homunculi are *bogeymen* only if they duplicate *entire* the talents they are rung in to explain. . . . If one can get a team or committee of *relatively* ignorant, narrow-minded, blind homunculi to produce the intelligent behavior of the whole, this is progress.[56]

The idea is to progress by top-down analysis, from the fully intentional through a simpler and simpler *intentional* division of labor until we reach a point where we have only to substitute some *non-intentional* (extensional: that is, non-intensional) operation for the utterly simple ("stupid") work of some ultimate homunculi – where, that is, it no longer makes any difference whether we call it intentional or extensional. Simplicity itself!

But there is a double difficulty. For one thing, as part of a top-down strategy, homuncular terms are entirely *relational* – that is, introduced only by factoring into sub-functions the functioning (already conceded) of a given molar system – and so are logically incapable of *replacing* the molar. Thus Dennett says, "the information or content an event within [a given] system has [it has] *for the system as a (biological) whole. . . .* The *content* (in this sense) of a particular vehicle of information, a particular information-bearing event or state, is and must be a function of its function in the system . . . of which it is a part."[57] Dennett also says, inconsistently with the above (or, at any rate, on independent grounds that he fails to supply), "Any psychology with undischarged homunculi [that is, 'agents' that manipulate 'internal representations', function informationally or intentionally, are not theoretically replaced (discharged) by 'agents' described in purely physical terms] is doomed to circularity or infinite regress, hence psychology is impossible."[58]

The fact is that, *if* homunculi are introduced relationally – as pertinent sub-functions *of* some complex function – there is no clear sense in which the *aggregated* homunculi (at the end of Dennett's story) can replace the *relationally factored* homunculi (introduced at the beginning). The maneuver is a complete *non sequitur*. It could only work if (the second and principal difficulty) it could be shown, on independent grounds, *that there is at hand a viable extensional treatment of intentionally complex phenomena.* That is, Dennett shows us (informally) *what* an extensional replacement would be like in information-theoretic terms (just as Quine had done, in syntactic terms); but Dennett never shows us *how* to work out the replacement, or *that* the replacement is genuinely viable. Notice that Dennett means to replace *persons*, not (as in Putnam's account) mere finite *segments* of the behavior of (intelligent) persons. To concede a rough simulation of the second sort goes no distance at all toward sketching the conditions for success of the first sort. Also, it does no good to invoke Church's Thesis (which Dennett paraphrases, "anything computable is Turing-machine computable"[59]). It is not that Church's Thesis is "false" or irrelevant. It is simply that *it can't be called into play* until Dennett shows that the intentional (or intelligent) capacities of persons *are* analyzable in computable terms; and that he has not yet done. (It may, by the way, be usefully remembered that, on Davidson's view, the mental is holistic – that is, conceptually such that an extensional decomposition of the mental is logically impossible.[60])

The second strategy (the strategy of the unpublished paper mentioned) is most revealing – and very simply disposed of. The strenuous argument we have been examining proceeds, as remarked, top-down. The alternative argument proceeds bottom-up by introducing, first, a vending machine (say) that functions in a way that may be fully described in extensional terms (it sorts US coins, or some such thing) – that may, that is, be fully described in terms of covering physical laws. *Then*, the extensionally described behavior is *assigned* a certain intentional import. But, of course, that strategy, meant to subvert the thesis of "original intentionality," is utterly irrelevant. We may ascribe intentionality where there was none before: then it is only an artifact of our intrusion. But how does that show that, where we begin with real intentional phenomena, it is as clear that we are only imputing intensionality to *extensionally accessible* phenomena? That calls for just the kind of answer Dennett has never managed to supply. So the first argument generates a dilemma that can only be resolved by an empirical demonstration that genuinely intentional phenomena can be analyzed in extensional terms – which, in effect, would show that top-down and bottom-up strategies differ only as a matter of convenience; and the second argument is quite beside the point.[61]

IV

Let us turn now to Davidson. Roughly, what Davidson recommends is a program of analysis that treats language essentially in *syntactic* terms: if he were right, meaning and reference could, in a sense, be entirely obviated; or else, they could be benignly conceded to color (inessentially) an otherwise (scientifically) sufficient explanation of the phenomenon of language. By parity of reasoning, the entire range of psychologically, culturally, historically, semiotically, intentionally rich characterizations of human selves are taken to be eliminable (or, decisively displaceable) in an adequate scientific psychology; alternatively, they are regarded as part of a colorful (even efficient) idiom in which we move among our own kind – which, in the eyes of a disciplined science (but only there) are entirely eliminable. This conceptual linkage between the analysis of languages and the analysis of psychological phenomena helps to explain Davidson's adherence to (what he calls) anomalous monism and Tarski's Convention T.[62] Their union, in fact, is quite characteristic of much recent analytic philosophy of psychology.

This may seem a strange way of boarding the question of the conceptual role of selves in the human sciences. But if we are looking for a sense of the entire range of possible strategies for excluding or diminishing that role – in the name of the normal constraints of science – we can hardly do better. Davidson's philosophical programs are certainly among the most influential in the current Anglo-American literature devotedly loyal to the unity of science program or to the more general bottom-up strategies.

The project is simplicity itself: "Words have no function," Davidson tells us, "save as they play a role in sentences: their semantic features are abstracted from the semantic features of sentences, just as the semantic features of sentences are abstracted from *their* part in helping people achieve goals or realize intentions."[63] But then, following Quine (even going beyond him), Davidson adds,

> ... a translation manual is only a method of going from sentences of one language to sentences of another, and we can infer from it nothing about the relations between words and objects. Of course we know, or think we know, what the words in our own language refer to, *but this is information no translation manual contains.* Translation is a purely syntactic notion. Questions of reference do not arise in syntax, much less get settled.[64]

There you have the heart of the argument. Words have their function in sentences; sentences, their function in the behavior of the members of human societies; *and it is possible to provide an adequate schema for translation without any attention at all to the actual meanings and references of determinate discourse.* The principal thing to understand is how *entire* languages function (or how the *entire* psychology of man is ordered), *not* the piecemeal meaning of this or that sentence (or the piecemeal intention and significance of this or that bit of behavior). Davidson hints at the possibility of a physics of psychology.[65] And he concludes that "The theory gives up reference, then, *as part of the cost of going empirical.*"[66] Correspondingly, we may suppose, we give up, in psychology, intentions and beliefs (and intentionality) as well, as the same price of "going empirical."[67] Hence, Davidson answers the *empirical* question of the competence of a science (of potentially closed systems empirically adequate as sciences) by demonstrating that the language of science is analyzable in terms of a completely extensional syntax – *without* reference at all to the distributed, semantic (apparently, empirically requisite) features that distinguish this would-be science from that; *and* that, thus subordinated, the special idioms of each particular science can thereupon be introduced, within the scope of that syntax, without fear of disturbing its assured extensionality considered *en bloc.*

The trouble with Davidson's program is extremely important to fix – because, in resisting it, we are led to see something of one of the deepest and most global intellectual contests that underlie the speculative quarrels of philosophy and science in our own day. More to the point, that contest is nearly invisible in the smaller projects of every discipline. Here, then, is Davidson's proposal: treat meaning (and reference) solely in terms of truth; treat truth entirely in *syntactic* terms, conformably with Tarski's satisfaction condition (which condition, Tarski believed, was apt only for suitably formalized languages[58]); and construe the *scientific* ("empirical") interest in the phenomenon of language as restricted to such matters only.[69]

No doubt, the bearing of these matters on the theory of selves will seem quite remote. But that is their attraction. The trick rests with a very simple maneuver – that, reciprocally, almost isomorphically, appears in theorizing about language and theorizing about human action and human psychological states. For one cannot interpret linguistic behavior without making ascriptions of beliefs and intentions and the like; and, in the human (the paradigm) case, one cannot ascribe beliefs and intentions without both modeling such ascriptions linguistically and construing them as naturally manifested in linguistic utterances. To anticipate our objection to Davidson: it is impossible to translate a language without attention to actual usage; and it is impossible to trace usage without attention to the actual history and practices of a people. Translation could never be "a purely syntactic notion." By the same reasoning, there cannot be a competent science that does not cleave to some initial, however gradually altered, sense of the actual saliences of a given domain. Davidson could not be more mistaken in his intended economy.

Davidson concedes quite straightforwardly that his "Tarski-like theory of truth *does not* analyze or explain either the *pre*-analytic concept of truth or the pre-analytic concept of reference."[70] He means by the "pre-analytic"

1   whatever may be the concepts of truth and reference that could be fitted to the linguistic and linguistically informed behavior of human agents normally acknowledged in the practices or forms of life of natural societies (whose members acquire their linguistic and cultural aptitudes merely by growing up in a community of apt adults); *and*
2   whatever may be the adequate story of the semantic and pragmatic function of words, sentences, utterances, acts generated by such agents *within* such practices.

In defending "a version of the holistic approach,"[71] Davidson means to exclude *such* factors from his linguistic and psychological theories. Alternatively put, he distinguishes sharply between explanation *within* the theory and explanation *of* the theory. Within the theory, the conditions of truth of a sentence are specified by adverting to postulated structure and semantic concepts like that of satisfaction or reference. But when it comes to interpreting the theory as a whole, it is the notion of truth, as applied to closed sentences, which must be connected with human ends and activities.[72]

Step back, now, and reconsider what's at stake. We are attempting to sort what may be called "simple" and "sophisticated" versions of bottom-up reductionism. Rudolf Carnap affords a particularly sanguine paradigm of all "simple" reductive programs, because Carnap believed that psychological predicates could actually be translated without remainder by a purely physicalist idiom. It is true that Carnap retreated from the required labor; but his more resilient followers realized that translation, predicate by predicate, was quite unlikely – and also *unnecessary*.[73] By contrast, Davidson's strategy serves as a paradigm of "sophisticated" reductive programs. It proposes to show that *the closure of science itself – the very condition for the adequacy of the scientific explanation of all empirical phenomena – does not require the regimentation of the idiom the simple reductionists had labored so fruitlessly to achieve.* Davidson concedes the irreducibility *and* useful function of psychological idioms: he "merely" disallows their having any essential function (as such) *within the bounds of science.*

If, however, there were (irreducibly) psychophysical laws, or if there were (irreducibly) psychophysical causal processes (even if we were unable to formulate their covering laws), *the physicalist closure of science would be empirically untenable*: we should then need to reinstate the concept of *selves* once again within the "space" of science. To meet the challenge, Davidson divides his strategy. He argues as a "simple" reductionist, in advancing his doctrine of anomalous monism; and he argues as a "sophisticated" reductionist, in drawing out the full import of his extension of Tarski's Convention T; *and* he supports each project by appeal to the work of the other.

The countermoves to Davidson's theory of psychology (anomalous monism) are extremely powerful. First of all, the usual simple theories (Davidson's, Carnap's, J. J. C. Smart's, Herbert Feigl's) are committed to a *realist* interpretation of psychology: they concede causal processes involving the psychological, and attempt to save the reductive undertaking by supporting *some* version of the identity thesis. *There is no other option* under the simple realist concession. The only other prospects are these:

1   treat psychology solely as an *idiom* that has had its inning, for historically contingent reasons, and that may now be retired, in principle, by a physicalist idiom with respect to whatever (neutrally identified) *is* empirically real;
2   treat psychological descriptions and explanations as *lacking any realist import at all*, as falsely appearing to have referred to what is actual.

Both alternatives have of course been pursued, and both (generously construed) may be taken as instances of "sophisticated" strategies replacing the "simple."

The first of these ("sophisticated" strategies) is notably associated with the views of Wilfrid Sellars and of such advocates of the unity-of-science orientation as May Brodbeck and Stephan Körner.[74] In a sense, it obviates the need for supporting an identity thesis, since (on the argument) what is real *is* physical (in whatever sense may be defended) and since the picturesque idiom of the mental proves scientifically otiose. The trouble is that this has *never* actually been shown to be true. The second option is notoriously associated with an early version of Paul Feyerabend's views, with the early views of Richard Rorty, and, more recently, with those (oscillating somewhat between the two options) of Paul Churchland and Stephen Stich.[75] Here again, the trouble is that there is no known or remotely promising way to specify *any* physicalistically real set of phenomena able to capture in an explanatorily adequate sense whatever may be salvaged *as* real (that the psychological idiom may be supposed to have luckily identified) and able, at the same time, to vindicate discarding as illusory or a mere artifact of the idiom whatever else that idiom mistakenly took to be real.

Dialectically, the prospect of psychophysical laws and psychophysical causal processes poses the most troublesome threat. Curiously, the proponents of a reductive psychology – notably, Davidson again – tend to be rather orthodox or vague about the nature of scientific laws.[76] (Davidson does not regard himself as a reductionist, of course, essentially because he opposes type-identity – not token-identity. But token-identity is also a form of reductionism. We shall return to the issue in a later chapter.) It is, however, a dogma of analytic philosophy of science that *causal contexts* (as and if distinct from contexts of causal explanation) invariably behave extensionally.[77] This is simply the nether side of the thesis of the (scientifically) closed physical universe: *if* there were psychophysical causal processes, and *if* the psychological were *not* reducible in physicalist terms, then it would be impossible (by any "simple" strategy) to avoid the peculiar intensional, semantic, pragmatic, semiotic, historical, and cultural complexities ineluctably introduced *by acknowledging the reality of human selves*. In that case, causal contexts would *not* behave extensionally only – which is to say there would be no operable basis on which to reidentify all causes extensionally (since, on any theory, causes remain self-identical). That would lead to a stalemate regarding efforts to reduce or eliminate selves or would oblige us (supporting the reductive objective) to seek subtler versions of the "sophisticated" strategy.

At this point Davidson shifts, effectively, to advocating his extension of Tarski's Convention T. The theory is somewhat involved and invites a bit of patience.

Concede Davidson's characteristic conditions: bracket all questions requiring explaining the truth of individual sentences by reference to "the semantic features of words" (psychologically descriptive natural-language terms) *within* the scope of the linguistic theory Davidson advances; consider only the explanation *of* the theory *en bloc*. On this maneuver, as Davidson rightly says, "words, meanings of words, reference, and satisfaction are *posits* we need to implement a theory of truth. They serve this purpose *without needing independent confirmation or empirical basis*."[78] What this means, evidently, what it can only mean, is that categories of such linguistic sorts are first introduced in a purely formal way – they are either syntactic, semantically uninterpreted categories (with respect to their distributed instantiations) or completely explicable (as to their function) in purely syntactic or formal terms. Anything else would smack of apriorism – and Davidson can hardly be accused of that. But, *if* it turned out that the maneuver was *not* semantically neutral as supposed (in the sense Davidson favors), the planned extension (or adjustment) of Tarski's account would fail – would be arbitrary, undefended, *a priori* (actually), or question-begging; and, as a result, we should find ourselves quite reasonably obliged to restore the full pivotal function of the concept of selves.

The beauty of proceeding in this way is just that it permits us to grasp, *without any partisan doctrine of human nature whatsoever*, that we simply cannot manage to describe and explain the psychologically real phenomena that all theoretical hands acknowledge "pre-analytically" – in Davidson's way. Davidson would have us believe that the function of words, meanings of words, reference, and satisfaction (or truth) may be systematically assigned to languages treated holistically – hence, in accord with the extensional syntax favored, without regard at all to the distributed semantic, pragmatic, or intentional complexities of actual discourse. Once that is conceded, the *distributed* use of those very notions (with respect to experience and intentional complexities) count *only* (on Davidson's view) as internal adjustments within the holistic language first identified. Hence, we are blocked from challenging its initially confirmed extensionality. *If*, however, the two issues cannot be treated, in any empirically pertinent way, as *ordered* so as to favor the independence or priority of the first with regard to the second (a claim that must recall certain cognate claims of Lévi-Strauss and Chomsky, though now in an entirely different spirit), then Davidson's thesis (like Lévi-Strauss's ánd Chomsky's) collapses at once. There is no empirical basis for confining extensionality holistically.

The issue grows murkier. Davidson admits that "a general and pre-analytic notion of truth *is* presupposed by the theory" itself, in order that "we can tell what counts as evidence for the truth of a T-sentence" (that is, a sentence cast in the extensional, syntactically canonical form the Tarskian theory affords – which, on the first condition, does not affect empirical questions at all). He warns us that such a notion is *not* required for merely introducing the concepts of *satisfaction* and *reference* (which *are*, when used distributively, systematically tied to the admission of selves): "Their [generic] role is theoretical, and so we know all there is to know about them when we know how they operate to characterize truth." It is in this sense – an entirely fair sense – that "we don't need the concept of reference."[79] *Consequently, we can't need the concept of selves*. Obviously, Davidson would expect to capture the pre-analytic consideration introduced for T-sentences by a higher-order application of the *same* theory – and would hope, in doing so, to exhaust all interesting challenges to his account. Perhaps. But it may be noted, in allowing the point, that the implied regress does *not* actually show how, in real-time terms, *to* escape relying on semantically rich notions of satisfaction (or truth) and reference. The very formality of Convention T, therefore, appears to confirm our ultimate dependence *on* (semantically pertinent applications of) satisfaction and reference, *whether* for the truth of our object-language sentences *or* for their T-sentence replacements. (That would effectively confirm our ultimate dependence on the concept of selves.) We must bear in mind that to postpone the application of the concept is hardly to attenuate its critical role.

Here, a third consideration of a most troublesome sort arises, which may be conveniently clarified by recalling some observations of Hilary Putnam's with which Davidson seems to agree. "'True,'" on Tarski's account (also, on Carnap's, Quine's, Ayer's and that of others), "is [says Putnam], amazingly, a *philosophically neutral* notion. 'True' is just a device for semantic ascent': for 'raising' assertions from the 'object language' to the 'metalanguage,' and the device does not commit one epistemologically or metaphysically."[80] If we put quotation marks around the sentence *Snow is white* and adjoin the words "is true," "The resulting sentence is itself one which is true if and only if the original sentence is true. It is, moreover, assertible if and only if the original sentence is assertible; it is probable to degree r if and only if the original sentence is probable to degree r; etc." So, "to understand *P is true*, where P is a sentence in quotes, just 'disquote' *P*: take off the quotation marks (and erase 'is true')."[81] This surely raises the puzzling question of how, on a *realist* reading of empirical assertions, satisfaction (or truth) and reference can be avoided; *and*, more significantly, it raises the question of how the *formal adequacy* of *any* would-be T-sentences (as replacements for natural language sentences) could be *decided* without invoking, at that very level, the original worries about satisfaction and reference. One may even invoke, here, the so-called anti-realist strategy introduced by Michael Dummett[82] – without subscribing to Dummett's would-be solution of the matter. For, in merely raising the decidability question, anti-realism poses once again the issue of the eliminability of personal agents or selves (the very agents *who* decide) and the intensionally complex world they appear to inhabit. Another way of putting the point is this. A T-sentence must be strictly constructed in an extensionally satisfactory way in accord with Tarski's view of a *suitably* formalized language.[83] Tarski believed that natural languages could not be completely regularized to this end. This is most important: for, if "true" is "philosophically neutral" as Putnam says (and Davidson obviously believes), then quotation and disquotation *are not conceptually linked with Tarski's account of the conditions on which "true" is a predicate of sentences admissible in his theory (T-sentences); and if they are so linked, then "true" is not in the least philosophically neutral.* Either way, Davidson *cannot* avail himself of *any* use of Tarski's theory to strengthen the empirical adequacy of *any* form of simple or sophisticated reductionism. (This is a precise analogue of our objection to Dennett's premature appeal to Church's Thesis and to Quine's extensional treatment of sentences within the scope of his attack on the analytic-synthetic dogma.) More pointedly put, the question of whether sentences describing or explaining psychologically real phenomena can be extensionally construed or are irreducibly complex in nonextensional ways is a substantive question that needs to be independently addressed. Either, then, Davidson must already have succeeded in his theory of *psychology* (anomalous monism or some other "simple" reductive account; in demonstrating an extensionally adequate analysis of the relevant *phenomena* (which he has merely chosen to cast in terms of T-sentences), or else the project has already been aborted by the failure of the antecedent psychology. Either the empirical psychology must precede the formal linguistics or the intended linguistics is itself empirically fitted to the data and cannot be known in advance to be congruent with Tarski's formal constraints. Here is the most salient version of the conceptual lacuna of all formal versions of extensionalism. The irony is that it was already anticipated by Tarski. Davidson is candid enough to say that the Tarski-like theory of truth he favors can "*at best*" give "the extension of the concept of truth for one or another language with a fixed primitive vocabulary."[84] It cannot explain the truth of semantically rich individual sentences. If there *were* an empirically available primitive vocabulary(that is, one that did behave extensionally), and if *all* complex concepts could be extensionally treated in terms of such a vocabulary, then either type or token identity might ultimately be empirically defended. As it happens, this is precisely the undertaking that Jerry Fodor favors.[85] Unfortunately, Fodor's theory is cast in an extremely unlikely platonist (or nativist) form and is utterly inexplicit about the actual extensional reduction *of* complex concepts (along Tarskian lines) to some original core of putatively primitive concepts.

Now, the originally intended overview threatens to overwhelm us with the intriguing details of an endless queue of detailed tactics. Let it suffice to say that there is no known empirical or operative resolution of the problem. There are, of course, ever so many voices that assure us that it can be managed.

Notes

1 See Joseph Margolis, "Animal and Human Minds," *Culture and Cultural Entities* (Dordrecht: D. Reidel, 1984).

2 See Stanley Milgrim, *Obedience to Authority* (London: Tavistock, 1974); also, Don Mixon, "Behavior Analysis Treating Subjects as Actors Rather than Organisms," *Journal for the Theory of Social Behavior*, I (1971).

3 The question is the essential pivot in J. J. Gibson, *The Senses Considered as Perceptual Systems* (Boston: Houghton Mifflin, 1966).

4 See Rudolf Carnap, "Psychology in Physical language," tr. George Shick, in A. J. Ayer (ed.), *Logical Positivism* (Glencoe, Ill.: Free Press, 1959).

5 Sigmund Freud, "Project for a Scientific Psychology," in *The Standard Edition of the Complete Psychological Works of Sigmund Freud*, vol. i: *1886–1899*, tr. J. Strachey et al. (London: Hogarth Press and the Institute of Psycho-analysis, 1966).

6 See Michael S. Gazzaniga (ed.), *Handbook of Behavioral Neurobiology*, vol. 2 (New York: Plenum, 1979) and *Handbook of Cognitive Neuroscience* (New York: Plenum, 1984).

7 See Noam Chomsky, *Rules and Representations* (New York: Columbia University Press, 1980).

8 See Jerry A. Fodor, *The Language of Thought* (New York: Thomas Y. Crowell, 1975) and *Representations* (Cambridge, Mass.: MIT Press, 1981).

9 See Zenon Pylyshyn, *Computation and Cognition* (Cambridge, Mass.: MIT Press, 1984).

10 See Daniel C. Dennett, *Brainstorms* (Montgomery, Vt: Bradford Books, 1978).

11 See Paul K. Feyerabend, *Against Method* (London: New Left Books, 1975).

12 See Stephen P. Stich, *From Folk Psychology to Cognitive Science* (Cambridge, Mass.: MIT Press, 1983).

13 See Paul M. Churchland, *Scientific Realism and the Plasticity of Mind* (Cambridge: Cambridge University Press, 1979) and *Matter and Consciousness* (Cambridge, Mass.: MIT Press, 1984); also Patricia S. Churchland, *Neurophilosophy* (Cambridge, Mass.: MIT Press, 1986).

14 The use of the term in its derogatory sense is more or less installed in Stich, *From Folk Psychology to Cognitive Science*; see, also Patricia Churchland, *Neurophilosophy*.

15 Peter Hacker has been gradually developing this theme, largely, of course, beyond the narrowly "therapeutic" vision of philosophy that Wittgenstein had espoused in a distinctive way, regarding the materials of *Philosophical Investigations* and related studies. There is apparently to appear fairly soon a revision of P. M. S. Hacker, *Insight and Illusion* (Oxford: Clarendon Press, 1972), that promises to present the theme in a fully ramified way.

16 See B. F. Skinner, *Science and Human Behavior* (New York: Macmillan, 1953); also, R. J. Nelson, "Behaviorism is False," *Journal of Philosophy*, LXVI (1969).

17 See D. M. Armstrong, *A Materialist Theory of the Mind* (London: Routledge and Kegan Paul, 1968) and *Belief, Truth and Knowledge* (Cambridge: Cambridge University Press, 1973).

18 See Chomsky, *Rules and Representations*, ch. 1; also Donald Hockney, "The Bifurcation of Scientific Theories and Indeterminacy of Translation," *Philosophy of Science*, XLII (1975).

19 See Andras Pellionisz and Rodolfo Llinás, "Brain Modeling by Tensor Network Theory and Computer Simulation. The Cerebellum: Distributed Processor for Predictive Coordination," *Neuroscience*, IV (1979), and "Tensor Network Theory of the Metaorganization of Functional Geometries in the Central Nervous System," *Neuroscience*, XVI (1985).

20 See Robert D. Hawkins and Eric R. Kandel, "Steps toward a Cell-Biological Alphabet for Elementary Forms of Learning," in G. Lynch et al. (eds), *Neurobiology of Learning and Memory* (New York: Guilford, 1984).

21 See ch. 4, above.

22 See Paul Oppenheim and Hilary Putnam, "Unity of Science as a Working Hypothesis," in Herbert Feigl et al. (eds), *Minnesota Studies in the Philosophy of Science*, vol. 1 (Minneapolis: University of Minnesota Press, 1958).

23 See Dennett, "Artificial Intelligence as Philosophy and as Psychology," *Brainstorms*, pp. 110–11.

24   See Thomas S. Kuhn, *The Structure of Scientific Revolutions*, 2nd enlarged edn (Chicago: University of Chicago Press, 1970); Karl R. Popper, *The Logic of Scientific Discovery*, 2nd, rev. edn (New York: Harper and Row, 1965); Feyerabend, *Against Method*. See also Joseph Margolis, *Pragmatism without Foundations; Reconciling Realism and Relativism* (Oxford: Basil Blackwell, 1986).

25   See L. R. Squire and N. J. Cohen, "Human Memory and Amnesia," in G. Lynch, J. L. McCaugh, and N. M. Weinberger (eds), *Neurobiology of Learning and Memory* (New York: Guilford Press, 1984).

26   See Karl R. Popper and John C. Eccles, *The Self and its Brain* (New York: Springer International, 1977), pt II (by Eccles).

27   See Joseph Margolis, *Philosophy of Psychology* (Englewood Cliffs, NJ: Prentice-Hall, 1984), ch. 4, and *Culture and Cultural Entities*, ch. 1.

28   George A. Miller, "Informavores," in F. Machlup and U. Mansfield (eds), *The Study of Information: Interdisciplinary Messages* (New York: John Wiley, 1984).

29   Patricia Churchland seems to think that "folk" or top-down views are peculiarly intransigent about such accommodation. This seems to be her principal reason for championing an exclusively bottom-up strategy; see *Neurophilosophy, passim*.

30   Robert Cummins, *The Explanation of Psychological Explanation* (Cambridge, Mass.: MIT Press, 1984), particularly chs 1–2. See also ch. 10, below.

31   Jerry Fodor rather exorbitantly treats thinking as natively linguistic in structure; but we need not follow him in this, in admitting the sentential modeling of thought (including animal mentation – whether we should call that thought or not). See Fodor, *Representations*.

32   See Donald Davidson, "Mental Events," *Essays on Actions and Events* (Oxford: Clarendon Press, 1980).

33   See Margolis, "Puzzles about the Causal Explanation of Human Actions," *Culture and Cultural Entities*.

34   See ch. 12, below.

35   See Joseph Margolis et al., *Psychology: Designing the Discipline* (Oxford: Basil Blackwell, 1986).

36   See Peter Marler, "The Evolution of Communication," in Thomas S. Seboek (ed.), *How Animals Communicate* (Bloomington: Indiana University Press, 1977).

37   See Maurice Merleau-Ponty, *The Visible and the Invisible*, ed. Claude Lefort, tr. Alphonso Lingis (Evanston, Ill.: Northwestern University Press, 1968); Pierre Bourdieu, *Outline of a Theory of Practice*, tr. Richard Nice (Cambridge: Cambridge University Press, 1977).

38   See Davidson, "Mental Events," *Essays on Actions and Events*.

39   See Chomsky, *Rules and Representations*, ch. 1, and *Knowledge of Language: Its Nature, Origin, and Use* (New York: Praeger, 1986); also, Fodor, *Representations*. See also, for a sense of the bottom-up theorist's

charge, Dennett, "A Cure for the Common Code?" and "Artificial Intelligence as Philosophy and as Psychology," *Brainstorms*.

40 John Haugeland, "Semantic Engines: An Introduction to Mind Design," in J. Haugeland (ed.), *Mind Design* (Montgomery, Vt: Bradford Books, 1981), pp. 32–3. Dennett cites the passage in the unpublished article mentioned, "Evolution, Error and Intentionality."

41 Of the published papers that Dennett cites, the following may be of particular interest: Saul A. Kripke, *Wittgenstein on Rules and Private Language* (Cambridge, Mass.: Harvard University Press, 1982), pp. 35–40; Tyler Burge, "Individualism and Psychology," *Philosophical Review*, XCV (1986); and Fred Dretske, "Machines and the Mental," *Proceedings and Addresses of the American Philosophical Association*, vol. 69 (Newark, Del.: American Philosophical Association, 1986).

42 Daniel C. Dennett, *Content and Consciousness* (London: Routledge and Kegan Paul, 1969), pp. 28–9.

43 Ruth Garrett Millikan, *Language, Thought, and Other Biological Categories* (Cambridge, Mass.: MIT Press, 1984).

44 Hilary Putnam, "Minds and Machines," *Philosophical Papers*, vol. 2 (Cambridge: Cambridge University Press, 1975), pp. 362, 364. See also Putnam's "Mind and Body," *Reason, Truth and History* (Cambridge: Cambridge University Press, 1981); and Ned Block, "Troubles with Functionalism," in C. Wade Savage (ed.), *Minnesota Studies in the Philosophy of Science*, vol. 9 (Minneapolis: University of Minnesota Press, 1978).

45 Hilary Putnam, "The Nature of Mental States," *Philosophical Papers*, vol. 1 (Cambridge: Cambridge University Press, 1975) p. 435.

46 Wilfrid Sellars, Introduction to "Intentionality and the Mental," *Minnesota Studies in the Philosophy of Science*, vol. 2 (Minneapolis: University of Minnesota Press, 1958), Appendix, p. 508.

47 Wilfrid Sellars, "Being and Being Known," *Science, Perception and Reality* (London: Routledge and Kegan Paul, 1963), p. 59 (cf. p. 57n), and "Philosophy and the Scientific Image of Man," in the same volume.

48 The quarrel seems to have exercised recent phenomenologists considerably. There is even a tendency to speak of different "schools" of phenomenology, viewed in terms of these competing interpretations of Husserl's notion of intentionality. See Aron Gurwitsch, *Studies in Phenomenology and Psychology* (Evanston, Ill.: Northwestern University Press, 1966); Dagfinn Føllesdal, "Husserl's Notion of Noema," *Journal of Philosophy*, LXVI (1969). Ischak Miller gives a clear sense of the quarrel. See his *Husserl, Perception, and Temporal Awareness* (Cambridge, Mass.: MIT Press, 1984), ch. 1. See also J. N. Mohanty, *Husserl and Frege* (Bloomington: Indiana University Press, 1982), ch. 3, section 6.

49 W. V. Quine, *Word and Object* (Cambridge, Mass.: MIT Press, 1960), p. 231.

50 Ibid.

51 See Donald Davidson, "Semantics for Natural Languages," *Inquiries into*

Truth and Interpretation (Oxford: Clarendon Press, 1984).

52 See Joseph Margolis, "The Stubborn Opacity of Belief Contexts," *Theoria*, XLIII (1977); and Quine, *Word and Object*, pp. 141–56.

53 Quine, *Word and Object*, p. 231.

54 Fred I. Dretske, *Knowledge and the Flow of Information* (Cambridge, Mass.: MIT Press, 1981), p. 75.

55 Dennett, *Content and Consciousness*, p. 190.

56 Dennett, "Artificial Intelligence as Philosophy and as Psychology," *Brainstorms*, p. 123.

57 Dennett, "Toward a Cognitive Theory of Consciousness," ibid., p. 163.

58 Dennett, "A Cure for the Common Code?", ibid., p. 101.

59 Dennett, "Why the Law of Effect Will Not Go Away," ibid., p. 83.

60 See Davidson, "Mental Events," *Essays on Actions and Events*.

61 I allow myself the luxury here of merely remarking that Ruth Millikan's account of intentionality, which Dennett (in the unpublished paper) suggests converges with his own (though it arrives by an entirely different trail), succeeds (as far as I can see) in showing only that, *if*, indeed, intentionality is what it is only if it has properties that can be extensionally mapped with respect to the real world, then an extensional treatment of intentionality is quite possible. I take this to be the upshot of chs 6–7 of Millikan's *Language, Thought, and Other Biological Categories*. If that is a fair reading, then Dennett is quite right, but for a reason that could hardly support his own position.

62 The issue is canvassed in ch. 1, above. Davidson's central papers bearing on the present issue include the following: "Mental Events" (in *Essays on Actions and Events*) and (in *Inquiries into Truth and Interpretation*) "In Defence of Convention T," "Reality without Reference," "Semantics for Natural Languages," and "The Inscrutability of Reference."

63 Davidson, "Reality without Reference," *Inquiries into Truth and Interpretation*, p. 220.

64 Ibid., p. 221 (italics added).

65 Ibid., p. 222.

66 Ibid., p. 223 (italics added).

67 For an analysis of Davidson's theory of psychology, see Margolis, *Philosophy of Psychology*, pp. 28–9.

68 See Alfred Tarski, "The Concept of Truth in Formalized Languages," *Logic, Semantics, Metamathematics*, 2nd edn, tr. J. H. Woodger, (ed.) John Corcoran (Indianapolis: Hackett, 1983).

69 For a view of standard challenges to Davidson's thesis, see Ian Hacking, *Why Does Language Matter to Philosophy?* (Cambridge: Cambridge University Press, 1975); Gilbert Harman, "Meaning and Semantics," in Milton I. Munitz and Peter K. Unger (eds), *Semantics and Philosophy* (New York: New York University Press, 1974).

70 Davidson, "The Inscrutability of Reference," *Inquiries into Truth and Interpretation*, p. 239.

71 Davidson, "Reality without Reference," ibid., p. 221 (italics added).
72 Ibid.
73 See Herbert Feigl, *The "Mental" and the "Physical": The Essay and a Postscript* (Minneapolis: University of Minnesota Press, 1967); J. J. C. Smart, "Sensations and Brain Processes," rev, in V. C. Chappell (ed.), *The Philosophy of Mind* (Englewood Cliffs, NJ: Prentice-Hall, 1962); and J. J. C. Smart, *Philosophy and Scientific Realism* (London: Routledge and Kegan Paul, 1963).
74 See Sellars, "Philosophy and the Scientific Image of Man," *Science, Perception and Reality*; May Brodbeck, "Mental and Physical: Identity versus Sameness," in P. K. Feyerabend and G. Maxwell (eds), *Mind, Matter, and Method* (Minneapolis: University of Minnesota Press, 1966); Stephan Körner, *Experience and Theory*, (London: Routledge and Kegan Paul, 1966).
75 See Paul K. Feyerabend, "Materialism and the Mind–Body Problem," *Review of Metaphysics*, XVII (1963); Richard Rorty, "In Defense of Eliminative Materialism," *Review of Metaphysics*, XXIV (1970); Paul M. Churchland, *Scientific Realism and the Plasticity of Mind*, and "Eliminative Materialism and Propositional Attitudes," *Journal of Philosophy*, LXXVIII (1981); Stich, *From Folk Psychology to Cognitive Science*.
76 See Davidson, "Mental Events," *Essays on Actions and Events*; also, Nancy Cartwright, *How the Laws of Physics Lie* (Oxford: Clarendon Press, 1983).
77 See Davidson, "Causal Relations," *Essays on Actions and Events*.
78 Davidson, "Reality without Reference," *Inquiries into Truth and Interpretation*, p. 222 (italics added).
79 Ibid., pp. 223, 224 (italics added). See also Hilary Putnam, "Beyond Historicism," *Philosophical Papers*, vol. 3 (Cambridge: Cambridge University Press, 1983).
80 Putnam, "Reference and Truth," *Philosophical Papers*, vol. 3, pp. 75, 76.
81 Ibid.
82 See Michael Dummett, *Truth and Other Enigmas* (Cambridge, Mass.: Harvard University Press, 1978).
83 A convenient summary appears in Putnam, "Reference and Truth," *Philosophical Papers*, vol. 3.
84 Davidson, "Reality without Reference," *Inquiries into Truth and Interpretation* p. 221 (italics added).
85 See Fodor, *The Language of Thought*.

6

Cognition, Representation, and Information

I

It is difficult to grasp what sort of theory of human life could be reasonably comprehensive and empirically responsible that would eliminate, in realist terms, any admission of man's cognitive aptitude. The puzzle deserves a naive reading, not because it is a primitive question but because it is so fundamental and because it arises so spontaneously everywhere. We cannot expect to anticipate its every appearance. When, for instance, Noam Chomsky, fiddling with the conceptual prejudices associated with the expression "know," offers "cognize" as a term of art suited to his own project, he means at one and the same time to preserve the sense of the functional capacity of humans at the molar level – functioning, that is, at the level of what we usually acknowledge to be powers only assigned "selves" or "persons" – and to reject the excessively unilevel, fully conscious modeling of such powers. So he proposes that, by "cognizing," we shall mean "tacit or implicit knowledge," which is well on its way to accommodating his ulterior notion of "innate knowledge" (of the deep structures of natural languages). He does not, here, wish to deny "conscious knowledge"; on the contrary, conscious and tacit knowledge are very much the same for Chomsky: "cognizing has the structure and character of knowledge, but may be and in the interesting cases is inaccessible to consciousness." For Chomsky, "The fundamental cognitive relation is knowing a grammar [which, in the sense crucial to his own theory, is known innately]; *knowing the language determined by it is derivative.*"[1] This means that, on Chomsky's theory, conscious knowledge of one's (natural) language (in particular, knowledge of its semantic and pragmatic features) presupposes and entails innate knowledge of that language's deep grammar.

One may quarrel with the thesis, of course. But it does not proceed bottom-up, in the sense that, on Chomsky's view, it is the molar self or person that, somehow, has access to grammatical structures: normally, the entailed cognition is not consciously accessible at the same molar level at which it is tacitly (in fact, innately) accessible. But, *if* the power to "cognize" such structures forms a modular *sub*-system for Chomsky, that system is, on the theory, a sub-system of the "self's" molar cognitive system, it is a form of the *sub*-functioning *of* the molar system. This is the reason why Daniel Dennett's blunderbuss criticism of numerous top-down psychologies – including Freud's, J. J. Gibson's, Jerry Fodor's, and Chomsky's – is off the mark when it charges each such theory with

> designing a system with component subsystems whose stipulated capacities are *miraculous* given the constraints one is accepting. (E.g., positing more information-processing in a component than the relevant time and matter will allow, or, at a more abstract level of engineering incoherence, positing a subsystem whose duties would require it to be more "intelligent" or "knowledgeable" than the supersystem of which it is to be a part.)[2]

Dennett's view would have us believe that, in spite of the fact that (top-down) the functioning of sub-systems is the sub-functioning of molar systems, the introduction of such sub-systems (homuncular systems) must (bottom-up) lead to the elimination of the molar level (*and* to that of themselves).[3] But, for one thing, Dennett never actually shows us how, empirically, to achieve this pretty objective; and, for another, he is quite mistaken in believing that a model such as Chomsky's (whatever its own drawbacks, or Freud's for that matter) must either be miraculous or incoherent. (Chomsky's and Freud's theories may simply be false, as both of those ingenious authors would acknowledge.) The upshot is that the *elimination* of molar cognitive agents (the centerpiece of so-called "folk psychology") cannot be adequately justified on charges of "miracle" and "more-than-human" powers. It can only be vindicated (if it can at all) on the basis of demonstrably better explanatory theories.

Chomsky sees at once that his account requires a positing of "interior mental objects" as "theoretical entities," in virtue of cognizing which the apt speakers of a language are able to perform as they do. He rejects, here, the argument against "the museum myth" (that is, the argument, not the myth), which putatively captures the objections of such different thinkers as Quine and Wittgenstein against interior or private mental objects. Chomsky's reasonable view is that the would-be argument accomplishes nothing – at least does not "until something is added to explain why positing interior mental objects gets in the way of explanation and papers over gaps in our understanding of the acquisition of language-mastery, and furthermore why this must be the case. In the absence of such additional steps, what we have is pseudo-argument against the theoretical entities."[4] This very neatly collects and confirms several important themes:

1   a theory of cognition (innatist, tacit, or conscious) is likely to entail a theory of representation, of interior mental computations of some sort on posited mental entities (image-like, sentence-like, propositional, or sign-like) in virtue of which molar behavior, molar states, molar dispositions are assigned some cognitively informed feature or function;

2   representation is, in this sense, a central-state capacity of molar agents posited on the strength of a theory designed to yield the best explanation of human cognitive life (or of the life or functioning of suitably cognate systems, animal or electronic);

3   the elimination of cognitive molar functioning (perhaps along David Marr's lines) eliminates the conceptual moorings (at least the usual moorings) for a theory of mental representation, and the rejection of a theory of mental representation effectively makes a mystery (at least a conditional mystery) of the intra- or infra-psychological processes of molar cognition.

Dennett's strategy (consistently with his ultimate eliminative purpose, inconsistently with his actual characterization of homunculi) would eliminate altogether mental representation.[5] Stephen Stich's program, more detailed than Dennett's but congruent with its final objective, is considerably more puzzling. Stich means *both* to eliminate molar agents ("folk psychology") *and* to retain a particularly spare theory that certainly appears (in some sense) to be a representational account, one that gives up both so-called "strong" and "weak" representational theories (since *they* require the admission of "mental content," semantic or intentional content, within the space of whatever "syntactic objects" mental processing operates on). The charge against representationalism is chiefly directed against Jerry Fodor, whom Stich finds equivocal on the question but whom he also favors.[6] The puzzle invites unraveling and promises an important measure of clarity.

Stich introduces what he calls the "syntactic theory of the mind" (STM). STM is not, Stich says, "a cognitive theory": "it is not sanguine about the use of folk psychological notions in cognitive science. It does not advocate cognitive theories whose generalizations appeal to the notion of content." True to its eliminative purpose, STM means to transform cognitive theory. "The basic idea," Stich declares,

> is that the cognitive states whose interaction is (in part) responsible for behavior can be systematically mapped to abstract syntactic objects in such a way that causal interactions among cognitive states, as well as causal links with stimuli and behavioral events, can be described in terms of the syntactic properties and relations of the abstract objects to which the cognitive states are mapped. More briefly, the idea is that causal relations among cognitive states mirror formal relations among syntactic objects. If this is right, then it will be natural to view cognitive state tokens as tokens of abstract syntactic objects.[7]

But these abstract objects do appear to be a set of sentences or codes, "complex syntactic objects," an interior language (in effect) that *has no semantics.*[8]

We must be careful here. Stich's account looks like a representational theory (a representational theory *manqué*) but it is not. Still, its empirical payoff *may* (ironically, against Stich's own intent) be accessible only if construed as a reform of (existing "folk-theoretic") representational theories of the mind. Whether his syntactic notion is empirically productive at all with regard to psychological or cognitive explanation, Stich has hardly any evidence for; and whether something like a Chomskyan thesis regarding the independence of grammar from semantic and pragmatic encumbrance is genuinely viable, or whether something like Davidson's intended extension of Tarski's formal semantics of truth would work, Stich does not actually discuss.[9] In any case, theories such as Chomsky's, Tarski's, Davidson's give priority to syntactic structures only *within* linguistic systems that *do* have semantic dimensions and are actually used in some recognizable way at the level of molar cognition – hence, representationally. What Stich means, therefore, by a comparable system that "has no semantics" is as yet unclear: whether in fact it would be a coherent notion (so construed) is also unclear. If, however, for the sake of argument, we allowed that reading, it would appear that Stich's "cognitive states" would still operate on a certain set of syntactically specified sentences; and that, in effect, would be representationalism. It might even be reasonable then to hold that, *qua* represented, syntactically restricted sentences would require semantic features – namely, the feature at least that they were syntactically restricted.

"*The virtue of STM theories,*" Stich avers, "*is that they eliminate the middleman.* The mental states postulated by an STM theory are not characterized by their content sentences but, rather, by the syntactic objects to which they are mapped."[10] Stich cannot have meant this seriously without intending to dismantle representational theories. So the suggested reading must be a false start. And yet, in spite of that, it is difficult to see how the theory actually works. (Of course, if it *merely revised* our folk psychology – for example, revised our notions of beliefs and desires – it would remain itself a folk psychology.) The question rests: *how* can we admit the processing of mental representations without admitting, at the level at which such processing obtains, something self-life and strongly analogous to what, in familiar top–down theories, is acknowledged as a cognitive agent? Or, alternatively, how can we admit the causal role of syntactic objects without admitting representational-ism?[11]

Now, Stich clearly adheres to a functionalist account. He holds (confining our sketch, for convenience, to biological species) that, although "mental" states ("state tokens ... causally implicated in the production of behavior") "are physical states of the brain," are indeed neurological states, causal regularities between relevant input stimuli and "hypothesized [token] neurological states" are not identified "by adverting to their essential neurological types but, rather, by adverting to the syntactic objects to which the neurological types are mapped"; the same, he ventures, will be true for "causal relations between neurological states and behavior."[12]

Here Stich is properly cautious about generalizing beyond biological species (to computers, say), about generalizing even within species and across species (which might exhibit "quite different neurological state types"), and about generalizing within species with regard to possibly unique mappings ranging over different state types and his (or any) preferred class of formal objects. All this is well and good and thoroughly functionist in tone. But it looks as if the *only* way in which Stich could possibly distinguish his STM from (what he calls) representational theories would require that the syntactic codes he postulates function somewhat in the manner of a closed, abstract program *empirically assignable* but not *representationally accessible* to the set of token neurological states of a token member of a given species, so that the regularities of stimulus–neurological connection and neurological–behavioral connection may be specified in functional terms only. Thus construed, Stich's program merely uses folk-theoretic descriptions and predictions as an interim test of the empirical promise of his thesis; his own thesis remains at bottom an eliminative rather than a revisionist project. It is *not* representational, in the sense that the pertinent neurological states are merely mapped with regard to postulated syntactic objects: the organism does not itself engage in processing tokens of any such syntactic system; furthermore, it is functionalist rather than neurobiological, in the sense that the regularities posited apply *to* neurobiological systems (token-wise at least) but are not expressible *in terms of* a neurobiological vocabulary. One possible way of characterizing Stich's program (which he does not consider and which doubtless he would not favor) suggests that his system admits a *heuristic use of an information-processing idiom for the abstractly functional organization of "cognitive" life* but *rejects a representational or "folk" interpretation of the real functionalist organization of such life*. In this sense, Stich's account is, finally, neutral as between bottom-up and top-down strategies; these strategies are, pretty well as Dennett supposes, symmetrical alternatives the selection of which depends only on a perceived advantage to a developing science. But whether there is an empirical basis for *any* such reduction Stich never demonstrates and hardly pursues. *All* the evidence he musters bears solely on the gradual (if radical) *revision* of folk-theoretic concepts. His eliminative *dream* (also Dennett's) remains an altogether independent concern. There is in fact no direct conceptual link between improvements of the first sort and achievements of the second. It was the apparent conflating of the two that symptomatically suggested a representational reading of Stich's program. But that there are law-like regularities of a purely functional sort (in psychological contexts) – either across all species or species-wide or across sub-species-wide "neurological types" or across the life of "token" instances of particular neurological types (individual organisms) – is a claim we have absolutely no reason to believe and no evidence about. In fact, the best evidence we have about the law-like regularities of sub-processes entailed in cognitive states (granting once again that folk-theoretic concepts are subject to considerable revision under the pressure of empirical findings) is that they are not likely to be formulable in terms that exclude fine-grained neurobiological structures.[13] And that means that a pure functionalism, whether top–down (Zenon Pylyshyn's, for instance[14]) or bottom–up (Stich's), is probably hopeless on empirical grounds.

We may strengthen this finding by a final consideration. Stich's contention is "that STM theories are a better choice for the cognitive theorist than those theories whose generalizations appeal to content [strong representational theories], since *syntactic theories can do justice to all* of the generalizations capturable by quantifying over content sentences while avoiding the limitations that the folk language of content impose."[15] Here, the *only* pertinent and reasonably explicit model for advancing such a claim corresponds more or less to either Davidson's or Fodor's use of Davidson's intended use of Tarski's semantic conception of truth (the so-called Convention T).[16] But Tarski himself was not convinced that his own model could be satisfactorily applied to natural languages (*a fortiori*, though only implicitly, he was not convinced that it could be applied to folk psychology); no one has shown that it does apply; there are serious questions about alleged neutrality;[17] and in any case Stich nowhere explains what reasons could lead us, if it did fit natural-language sentences (or "content sentences"), to construe it as detachable in such a way as to disallow representationalism.

There is only one possible strategy, and that one is just the one Stich admits is not empirically available at the present time or "for the foreseeable future": "So," Stich says, "if theory [STM] is to confront data, the syntactic theorist will have to make a significant number of ad hoc assumptions about causal links between [say, for illustration] B- and D-states [that is, states 'roughly analogous to beliefs (and) desires'] on the one hand, and stimuli and behavior on the other." The syntactic theorist would in fact "have to view B- and D-states as having a structure which parallels the structure of wffs ['well-formed formulas, that is, having the simple underlying syntax of first-order quantificational theory,' not really a language, 'no more than an infinite class of complex syntactic objects']. But," Stich adds, "an STM theorist need say very little on all this, *since he is merely postulating the existence of B- and D-states and the existence of the mappings.*"[18] In short, Stich probably has no empirical reason for believing that such a psychological model would work at all.

The single most troublesome feature of the "B- and D-state" model – which explains at once both why Stich does not seriously address any particular syntactic schema such as Tarski's and why his notion appears somewhat in the guise of a representational theory – is that he introduces in his eliminative theory distinctions tantamount to what would be that theory's "cognitive" states, functioning there in a way utterly unlike the functioning of the full-blooded cognitive states of a folk psychology. This *ad hoc* invention exhibits two particular features: first, it introduces "content sentences" relative to B- and D-states (and, therefore, does function provisionally in a representational way); but second, it does so in accord with the standard model of wffs in first-order quantificational logic (and, therefore, anticipates eliminative options). (If, then, Stich is not directly appealing to Tarski's explicit model, he is certainly appealing to something like Quine's or Davidson's thoroughgoing but *empirically* quite incomplete extensionalism.) No ground, however, has been laid for the crucial presumption on which an *empirically* motivated replacement of belief- and desire-states by B- and D-states is to be made: namely, the presumption that, relative to psychologically pertinent stimuli and psychologically significant behavior (*whatever* we may plausibly mean by that), B- and D-states may be identified as *discrete* or atomic states (*not* holistically and intentionally interrelated with one another, as for instance on Davidson's reading of folk psychology – which Stich is prepared to follow[19]). It is central to Stich's program and to *all* similar-minded programs that

> the characterization of a B-state [say] *does not depend* on the other B-states that the subject happens to have [whereas the belief-states of a normal subject *do* inextricably, holistically, depend on the subject's other belief-states, desire-states and the like]. A B-state will count as a token of a wff if its potential causal links fit the pattern detailed in the theorist's generalizations, *regardless* of the further B-states the subject may have or lack.[20]

But this only shows that the proposal is internally coherent; it does not show that it is an empirically plausible replacement for a folk psychology or that, if we could even get off the ground with such a model, there is any prospect that it would yield law-like regularities or that such regularities could be formulated in functionally abstract terms. There is absolutely nothing of an empirically salient sort in its favor.

To break off here, however, would be to risk missing the essential point. Every bottom-up psychology that means to be empirically responsible *must* justify in empirical terms *just how* it proposes to replace a holistic folk psychology with an array of discrete states (such as B- and D-states or homuncular states or the like). To neglect the issue is philosophically and scientifically irresponsible; and to fail in repeated attempts is to concede, possibly against the grain, the continued salience and near-hegemony of the folk model. It does no good to insist that the folk model's time is up: a fair request requires some evidence at least; *and* the evidence – at least the neurobiological evidence – shows no more than that there are developing reasons for drastic revisions in our folk-theoretic concepts, *not* that they can be successfully superseded. At the present stage of theorizing, then, folk psychologies are distinctly hospitable to, and are even thought to entail, representational theories; and representational theories are usually folk psychologies or top-down psychologies that posit suitably complex molar agents.

It is true that a bottom-up psychology *may* seek to preserve folk-psychological concepts, for instance as designating "emergent" functions with respect to admissible bottom-up theories. Stich will have none of this. Dennett appears to mediate irenically between the two extremes by introducing homunculi; but finally (as in insisting on "discharging" homunculi[21]), Dennett clearly sides with (or anticipates) Stich. And Zenon Pylyshyn, consistently at least with the usual approach of a realist cognitive science, would preserve the top-down functions of top-down agents but *only* as "emergent" with respect to a bottom-up strategy (that is, on the view that emergent functions are not capable of being explained within the closure of such "lower-level" disciplines as physics and biology *and* noncognitive computer science[22]).

So there is an inning for bottom-up representational theories. We shall in a moment turn to Pylyshyn's account. But what we may conclude thus far is this:

(i)   functionalist psychologies need not be representational;
(ii)   representational theories can take both bottom-up and top-down forms, though such alternatives are not mere equivalents in explanatory or ontological respects; and
(iii) whether representational theories should be construed in abstract functional terms – heuristically or realistically – or whether they should be construed in some more complex way is an empirically open question.

Stich apparently treats the provisional representational role of his B- and D-states as a mere *façon de parler* for a functionalist psychology that concedes the irreducibility of functional sciences to physics and biology (as those disciplines are canonically construed). (This is not to say that Stich concedes the ultimate irreducibility of the functional to the physical.) It *is* the "physical" that exhibits functional properties: the functional confirms the complexity of the physical itself. As we have already remarked, this (suitably isolated theme) is just the theme of Chomsky's linguistic nativism (also, of Fodor's nativism). The trouble is that Stich's counterarguments directed not only against preserving folk-psychological-like concepts or functions (Pylyshyn's project) but also against a fully top-down folk psychology (one that concedes real molar agents whose mental states and behavior are causally efficacious *in* the physical and biological world) are essentially driven by a favored conception of science (a version of the unity-of-science model) and not by any sustained empirical examination of the explanatory powers of competitors to the folk-psychological approach. Once we admit that our theories of what a science is are themselves empirically testable (in some suitably generous sense), it is difficult to assign much promise to Stich's program; for, on his own view, the folk-theoretic approach is distinctly more successful empirically than the one he himself advocates. The matter, of course, is hardly settled.

II

When David Marr, introducing his well-known account of vision, straightforwardly affirms that "vision is the *process* of discovering from images what is present in the world, and where it is," *and* then moves directly from that to treat vision as "an information-processing task," he is entirely aware that he is proposing a representational theory of perception: "if we are capable of knowing what is where in the world, our brains must somehow be capable of *representing* this information."[23] Marr's formulation certainly suggests a top-down model, however radically it may revise our folk psychology, though we may be premature in committing Marr to a thoroughly top-down view. But it also draws attention to two further features of a well-formed theory of cognitive representation (to be added to our earlier tally):

4   representational theories are motivated in realist terms, with regard to molar agents, though they need not be committed, for that reason alone, to any particular epistemology; and

5   the precise nature of the theoretical mental objects that we process as representations (in cognitive contexts) is an empirical question (both as regards whether we process them by means of propositions or images in different processing sectors or even by other means; and as regards "where" such processing obtains – "in" the brain or elsewhere).

A fair way of putting the point is to concede that a robust representationalism is meant as a realist theory of cognition in accord at least with the pragmatist sense of such a theory (holistically and non-transparently); but it need not be of a top-down or molar sort. This marks the essential difference between Fodor and Marr. Locke's sort of representationalism is also hardly required, both because the requisite processing need not obtain at the level of conscious functioning (hence, may be encumbered in all sorts of ways by the interior processing mechanisms of the organism) and because, at the level of molar cognition, our theories of knowledge and reality may consistently (and variously) yield in the

direction of the indissolubility of realist and idealist factors, when
distributively applied to particular claims.

One thing is clear, however. Mental representation may be construed
heuristically or in realist terms; also, in realist contexts, either directly
as a theory of cognition itself (however manifested in the neurobiology
of living organisms or in the electronics of actual computers) or indirectly
in conceptually parasitic terms (as in tracing cerebellar contributions to
sensorimotor coordination); and, finally, as emergent in either a top-
down or bottom-up sense. The reason why the distinction is important
is that it obliges us to concede, in the sense of a conceptual analysis of
molar cognition, that there can be no significant contrast between
propositions and *images* (or any other representata), simply because every
robust theory of cognition or cognitive states must concede that,
whatever psychological states count as cognitive states, they do so in
virtue of being assignable propositions or proposition-like intentional
functions. (In this sense, "images," if cognized, have, must have,
proposition-like features; although how any information is "stored" in
organisms is clearly a mystery.) This, at least, is the salient "folk"-
psychological theory without which we seem unable to begin pertinently
– and with which (to remind ourselves once again) Stich's theory has
not really come to grips.[24] Brusquely affirmed, if knowledge (however
attenuated) is rightly assigned propositional features or functions – if to
know is to know *that p* – then "cognitive states" of whatever description
or involving whatever "mental objects" ("images" for instance viewed
as a distinctive form of information-storage) must, consistently, be
assigned propositional features. Quarrels, therefore, about whether
perception processes images or propositions raise distinctly subordinate
issues and are potentially confused about a misplaced disjunction. It
does remain an empirical question, of course, whether, in some sense,
information is stored in "images" or "propositions."[25] But the answer
to that question presupposes a resolution of the top-down/bottom-up
quarrel. Once the propositional model is in place at the molar level it
is entirely possible to concede that, at different points in the actual
processing of representations (or of processing what contributes converg-
ently to the processing of representations), multiform "mental objects"
may be involved *and* some facilitating processing may not be representational at all, *a fortiori* may be neither propositional nor imagistic.

Zenon Pylyshyn's theory is, therefore, a most instructive middle-
range account, because it seeks to take a neutral stance (*contra* Stich) on
the adequacy of a folk-psychological model[26] while at the same time it
sketches the essential lines of an adequate cognitive science. In the terms
we have been favoring, Pylyshyn wishes to preserve some full and

ineliminable explanatory role for folk-psychological functions – those in fact required in admitting the reality of *cognition* – *within* the canonical framework of the physical and biological sciences. Pylyshyn's only option is to treat the *representational* (which he pretty well equates with the "cognitive," the "computational," and the "semantic," also with "intentional content") as *emergent* with respect to the "lower-level" phenomena of physics, biology, and (noncognitive) computer science. In this sense, he is a bottom-up theorist (as much as Stich). He simply is not convinced that psychology can be pursued without admitting cognition, or that cognition can be analyzed without admitting the internal processing of represented information, meanings, or "semantic content"; *and* he is convinced that a relatively straightforward theory can be formulated by which to reconcile the emergent phenomena at the cognitive level – with the requirements of a conventional unity-of-science model and with the actual saliences of empirical psychology as it is now practiced.

We may save ourselves some labor, here, by anticipating that there may be a fatal inconsistency in Pylyshyn's program – or, if "inconsistency" is too strong a term, that Pylyshyn imposes too stringent or too unpromising an empirical requirement on his theory – and that this adversely affects every possible version of his sort of strategy. The weakness in question has to do with *reconciling the emergence of the cognitive with the law-like regularities of the physical and biological, within the terms of closure of such sciences as the latter.* Roughly, this identifies the essential difficulty of a unified-science reading of a cognitive folk psychology (and of every social and human science that presupposes cognition). Pylyshyn's theory, therefore, serves as a touchstone of the prospects of bottom-up psychologies that mean to preserve (emergently) top-down functions. The charge here is that such theories (Pylyshyn's, for example) are either self-contradictory or empirically attenuated so radically (to escape contradiction) as to be effectively unworkable. If the case can be made, then we are left with a very well-defined choice between (1) *empirically* unmotivated (or, all but unmotivated) bottom-up theories that eliminate molar cognitive states (though not necessarily functionalist psychologies) as incompatible with the unity-of-science model and (2) top-down folk or molar psychologies that construe real cognitive states as empirically undeniable but also as empirically and conceptually irreconcilable with canonical unity-of-science views. The strategic importance of this forced choice is perfectly plain. Perhaps it is also sufficient to justify an extended and particularly close examination of Pylyshyn's argument. One is not likely to find a comparable measure of candor joined to such a transparently desperate effort to save the cognitive scientist's mode of resolution.

The success of Pylyshyn's strategy requires that we "identify basic, transducible, cognitive states."[27] A *transducer* is not a computational (or cognitive) process, it is "fundamentally ... a physical process"; "its behavior is explainable (its regularities can be captured) in terms of its intrinsic properties – physical, chemical, biological, and so on."[28] A transducer is a theoretically posited transition device systematically linking (in the only way deemed possible) physical or biological phenomena and cognitive phenomena – *each analyzed on a "level" necessarily independent of the other's* – such that the emergent cognitive phenomena (intrinsically not explicable in law-like terms) can be made congruent with the covering law explanations required by the unity model. A moment's reflection confirms that Pylyshyn has managed to construe the resolution of the mind/body problem in a way very much like that in which Descartes appeals to the functioning of the pineal gland: except that Pylyshyn's dualism is attributive and methodological rather than substantive (distinguishing the *law-like* and the *rule-like*) and except that Pylyshyn is committed to a thoroughgoing materialism (albeit one hospitable to genuine cognitive emergence).

The key constraint on transducers Pylyshyn formulates as follows:

> A transducer can occur at any energy transfer locus where, in order to capture the regularities of the system, it is necessary to change descriptions from physical to symbolic. The only requirement is that the transformation from the input, described physically, to the ensuing token computational [cognitive, representational, or semantic] event, *also described physically, follow from physical principles.*[29]

The transducer maps "certain classes of physical states of the environment into computationally relevant states of a device [described at an emergent level of analysis as a computational or cognitive device, thereby performing] a rather special conversion: converting *computationally arbitrary* physical events into computational events."[30]

The question remains, of course, how the feat may be accomplished. Any deviation from the constraint imposed Pylyshyn takes to be fatal to the prospects of a genuine cognitive *science*: he means, of course, fatal to a cognitive science strictly reconciled with a unity-of-science conception of the supposed discipline. It is after all to be a discipline addressed to a range of phenomena that are real, emergent, and intrinsically *not* definable in terms of physical laws or law-like regularities.

The solution requires a suitable "isomorphism" between the elements of the system at the psychological or computational level (described *independently* of any physical implementation) and the elements of the transducing system described entirely in physical terms, altogether independently of its psychological or "symbolic" realization. "[A] psychological transducer," Pylyshyn rather neatly puts it, "is considered nonsymbolic in its internal operation"[31] (that is, physical or biological and neither abstractly functional nor semantic or mental). Its processes must be "computationally primitive in the sense that their internal operation is not considered a rule-governed computation; they are simply performed, or, 'instantiated, by properties of the biological [or physical] substrate in a manner not requiring the postulation of internal representations'."[32]

There are complex questions to be answered, Pylyshyn feels, affecting the correct selection of psychological transducers; but the entire doctrine comes down to the claim that cognitive phenomena – the domain of psychology or of an even more inclusive science – form and must form a "natural domain." By this Pylyshyn means that "there will be principles [genuine laws; that is, principles that support relevant counterfactuals] that apply to all entities in that domain and only entities in that domain"; there will be *natural laws of cognition*.[33]

Nevertheless, it seems that there cannot be such laws. First of all, the "principles" that govern operations at the cognitive level are fundamentally different from those that govern physical processes: physical systems are causally "closed" with respect to their covering laws, whereas "semantic level principles ... apply to open-ended classes of physical properties." A physical "system" is rightly thus designated, Pylyshyn argues, if "no causal connections [obtain in it] other than [those expressible in terms of] the set of variables that define the system"; but, he adds, "there is no limit on the combinations of physical properties that can be used to instantiate [a semantic-level operation, say *modus ponens*, and such operations are not even specified in physical terms]. The two sets of principles have nothing in common; they are merely related by the happenstance of design or instantiation."[34] This seems to entail that there *could be no law-like regularity* ascribed to semantic-level operations – in which case the very notion of a psychological transducer would be self-contradictory. (Pylyshyn is not put off by this objection.)

Secondly, there can only be an "interface" between the physical and the semantic (which transducers instantiate in instantiating cognitive processing) – there cannot be "mixed-vocabulary principles" (that is, *prior* to the introduction of transducers), *laws* of a psychophysical or psychobiological sort, simply because the regularities of the conceptually emergent, independent level of cognitive processing are physically open-ended, not subject to physical closure in any way, and not directly expressible nomologically. Correspondingly, physical or biological processes – in particular, transducing processes – are not and cannot be "cognitively penetrable" (that is, open to direct causal influence by the mental or semantic): else, the entire structure of a *cognitive science* modeled along unity lines would be completely subverted. The semantic or intentional or representational or inferential or cognitive processes, after all, are not intrinsically law-like but only "rule-governed." Following Allan Newell, Pylyshyn proposes that "there are at least three distinct, *independent* levels at which we can find explanatory principles in cognitive psychology ... biological, functional, and intentional."[35] Closure is to be maintained at *each level of analysis*; the entire set of such closed systems admits the physical and biological as the basic level of analysis;[36] and the emergent cognitive level, which is not "causally *definable*, at least not in any direct way," which (because of closure and the independence of levels) yields explanations "*without using* physical or biological terms ... *without* [making] *reference* to structural material properties of the organism or its environment" and which remains a "natural domain" nevertheless,[37] is brought into accord with lower-level physical and biological regularities *by means (only) of mediating transducers*.

These conditions, however, are just the requirements of a relatively disciplined unity-of-science model accommodating emergence – on a theory that radically distinguishes cognitive science from, say, neurobiology. (We shall return to the issue.) On Pylyshyn's view, not unreasonably, "the main argument for the existence of levels" of natural phenomena amenable to scientific explanation is just that "there be valid generalizations at one level that are not expressible at a lower level": the need for positing a "new level can often be recognized by the presence of constraints on the behavior of a [given] system over and above the constraints that can be expressed in terms of principles currently available [at that level, presuming closure]."[38] For Pylyshyn, therefore, psychology and cognitive science are empirically needed at a level emergent with respect to physics, biology, and the functional (but not representational) aspects of computer science (which, following Newell,[39] Pylyshyn often calls "symbolic").

Finally, to fix once again Pylyshyn's (likely) difficulty: either it is logically impossible that there should be empirically formulable transducers or it is empirically impossible to specify them. In either case, a bottom-up psychology (or a bottom-up cognitive science) committed to *both* a reasonably strong version of the unity-of-science program *and* the real emergence of cognitive processes would be defeated; and that would mean, as already remarked, that we should be obliged to opt for either an eliminative bottom-up psychology or a top-down psychology that featured folk-theoretic conceptions of cognition irreconcilable with the unity model.

In fact, Pylyshyn's difficulty suggests an instructive *passage* of theory stretching all the way from Frege and Tarski to current speculations of cognitive scientists such as Pylyshyn and theoretical neurobiologists such as those reviewed by Patricia Churchland. For Tarski shows how to construe the adequacy conditions for truth applied to certain interpreted formal languages exhibiting a thoroughly extensional structure in accord with canonical first-order calculi; but Tarski was also very explicit in expressing his doubts about the application of his notion to language used "in everyday life."[40] Quine claimed that natural languages may be suitably "regimented" in the extensional terms of canonical logic, and he demonstrated a large number of ways in which recalcitrant sentences might be so interpreted; but he never attempted the project under the actual conditions of cognitive inquiry and empirical science – he offered only formal resolutions regarding how favorable paraphrases would look.[41] Davidson explicitly attempted to combine Tarski's conception and Quine's (with modifications) in an effort to demonstrate, among other things, how psychology *could* be paraphrased in accord with the strong extensionalism of the unity-of-science model; but, again, the empirical question of the relationship between (what in effect are) folk-theoretic notions and psychological notions suitably thus regimented is never actually pursued by Davidson.[42] He merely assumes that the paraphrase can be accomplished, though it implicates him in an inconsistent triad;[43] his project remains formally very much the same as Quine's (though it is actually less explicit).

Daniel Dennett has attempted, by way of introducing gradually more and more "stupid" homunculi, to analyze the molar, intentionally complex functions of folk-theoretic cognition in such a way that, ultimately, homuncular functions may be entirely replaced by discrete, extensionally well-behaved physical phenomena – in accord with the extensionalism demanded by Quine and Davidson and by generic versions of the unity of science program.[44] But Dennett never filled the essential lacuna of Quine's and Davidson's programs (with respect to psychology), in the sense that, as homunculi are (on his own view) only introduced *relationally* as the sub-functions of molar functioning, the problem of replacing intensionally complex characterizations (of molar or sub-molar states) by extensionally suitable ones remains entirely unaffected by the assumption – the empirically unmotivated assumption – that cognitive functions may be counted on to yield progressively to the reductive paraphrase.[45] Jerry Fodor holds (somewhat ambiguously) to a top-down functional psychology in terms of which he is prepared to countenance functionally distinctive psychological laws not reducible to physical laws but congruent nevertheless with the general requirements of the unity model – on the assumption (nowhere empirically supported) that the molar functions of cognitive agents are straightforwardly expressible in the vocabulary of an innate "language of thought" in complete accord with the extensionalism featured in Davidson's version of Tarski's account.[46] Stich's view we have already considered; but it is clear that, in "elminating the middleman," Stich simply retreats by way of a promissory note to a very strongly extensionalized mapping of law-like functional regularities (that are to replace "folk" notions but that are also entirely unspecified). There is very little difference – as far as the essential empirical puzzle is concerned – between Quine, Davidson, Dennett, and Stich. There are, to be sure, different economies and variations among them, but the core difficulty remains unaffected and unresolved. Chomsky's and Fodor's accounts *may* also converge in the same bottom-up spirit; but Chomsky nowhere discusses the actual complexities of molar linguistic behavior apart from getting on very rapidly to his idealized account of innate grammars; and Fodor pursues only the alleged requirement of an innate language of thought. Pylyshyn, for his part, attempts to recast the principal line of argument of this entire *passage* within the terms of cognitive science; but as we have seen, he simply brings us back again to the same apparent lacuna. It remains to report what Pylyshyn adds regarding this vexed question – which (in Pylyshyn's own idiom) is just the question of transducers.

III

So back once again to Pylyshyn. It is Pylyshyn's stated view that "semantic level principles differ considerably from physical laws," since physical laws entail closure over the domain of "bona fide physical properties" (to some fair first approximation) whereas semantic principles "apply to open-ended classes of physical properties" for which there is "*no limit* on the combinations of physical properties that can be used to instantiate" semantically relevant items. In fact, "the two sets of principles have nothing in common; they are merely related by the happenstance of design or instantiation."[47] (The correspondence, then, is at best taken token–token, never type–type.) Pylyshyn adds that even if "Every mental-event type (for example, thinking that $2 + 3 = 5$) is realized by a corresponding, token brain event ... there is no a priori reason for assuming that the entire class of brain events that could ever conceivably constitute the same mental event ... *has any physical property in common.*"[48] There is, therefore, a fundamental difficulty in accounting nomologically for the causal influence of mental events. *If* they are to have causal effects on behavior – as apparently they do in a variety of ways (as in producing effects under voluntary control or as in the indirect influence of goals and beliefs[49]) – they must fall under covering laws. Yet either they cannot be made to do so (on any known strategy) or the condition on which they might do so, empirically, must escape all finite inquiry. If, say, (1) an infinite correlation between semantic-level and physical-level events would be required in order to justify introducing nomological speculations about otherwise utterly unrelated phenomena, or if (2) semantic-level events (mental tokens) were so interrelated, intensionally, that they could not be treated, in *realist* terms, as discrete instances of rule-governed types and, as such, could not be treated as finitely generable computationally (no matter how "effective" extensional approximations over significantly idealized finite segments of the entire range of such events might be[50]), then Pylyshyn's project would fail – for principled or empirical reasons. (This sketches, incidentally, a sense in which achievements along the lines explored in a more sanguine spirit by Quine, Davidson, Chomsky, Fodor, and Stich need never be utterly discarded; they would merely be devalued in terms of psychological or cognitive realism.)

One may speculate here that, on such a reading, human cognition could easily remain representational without being computational or without being completely computational and without being computational at every level at which it operates with representations. For example, at the molar level of societal communication, human agents could be said to be using representations in using the sentences of a natural language. Perhaps only a part of the language they have internalized is computable: either because, at some tacit modular level of sub-molar functioning, socially induced habits have regularized or reconciled their speech in extensionally computable ways in tandem with the relatively unchanging core practices of their society; or because selected segments of their actual communicative practice may be described, retrospectively but with some predictive power (within slowly changing historical societies), in *computationally idealized terms*, without denying that molar communication in natural societies works primarily by interpretive consensus and consensual tolerance (particularly where intensional complexity, conceptual invention and linguistic novelty are paramount) and without at all denying that such idealized computational regularities are inevitably subject to being overridden by molar behavior anticipating and responding to such societal consensus. We may dub the theme a mixture of the Wittgensteinian and the phenomenological. It obviously deserves attention. To fix it somewhat hastily for the sake of convenient reference: it is perhaps suggestively sketched by Pierre Bourdieu, in his account of what he terms *habitus*.[51] In any case, it is quite remarkable that the possibility that a cognitive representationalism *may depend constitutively on societal practices – and therefore may not be fully computational in realist terms* – has never been directly explored (whether favorably or unfavorably) by bottom-up theorists committed to a computational model or to a deeper extensionalism. If something like consensual tolerance *at the societal level* obtains in human communication, across diachronically open-ended change, then it is unlikely that one could ever show that human languages were effectively computable in all important respects or that they were recursively analyzable and therefore computable.[52] It is not unimportant to observe that Pylyshyn's emphasis on the emergence of an open-ended semantic level invites an inquiry of the sort here sketched; but his own claims about transducers are clearly intended to obviate the need for such an inquiry.

Pylyshyn has the solution all right, the formal solution – that is, the solution under the narrow conditions he himself imposes: "If," as he explains, "you take the inputs and outputs of a system under a biological description, then, by definition, its regularities will be subsumed under biological generalizations. No law-like (counterfactual supporting) generalizations contain both cognitive and biological descriptions, not, that is, *unless the cognitive categories happen to be coextensive with the biological categories.*"[53] Cognitive states may produce changes in biological states, then, only insofar as they themselves are biological states. They can belong to both systems if and only if there are actual transducers; a psychological transducer must, therefore, be "an element of both a cognitive- and a biological-equivalence class";[54] the specification of (such a) "strong equivalence" is, in effect, the specification of a transducer. So a transducer fixes a functional algorithm (hypothesized to be real for the physical or biological system identified) in virtue of which the required equivalence can be accounted for. A transducer, therefore, is a "functional" or "symbolic" or "syntactic" structure (a structure specifying "the basic computational resources" of that system – what Pylyshyn terms the relatively fixed, possibly even species-specific, "functional architecture" of the system in question) through which the physical and the semantic (or mental) are seen to be isomorphic.[55] There you have the true Cartesian force of Pylyshyn's argument. "According to the strong realism view I advocate," Pylyshyn declares, "a valid cognitive model must execute the *same* algorithm *as the one executed by subjects*" – that is, the molar subjects of top-down folk psychology.[56] The transducers assigned to the functional architecture of the "cognitive virtual machine" that *is* the organism or computer (on Pylyshyn's theory) must be the actual transducers of that organism or computer. So Pylyshyn's constraints are admirably clear but terribly difficult to satisfy. Also, of course, it is entirely possible (for instance, on the consensual model sketched just above) that "subjects," human cognizers functioning communicatively within their natural societies, may actually employ representational schemata that are *not* extensionally algorithmic (or, not such invariably – or, not such in certain decisive contexts). Transducers, in Pylyshyn's sense, could not then account in one respect or another for actual cognitive processing at the molar or folk-theoretic level.[57] This draws attention once again to an important gap in the conceptual options considered.

How is the required isomorphism to be established? If there are no (psychological) transducers, then, on the argument, there is no way of "systematically relating behavior to properties of the physical world." Pylyshyn concedes "the relative scarcity" of promising candidates for the role of "primitive" or minimally adequate transducer.[58] But he also admits the explanatory importance in molar psychological contexts of the principle of "rationality":[59] on familiar arguments, for instance Davidson's,[60] it is precisely the intensionally complex holistic nature of rationality in molar settings that makes the extensional treatment of folk psychology conceptually impossible and that actually precludes, in principle, psychophysical and psychobiological laws, regardless of *any* seemingly promising empirical regularities. If Pylyshyn were to admit intensional complications due to psychological holism, then his search for psychological transducers could not avoid being self-contradictory and self-defeating. So Pylyshyn has impoverished from the start his chances of sustaining his own thesis. For he disqualifies the resources of the neurobiologist who would radically replace folk-psychological notions in order to bring *them* into accord with unity-of-science constraints;[61] he is not prepared to speculate in the empirically blind eliminative manner that Stich favors; he admits that productive psychological studies of cognition (for instance, those of J. J. Gibson and Julian Hochberg) do not usually employ descriptive and explanatory categories likely to accord with the nomological regularities of any of the fundamental sciences;[62] and he does not himself offer any favorable analysis of the actual mental concepts he wishes to provide transducers for. There could not be a less unlikely project.

What Pylyshyn finally offers on empirical grounds is also disappointing – but most scrupulous. First of all, he distinguishes very sharply between physical stimuli defined neurophysiologically and stimuli defined as "cognitively or computationally relevant": "only a small fraction of the physically discriminable states of a system are computationally, or cognitively, distinct." Computationally defined stimuli must be available as instantiations of "discrete atomic symbols" (which is to say, they must be definable in a strongly extensional, modular way, and must be systematizable as such in terms of a finite "program" or "encoded algorithm"); but they must also thereby provide the matching syntactic types for *independently* identified mental or semantic events. No wonder the admissible stimuli are scarce.[63]

It is a notable scruple of Pylyshyn's argument, however, that he pinpoints the essential weakness of *any* reductive, bottom-up neurophysiological account of cognition; for, on the argument regarding ascending levels of analysis ("biological," "symbolic," "semantic"),[64] *the cognitive relevance of neurophysiology cannot be assumed at any relatively fundamental level of functioning – say, the neuronal*. This is precisely the point (thus far considered) that David Marr also concedes in his overview of the recent history of *reductive* neurobiological accounts of visual perception – in particular regarding Horace Barlow's classic (but overzealous) claims enthroning single neurons. Barlow affirmed, on the strength of breakthrough experiments, that "It is now quite inappropriate to regard unit [neural] activity as a noisy indication of more basic and reliable processes involved in mental operations: instead, we must regard single neurons as the prime movers of these mechanisms . . . the activities of neurons, quite simply, are thought processes."[65] Marr himself remarks that, though he was strongly drawn to the reductive argument, it eventually became clear that "without an analysis of the . . . 'information processing task' [effectively, the cognitive task] there can be no real understanding of the function of all those neurons."[66] There is a double lesson here: first, the general outcome of the top-down/bottom-up quarrel is as yet hardly settled on empirical grounds; and, second, there are excellent reasons for pursuing the analysis of cognition in terms of emergent levels of explanation, even if one means to proceed in a way that might accommodate bottom-up strategies as far as they may take us.

Marr is most instructive here. For he shows quite convincingly that, in accord with an information-processing model of vision, there is good reason to suppose that, although there is a hierarchy of levels of relatively fixed modular processing, the modes of functioning of the highest levels cannot be inferred bottom-up from that of the lower constitutive levels and that the full assignment of lower-level functions depends on the definition of the informational "problems" of higher-level processes. This is already clear, at a relatively low level, from the evidence of stereopsis, against Herbert Barlow's extreme bottom-up reductionism.[67] But is only a metonym for Marr's larger thesis. Marr warns, "Remember the moral from the early stereopsis networks discussed [earlier]. None of them formulated the computational problem precisely at the top level, and almost all the proposed networks actually computed the wrong thing."[68] "Explanation of a given phenomenon," he adds, "must be sought at the appropriate level."[69]

Strictly construed, this means that, if vision employs representations, the modular assignment of *what* such representations represent can only be specified by reference to the hierarchical processing of them – that is, by reference both to instantiated representations and to the (relative) modular fixity of particular levels of representation. The cognizing system must in some sense possess a theory of the role particular information contributes to the relatively unified and steady representation *of the actual world* at the highest level of functioning. All in all, on Marr's thesis

(a)   we cannot ensure modularity at all levels;
(b)   we must guard against the mere mimicry of what we do at a high level, by simulations generated from lower-level modules;
(c)   higher-level functions cannot be compositionally inferred from lower-level functions either within the competences of a given species or in evolutionary terms;
(d)   the recognition of real objects cannot be generated solely on the basis of visual images; and
(e)   the most important implication (a matter Marr does not actually examine) – *the theory of highest-level functions is itself an artifact of reflexive functioning at that level.*[70]

Two observations of Marr's fix the sense in which these things must be so. First, as Marr reminds us, "Almost no single feature is necessary for any numeral. The visual descriptions necessary to solve this problem [recognition of a numeral] have to be more complex and less directly related to what we naturally think of as their representation as a string of motor strokes."[71] Secondly, considering the possibility that information about the identity of perceivable objects might be computed from features of visual images, Marr drily observes, "A black blob at desk level has a high probability of being a telephone."[72] These two remarks capture a volume of arguments that confirm the effective ineliminabiity, *for* an information-processing model of cognition, of the molar functioning of folk-theoretic cognitive agents (however algorithmically, at some level, we suppose – so far at least – such functioning may be reduced); they also identify the essential point of convergence between Marr's and Pylyshyn's projects.

If, however, the prevailing arguments against cognitive transparency – converging from both Anglo-American and Continental European philosophies – were accepted, for instance along such rather different lines as those of Quine and Nelson Goodman or Heidegger and Maurice Merleau-Ponty,[73] then a fair reading of item (e) in the above tally would confirm that Marr's own insistence on an "object-centered" (rather than "data-driven") framework for vision, as opposed to an "image-centered" or "viewer-centered" account,[74] remains relatively inattentive to complexities at the highest-level molar processing.[75] For, that philosophical concession – along the lines of ontic indeterminacy, diachronic forestructuring, reflexive prejudice and interest, and relativistic tolerances – would inevitably color our reading of so-called "object-centered" vision.

More than that, it would confirm the sense in which – without denying relatively fixed, low-level, modular processing of visual information through an ascending hierarchy of processors, congruent with a strongly computational model – such information, in man, may well be subject to and subserve intensionally quite complex, decidedly interest-driven theories of the objective world that cannot be entirely "decomposed" in the strongly extensionalized modular terms suited to the lowest levels. Marr himself makes the point very neatly and with due care in risking a generalization regarding the strengths and limitations of Chomskyan linguistics:

> it is easy to construct examples in which syntax cannot be analyzed without some concurrent semantical analysis. Thus, *syntax is not a truly isolated module*, and this fact led the artificial intelligence people to jump to the opposite conclusion, that syntax is not a module at all. This is incorrect – the true situation seems to be that syntax is almost a module, *requiring some interactions with semantics* but only a very small number of types of interaction. ... Noam Chomsky's transformational grammar ... is in no way concerned with *how* syntactical recognition should be implemented. It merely gives rules for stating *what* the decomposition of an arbitrary sentence should be. Chomsky's description of it as a competence theory was his way of saying this.[76]

Marr is quite correct, here, in criticizing "the artificial intelligence people" (he has in mind Terry Winograd, in particular); but he has completely misgauged (as Winograd himself may have done) the full significance of Winograd's complaint – and so betrays a weak grasp of the very sense, under the terms of his own theory, in which would-be modules in place can be said to be safely in place within an adequate theory of the higher-level "problems" and processing of molar systems that the modules in question are to serve.

As Winograd reasonably observes,

> In implementing a systemic grammar for a computer program for understanding language, we are concerned with the process of recognition rather than that of generation. We do not begin with choices of features and try to produce a sentence. Instead we are faced with a string of letters, and the job is to recognize the patterns and features in it. We need the inverse of realization rules – interpretation rules which look at a pattern, identify its structure, and recognize its relevant features. ... A program for parsing language is as much a "generative" description of the language as is a set of rules for producing sentences. The meaning of "generative" in Chomsky's original sense, is that the grammar should associate a structural description to each permissible sentence in the language. A parsing program does just that.[77]

Marr criticizes Winograd for confusing "level one (what and why) and [level] two (how) [analyses]" – that is, the difference between Chomsky's and Winograd's own projects.[78] But there are really two errors conflated here, both Marr's: first, if (as Winograd intends) the computer program is to provide a realist analysis of natural language, then the "inverse" of the generative grammar is what is required (and is what Chomsky originally took himself to be pursuing); second, if a theory of linguistic competence is distinguished, in a realist or heuristic sense, from the variety of modular powers collected in actual linguistic performance (so that the two are not symmetrical), the analysis of competence remains conceptually dependent on a theory of apt performance (a thesis which Chomsky latterly believes he has superseded). Marr does not notice the difficulty for his own view, because he favors sensory "performance"; and Winograd does not notice the difficulty for his own view, because he, too, is attracted to algorithmic reduction.

The essential point is that there is no convincing reason for believing that the processing of "semantic-level" distinctions – the level of molar psychological experience and interest – is decomposable in any modular way resembling either Chomsky's account of deep syntax or Marr's account of "early vision" or is analyzable algorithmically if not modularly. In a sense, it is the common theme (and admission) shared by Marr, Pylyshyn, Chomsky,[79] and Davidson, and it is the one that explains the desperate eliminative measures recommended by Stich and Paul Churchland.[80] In short, *modular and nonmodular, molar and sub-molar, extensional and intensional, and emergent "hierarchies" and decomposable hierarchies of functioning are entirely capable of being coherently coordinated in a unified theory of human cognition.* On the evidence, the rejection of this option (opposed to the one pursued in the "compositional" model) is not in itself incoherent. To pursue it in a sanguine way, however, *does* entail rejecting or at least neutralizing the unity-of-science program and an unqualified extensionalism. But why not?

We may now savor the essential difficulty of Pylyshyn's attempt to give transducers a determinate empirical reading. Pylyshyn had admitted, remember, that "there are few ... basic, transducible, cognitive states," few *types* of belief-states that could provide the required connection. Nevertheless, *if* there are some,

> then beliefs have their physiological effect in the same way percepts are generated from optical inputs (except conversely). They take part in inferences, the end product of which is a special, cognitive state that happens to be causally connected to a physiological state of special interest (because, say, it causes ulcers). *The last link is completely reliable, since the cognitive state in question happens to be type-equivalent to a physiologically described state.*[81]

If, however, as we have already remarked and as many are prepared to concede, cognitively pertinent phenomena exhibit an ineliminable holism and intensional interconnection, then (contrary to Pylyshyn's objective) it would be quite impossible to "discover a relevant biological property that happens to be coextensive with a certain *class* of cognitive descriptions."[82] The required cognitive "class" could not then be specified.

Hear the argument out, however. In Pylyshyn's view,

> There is one class of psychophysical experiment that requires *a minimum* of cognitive intervention, and which therefore provides relevant data on transducer functions. The most sensitive possible measure of transducer operation is a *null* measure. If we can demonstrate that two physical events are perceptually equivalent over a range of contextual variation, we can be reasonably confident

that the decriptions of the events are mapped onto the same symbolic structures *at some extremely low level* of the perceptual system – possibly, the transducer output. Although the question of which physical descriptions are relevant is not solved, the null-detection task minimizes the problem, since the fact that two events are indistinguishable (in the special sense that they do not result in a perceptual discontinuity) implies that *all* physical descriptions of one event are cognitively equivalent to all physical descriptions of the other event. Thus they will be true of the particular description we choose.[83]

There is no conceptual objection to Pylyshyn's proposal if perceptual discriminations at a reportable *molar* level can be reasonably disciplined. It is, however, more likely that (something like) Pylyshyn's transducers would be admitted at selected levels of the *sub*-molar functioning of molar cognition (that is, *without* first being able to confirm empirically, transduction regularities directly linking cognitive processes and macro-scopically perceivable physical phenomena) than it would be to establish transduction regularities at the full level of molar cognition. This is not because the idea (adjusted as recommended) is incoherent. It is rather that the evidence at best can be expected to show only that the empirical correlations collected are not actually inhospitable to transducers. In short, the procedure Pylyshyn recommends is no more than an advertisement for the theory, not a ramified confirmation of it. But that much we can see in advance: "null measure" tests will not frighten the thesis away.

Pylyshyn's remark that "class A" experiments (the experiments in question) involve only "some extremely low level of the perceptual system" is clearly meant to assure us that, at that "level," something like a Lockean account of representations of primary qualities must obtain. Of course, *if that is true*, then the heart of the transduction thesis is confirmed. But, for one thing, it is not confirmed by the correlations empirically estabished; and for another, it is nowhere explained why such regularities should not be understood in some other way. What we must grasp, first, is what the rationale is for distinguishing between a "low level" of the perceptual system and a level that is not so "low." Presumably, the answer lies with Pylyshyn's notion of "cognitive penetrability": transducers are supposed to be cognitively impenetrable, in the sense that their behavior is law-like and not alterable "in response to changes in goals and beliefs" or is capable of being explained "without appeal to principles and properties at the symbolic or the semantic level."[84] The trouble is, Pylyshyn utterly rejects the notion that the cognitive *may be directly efficacious in causal terms* – that is, without the mediation of transducers. This is the point of Pylyshyn's disjunction between "law-like" and "rule-governed" processes (the physical and the biological, on the one hand, and the computational and the semantic, on the other).[85] But if this rather Cartesian disjunction were disallowed, as any strongly neurobiological theory would be bound to favor, then Pylyshyn's criterion of "cognitive impenetrability" would have to be rejected[86] – and with it his theory of transducers.

The double lesson is this: *if* the "functional architecture" of cognition, physical and biological structures (including what will prove to be transducers), are described at a level "independent" of the abstract functional properties of cognition (and computation), then (for essentially Cartesian reasons) no mere correlation *at the level of molar ("folk") cognition can* have sufficient theoretical significance to justify admitting transducers; on the other hand, *if cognition (as opposed to sub-cognitive information)* is in some sense already decribable (at least "in part") in physical and biological terms, then it becomes impossible (theoretically contraindicated) to insist that transducers are cognitively impenetrable. The first option is a Cartesian option that (apart from a preestablished harmony) makes transduction unintelligible; the second rejects the disjunction of the mental and the physical, characteristically along biological or neurobiological lines, thus making interaction (and "penetrability") logically trivial. It is true enough that genetic structures are cognitively impenetrable, but it is not clear that neurobiological structures involved in cognitive processing – for example, in forms of learning – are not causally affected by such learning. Certainly, the studies of cerebellar processes pursued by Rodolpho Llinás and Andras Pellionisz, reported very thoroughly by Patricia Churchland (also by her husband, Paul Churchland), show very effectively how reasonable it is to pursue a modular (but not cognitively impenetrable) thesis of cognitive sub-processing in neurobiological terms.[87] Transducers must be *sub*-cognitive, "informational," not cognitive. In short, the inescapable dilemma of Pylyshyn's completely abstract functionalist account of information-processing palpably stalemates the usual notions of cognitive science.

IV

The upshot is that theories of cognition must go biological – must go *at least* biological if they are to avoid dualism while preserving the ineliminability of those features of the psychological that are usually (but misleadingly) characterized in purely functionalist terms.[88] The charm of Pylyshyn's account lies largely with its having attempted to preserve (at whatever cost) a full-bodied conception of molar cognitive functioning – and, therefore, of representations and information-processing suited to that level of discourse. The same is true of Marr's work, though Marr is clearly more phenomenologically oriented than Pylyshyn. The question that remains concerns the possibility of combining the elimination of the folk-theoretic idiom that Pylyshyn and Marr (in different ways) wish to preserve with a strongly biologized or neurobiological account of cognition. This is in fact the project of the Churchlands; its antithesis is probably best sketched in the philosophy of Merleau-Ponty, for it is Merleau-Ponty who, in our own time, rejects most comprehensively all forms of Cartesianism while preserving the *sui generis* emergence of a biologized self.[89] What we see, therefore, is that with every variation in the analysis of cognitive processing the top-down/bottom-up contest persists. But, with that thread in hand and the other speculative corridors already cleared, we can move rather quickly to the single exit that the maze affords.

There is a general adjustment that Patricia Churchland favors in sorting reductive strategies without yet broaching her own substantive thesis. She is attracted to Paul Feyerabend's subversive assessment of the philosophy of science and philosophy of mind – though not anarchically.[90] Her project is "to inquire into the possibility of a unified theory of the *mind–brain*, wherein psychological states and processes are explained in terms of neuronal states and processes."[91] She is a reductionist but not in the familiarly frontal ontological sense. Her adjustment in conceptual strategy rests with the following: "reduction involves *explanation* and ... typically reductions involve *corrections* of those levels of description that supply the explananda"; a new theory "displaces the old, there is a reconfiguring of *what needs to be explained*"; or, again, "What gets reduced are theories, and the stuff in the universe keeps doing what it is doing while we theorize and theories come and go. So the question 'Does mass itself get reduced?' does not make any sense."[92] "Ontological simplification" (as well as "explanatory unification") falls out either by the reduction or elimination of *theories* – which, in effect, signifies a suppler methodological policy (still strongly wedded to the "deductive–nomological theory of explanation") than that of the logical empiricist's insistence on treating reductionism in terms of a straightforward deducibility-relation between reducing and reduced theories: the correction and modification of theories through the history of science requires an improvement on the empiricist's improvement upon earlier methodological disorder.[93]

To draw further on our prepared thread, however: it remains just as possible to adjust anti-reductive strategies such as the folk-psychological in terms of Patricia Churchland's correctional constraint as it is to favor the neurobiologically reductive or eliminative strategies Churchland herself prefers. There is a sense, most noticeable in Stich's account, that top-down folk-theoretic approaches are not merely committed to the notion of molar selves and its inconvenient (and allegedly unproductive) categories but are committed as well to the timeless fixity of those categories. If so, that would mark an extravagance its champions can ill afford. The implication of Churchland's policy, read neutrally (thus far), is that "the candidate for reduction to neurobiological theory is some *future* theory," since the corrections that will be required are likely to entail the replacement of current theories, both folk-psychological *and* provisionally reductive; also, that a replacing theory can, despite the intuitively compelling features of a folk psychology, come to "be as routinely and casually used as the old folk theory." An analogue is provided from "folk physics" – in effect, the Feyerabendian move.[94]

We come then to the substantive argument that folk theories will be replaced not merely by other, more adequate *folk* theories but by neurobiological accounts that eliminate all versions of the other. We must be forewarned, however, that, in subscribing to a *folk* psychology, one need not remain hostage to all the admitted errors and untenable and unconvincing claims that have been made in its name – any more than the reductionist need be with respect to earlier conjectures of his kind. For example, it is entirely reasonable to hold that reasons (or having reasons) and desires and beliefs may be causes; that present folk theories are not or may not be essentially correct as they stand; that reductive policies need not proceed (in the order of discovery) "compositionally" (from "micro properties" first); that introspective appeals are not theory-free; that folk psychology is not immune to reduction and is not more privileged epistemologically than folk physics; that dualisms of substances are conceptually inadmissible; and (perhaps most important) that folk-psychological concepts such as consciousness, thought, learning, memory, emotion, belief and desire are not phenomena of "*a natural kind* . . . in the sense that there are natural laws about their properties and behavior [thus categorized]."[95] There is perhaps one further *caveat* to be singled out: dualisms of a "functionalist" sort are also conceptually disqualified – that is, hypotheses that hold that "the generalizations of psychology are *emergent* with respect to the generalizations of neuroscience and that mental states and processes constitute a domain of study *autonomous* with respect to neuroscience."[96] The last objection seems clearly directed against Pylyshyn and Fodor.[97]

The *caveat* cuts both ways, however: it is reasonable to deny the autonomy of psychology with regard to neurobiological considerations (this, after all, was the upshot of our argument against Pylyshyn); but it remains an empirical question whether psychological phenomena are or are not *emergent* with respect to neurobiology, in the specific sense that they cannot (as emergent) be explained in terms of the properties and laws of the neurobiological domain. Emergent properties need not be "nonphysical properties" – *only properties not reducible to physical or biological properties*, incarnate properties for instance.[98] Churchland is entirely unclear about this matter, and characteristically conflates the two issues (as does also Paul Churchland) in her objections against a mix of dualisms and against extravagant autonomy claims (where the latter claims sometimes require the former claims). It is obvious that *folk* psychologies oppose reduction, in the emergentist sense – not in the sense favoring autonomy.[99]

All this may be easily granted. What follows? There is only one consideration that Churchland introduces to offset the emergentist claim. She admits the use of the expression "emergent property" in the neurosciences, in the sense "roughly equivalent to 'network property'": "The functional property [for instance] of being a movement detector [in the retina] may understandably be described as 'emergent' relative to the individual neurons in the circuit. However, the functional property is certainly and obviously reducible to the neurophysiological properties of the network." The entire pattern of such possibilities Churchland labels (following Dennett) an "innocent emergence."[100] But her argument is merely a formal argument about the coherence of "innocent emergence"; it is hardly a demonstration that *emergence* at the putative level of the functioning of molar selves is "innocently" emergent, in the sense that *it is invariably reducible* – in being invariably concerned with "network properties." That needs to be empirically demonstrated. *No one, Churchland included, has actually provided even a promising sketch of such a reduction*. It is entirely possible, even plausible, to hold that there are many modular, *sub*-functional processes of molar cognition that are reducible in something like Churchland's sense *and* to insist at the same time that functioning at the highest level – linguistic, rational, reflexive, culturally informed, deliberate and the like – are not similarly reducible, that *all* sub-processing of the kind Churchland surveys subserve molar functions of these higher and higher sorts. The point is that it is an entirely coherent option. Admittedly, it is an option that conflicts with the (generally reductive) conceptual preferences of the unity-of-science program. But the adequacy of that program is as much an empirical issue as is the occurrence of emergent phenomena that are not "innocently" emergent. These are, in fact, obverse sides of the same issue. The trouble, here, with Churchland's countermove is that she believes (without the least evidence or argument) that no one *can* oppose the reductive theme of "innocent emergence" without becoming committed to a functionalist dualism (for instance, of, say, Pylyshyn's or R. W. Sperry's sort[101]).

That bias is already evident in Churchland's reading of the *functionalism* she herself favors. "Functionalism, in its most general rendition," she says, "is just the thesis that the nature of our psychological states is a function of the high-level causal roles *they* play in the internal system of states that mediate between sensory input and behavioral output."[102] The statement as it stands is entirely neutral as between reductive and emergentist readings. Churchland, however, does not read it neutrally.

The important distinction, here, is that the issue of empirically identifying "the high-level causal roles" that psychological states play in internal systems mediating between input and output is a quite different matter from that of reducing such functional roles to "network properties." That's all that's at stake. Churchland *never* addresses the distinction. She is indeed admirably clear about what she *takes* to be model instances of the latter, especially in the pioneer work of Pellionisz and Llinás. But their work is entirely open to an emergentist reading: it certainly does not demonstrate the conceptual adequacy of the reductive reading. She has in fact utterly neglected the import of the executive theme – something that both Dennett (whom she professes to follow in this matter) and Marr (as we have seen) warn about: namely, that modular or homuncular functioning is, at best, only *relationally* isolated as ingredient in the *sub*-functioning of some form of molar functioning: once that theme is preserved, it must be seen to be an entirely independent question whether modular or homuncular analyses of this kind (which, effectively, are top-down analyses, *not* yet bottom-up) *can also be replaced (reduced) in some neurobiological way.* Churchland has nothing to say about the latter issue, and no one really has. Her own master claim about Pellionisz and Llinás's cerebellar studies converts *relational* findings regarding transformations between *modular* "phase spaces" into *reductive* claims that treat those *relata* as no longer functionally factored *within* some higher-level (molar) functioning. This is the same *non sequitur* as Dennett had originally – one may almost say, classically – produced, and it is a mistake that Paul Churchland compounds.[103]

Finally, Patricia Churchland appeals to a unity-of-science policy, but it cannot possibly do the empirical work required: "The unity of science is advocated," she says, "as a working hypothesis not for the sake of puritanical neatness of ideological hegemony or old positivistic tub-thumping, not because theoretical coherence is the 'principal criterion of belief-worthiness for epistemic units of all sizes from sentences on up.'"[104] Clearly, this hardly shows that superseding (without rejecting where relevant) innocent emergence – or rejecting a fully comprehensive unity-of-science program – is either incoherent or beneath being "belief-worthy." On the contrary, it is an essential part of a longstanding philosophical contest.

Stalemate is what we wanted and provisional stalemate is what we have achieved. For there is an enormously widespread conviction that the mere advocacy of a folk-theoretic psychology is hopelessly opposed to science, scientific progress, scientific coherence, scientific evidence, the rationality of science and who knows what else; whereas the modest truth is that it is opposed to radically reductive and eliminativist programs and to the unity of science as that is canonically understood. Still, the stalemate is tipped somewhat in our favor because representational theories of cognition (which Pylyshyn and Marr and Churchland and Fodor wish to preserve) are, at least at the present stage of development of the psychological sciences and for the foreseeable future, both hospitable to and most clearly congruent with a top-down folk-theoretic approach; and eliminative treatments of the representational (for instance, Stich's and Dennett's) are plainly flawed as they stand and little more than promissory notes.

Notes

1   Noam Chomsky, *Rules and Representions* (New York: Columbia University Press, 1980), p. 70 (italics added).
2   Daniel C. Dennett, "Artificial Intelligence as Philosophy and as Psychology," *Brainstorms* (Montgomery, Vt: Bradford Books, 1978), p. 112.
3   Ibid., pp. 110, 124.
4   Chomsky, *Rules and Representations*, pp. 12–13. See also Jonathan Lear, "Going Native," *Daedalus*, Fall 1978; and Hilary Putnam, *Meaning and the Moral Sciences* (London: Routledge and Kegan Paul, 1978), pp. 49–50. Both titles are cited by Chomsky.
5   Dennett, "Brain Writing and Mind Reading," *Brainstorms*.
6   See Jerry A. Fodor, Introduction to *Representations* (Cambridge, Mass.: MIT Press, 1981).
7   Stephen P. Stich, *From Folk Psychology to Cognitive Science* (Cambridge: Cambridge University Press, 1983), p. 149.
8   Ibid., p. 153.
9   Ibid., chs 9–10.
10 Ibid., pp. 157–8.
11 See ibid., ch. 11.

12 Ibid., p. 151.
13 The most thorough philosophical canvass of the pertinent neurobiological literature – cast, it must be said, largely in eliminative terms without wishing to eliminate representationalism – appears in Patricia S. Churchland, *Neurophilosophy* (Cambridge, Mass.: MIT Press, 1986).
14 See Zenon W. Pylyshyn, *Computation and Cognition* (Cambridge, Mass.: MIT Press, 1985).
15 Stich, *From Folk Psychology to Cognitive Science*, pp. 157–8 (italics added).
16 See Donald Davidson, "In Defense of Convention T" and "Reality without Reference," *Inquiries into Truth and Interpretation* (Oxford: Clarendon Press, 1984); and Jerry A. Fodor, *The Language of Thought* (New York: Thomas Y. Crowell, 1975), pp. 79–97.
17 See ch. 1, above.
18 Stich, *From Folk Psychology to Cognitive Science*, pp. 153–6 (italics added).
19 Donald Davidson, "Mental Events," *Essays on Actions and Events* (Oxford: Clarendon Press, 1980); see Stich, *From Folk Psychology to Cognitive Science*, pp. 53–60.
20 Stich, *From Folk Psychology to Cognitive Science*, p. 158 (italics added).
21 Dennett, "A Cure for the Common Code," *Brainstorms*, p. 101, and "Artificial Intelligence as Philosophy and as Psychology," ibid., p. 132.
22 Pylyshyn, *Computation and Cognition*, ch. 2, particularly pp. 32–8.
23 David Marr, *Vision: A Computational Investigation into the Human Representation and Processing of Visual Information* (New York: W. H. Freeman, 1982), p. 3.
24 For a thorough canvass of views of knowledge of the "folk"-theoretical sort – cast however in a strongly foundationalist spirit – see Roderick M. Chisholm, *The Foundations of Knowing* (Minneapolis: University of Minnesota Press, 1982).
25 See Stephen Michael Kosslyn, *Image and Mind* (Cambridge, Mass.: Harvard University Press, 1980); Zenon W. Pylyshyn, "What the Mind's Eye Tells the Mind's Brain: A Critique of Mental Imagery," *Psychological Bulletin*, LXXX (1973), and "The Imagery Debate: Analogue Media versus Tacit Knowledge," *Psychological Review*, LXXXVIII (1981).
26 Pylyshyn, *Computation and Cognition*, p. xx.
27 Ibid., p. 142.
28 Ibid., p. 148.
29 Ibid., p. 154 (italics added).
30 Ibid., p. 152 (italics added).
31 Ibid., p. 154; also p. 155.
32 Ibid.
33 Ibid., pp. 155, 35. See also pp. xi–xii: "Many feel, as I do, that there may well exist a natural domain corresponding roughly to what has been called 'cognition' ... that cognitive science is a genuine scientific domain like the domains of chemistry, biology, economics, or geology ... that included in this category are the higher vertebrates and certain computer systems." See,

also, George A. Miller, "Informavores," in F. Machlup and U. Mansfield (eds), *The Study of Information: Interdisciplinary Messages* (New York: John Wiley, 1984).

34 Pylyshyn, *Computation and Cognition*, pp. 140–1, 24–5n.
35 Ibid., pp. 141, 130–1 (italics added). See Allan Newell, "The Knowledge Level," *Artificial Intelligence*, XVIII (1982).
36 Ibid., pp. 210–11.
37 Ibid., pp. 42, 147 (I have italicized "without using" and "without making"). See also Jerry A. Fodor, "Methodological Solipsism Considered as a Research Strategy for Cognitive Psychology," *Behavioral and Brain Sciences*, III (1980).
38 Pylyshyn, *Computation and Cognition*, p. 35.
39 Newell, "The Knowledge Level," *Artificial Intelligence*, XVII.
40 Alfred Tarski, "The Concept of Truth in Formalized Languages," *Logic, Semantics, Metamathematics*, 2nd edn, tr. J. H. Woodger, ed. John Corcoran (Indianapolis: Hackett, 1983).
41 W. V. Quine, *Word and Object* (Cambridge, Mass.: MIT Press, 1960). See Joseph Margolis, "The Stubborn Opacity of Belief Contexts," *Theoria*, XLIII (1977).
42 Davidson, "Mental Events," *Essays on Actions and Events*, and "Reality without Reference," *Inquiries into Truth and Interpretation*.
43 See Joseph Margolis, *Culture and Cultural Entities* (Dordrecht: D. Reidel, 1984), ch. 4.
44 Daniel C. Dennett, *Content and Consciousness* (London: Routledge and Kegan Paul, 1969) and *Brainstorms*.
45 See Margolis, *Culture and Cultural Entities*, ch. 6; also Joseph Margolis, *Philosophy of Psychology* (Englewood Cliffs, NJ: Prentice-Hall, 1984), pp. 73–7.
46 Fodor, *The Language of Thought* and *Representations*. See also Jerry A. Fodor, "On the Impossibility of Acquiring 'More Powerful' Structures," in Massimo Piattelli-Palmarini (ed.), *Language and Learning: The Debate between Jean Piaget and Noam Chomsky* (Cambridge, Mass.: Harvard University Press, 1980).
47 Pylyshyn, *Computation and Cognition*, pp. 140, 141 (italics added).
48 Ibid., p. 125.
49 ibid., p. 138.
50 Ibid., pp. 70–1, 154.
51 See Pierre Bourdieu, *Outline of a Theory of Practice*, tr. Richard Nice (Cambridge: Cambridge University Press, 1977), especially ch. 2.
52 The thesis that effectively computable processes are formulable in terms of recursive functions is what is known as Church's Thesis. The thesis is not in dispute here, only its full application to natural-language communication. See Alonzo Church, "An Unsolvable Problem of Elementary Number Theory," *American Journal of Mathematics*, LVIII (1936); also R. J. Nelson, *The Logic of Mind* (Dordrecht: D. Reidel, 1982), ch. 2. There is in fact a

very nice parallel to be drawn between challenging the application of Church's thesis and challenging the application of Tarski's semantic conception of truth in natural-language contexts, without challenging either formal thesis.

53 Pylyshyn, *Computation and Cognition*, p. 143 (italics added).

54 Ibid.

55 Ibid., pp. 95–6, 259.

56 Ibid., p. 95 (I have italicized the final phrase).

57 If I understand his theory, this much at least is fairly congruent with the argument advanced, largely against Marr and John Frisby, by Patrick Heelan. See Patrick Heelan, "Machine Perception," in Carl Mitcham and Alois Huning (eds), *Philosophy and Technology*, vol. II (Dordrecht: D. Reidel, 1986). See also John P. Frisby, *Seeing: Illusion, Brain, and Mind* (New York: Oxford University Press, 1980). Heelan develops his own theory of "hyperbolic visual space" and its relation to the philosophy of science, but without reference to Marr, in Patrick A. Heelan, *Space-Perception and Philosophy of Science* (Berkeley, Calif.: University of California Press, 1983).

58 Pylyshyn, *Computation and Cognition*, p. 168.

59 Ibid., p. 136.

60 Davidson, "Mental Events," *Essays on Actions and Events*. Pylyshyn does not discuss Davidson, nor the difficulties posed by rationality for that matter. See, also, Stich, *From Folk Psychology to Cognitive Science*, ch. 4.

61 See Patricia Churchland, *Neurophilosophy*.

62 Pylyshyn, *Computation and Cognition*, pp. 166–8. See also James J. Gibson, *An Ecological Approach to Visual Perception* (Boston, Mass.: Houghton Mifflin, 1979); Julian Hochberg, "In the Mind's Eye," in R. N. Haber (ed.), *Contemporary Theory and Research in Visual Perception* (New York: Holt, Rinehart and Winston, 1968); and Jerry A. Fodor and Zenon W. Pylyshyn, " How Direct is Visual Perception? Some Reflections on Gibson's 'Ecological Approach,'" *Cognition*, IX (1981).

63 Pylyshyn, *Computation and Cognition*, pp. 88–9, 154, 172.

64 Ibid., p. 259.

65 Horace B. Barlow, "Single Units and Sensations: A Neuron Doctrine for Perceptual Psychology?" *Perception*, I (1972); also Horace B. Barlow, "Summation and Inhibition in the Frog's Retina," *Journal of Physiology*, CXIX (1953). The passage from Barlow is cited in Marr, *Vision*, pp. 12–13.

66 Marr, *Vision*, p. 19.

67 Ibid., pp. 111–59. See also Bela Julesz, *Foundations of Cyclopian Perception* (Chicago: University of Chicago Press, 1971).

68 Marr, *Vision*, p. 336 (italics added).

69 Ibid., p. 317; cf. p. 349.

70 All of these points except the last are clearly aired in Marr's idealized "Conversation" at the end of *Vision*, ch. 7. Note, in particular, his sympathy

with Joseph Weizenbaum's assessment of his own program ELIZA as a program mimicking a part of human performance in a relatively unenlightened way; see Joseph Weizenbaum, *Computer Thought and Human Reason* (San Francisco: W. H. Freeman, 1976). The last issue is the one Heelan presses particularly, along the lines of what he calls a "hermeneutic phenomenology," largely indebted to Heidegger, Merleau-Ponty, and Gadamer. Heelan's general criticism does not, however, commit us to his own particular view of natural vision.

71 Marr, *Vision*, p. 341.
72 Ibid., p. 271.
73 See Joseph Margolis, *Pragmatism without Foundations: Reconciling Realism and Relativism* (Oxford: Basil Blackwell, 1986), pt II.
74 See Marr, *Vision*, pp. 56–7, for a compendious sketch of Marr's conceptual strategy.
75 This, rightly, is the principal criticism Heelan (in "Machine Perception") directs against Marr and his associates. Although they are extremely interesting, Heelan's own preferences, in particular Heelan's arguments in favor of a natural "hyperbolic vision," to which Marr's "2 1/2 sketch" of "viewer-centered" vision converges, cannot be pursued here. See Heelan, *Space-Perception and the Philosophy of Science*; also Rudolf Arnheim, *Art and Visual Perception* (Berkeley, Calif.: University of California Press, 1974).
76 Marr, *Vision*, p. 357. Marr has italicized the words "how" and "what"; the other expressions I have italicized.
77 Terry Winograd, *Understanding Natural Language* (New York: Academic Press, 1976) [reprinted from *Cognitive Psychology*, III (1972)], pp. 21–2. See also Noam Chomsky, *Syntactic Structure* (The Hague: Mouton, 1957) and *Aspects of the Theory of Syntax* (Cambridge, Mass.: MIT Press, 1965), both cited by Winograd.
78 Marr, *Vision*, p. 357.
79 See Noam Chomsky, *Language and Responsibility*, tr. John Viertel (New York: Pantheon, 1977), pp. 152–3.
80 See Paul M. Churchland, *Scientific Realism and the Plasticity of Mind* (Cambridge: Cambridge University Press, 1979).
81 Pylyshyn, *Computation and Cognition*, p. 142 (italics added).
82 Ibid., p. 144 (italics added).
83 Ibid., p. 175 (I have italicized "minimum" and "at some extremely low level"). See also G. S. Brindley, *Physiology of the Retina and the Visual Pathways*, 2nd edn (Baltimore: Williams and Wilkins, 1970), whom Pylyshyn follows here.
84 Pylyshyn, *Computation and Cognition*, pp. 132–3.
85 Ibid., pp. 49–50.
86 See for instance the extremely penetrating criticism of Pylyshyn's theory offered by Patricia S. Churchland, "Neuroscience and Psychology: Should

the Labor be Divided?" Open Commentary to Zenon W. Pylyshyn, "Computation and Cognition: Issues in the Foundation of Cognitive Science," *The Behavioral and Brain Sciences*, III (1980). Pylyshyn does not seem to have grasped the full import of Churchland's criticism, despite the impression of fending off irrelevancies.

87 See Patricia Churchland, *Neurophilosophy*, ch. 10; also Andras Pellionisz and Rodolpho Llinás, "Tensor Network Theory of the Metaorganization of Functional Geometries in the Central Nervous System," *Neuroscience*, XVI (1985), and Paul M. Churchland, "Cognitive Neurobiology: A Computational Hypothesis for Laminar Cortex," *Biology and Philosophy*, I (1986).

88 On functionalism, see Margolis, *Philosophy of Psychology*, ch. 4.

89 Maurice Merleau-Ponty, *The Visible and the Invisible*, tr. Alphonso Lingis, ed. Claude Lefort (Evanston, Ill.: Northwestern University Press, 1968), and *Phenomenology of Perception*, tr. Colin Smith (London: Routledge and Kegan Paul, 1962).

90 See Paul K. Feyerabend, "How to be a Good Empiricist," repr. in Harold Morick (ed.), *Challenges to Empiricism* (Belmont, Calif.: Wadsworth, 1972).

91 Patricia Churchland, *Neurophilosophy*, p. 277 (italics added).

92 Ibid., pp. 288, 292, 294.

93 Ibid., pp. 280–1, 294.

94 Ibid., pp. 288–91, 295.

95 Ibid., pp. 152, 295–312, 321. See also Paul M. Churchland, *Scientific Realism and the Plasticity of Mind*; and "Eliminative Materialism and the Propositional Attitudes," *Journal of Philosophy*, LXXVIII (1981).

96 Patricia Churchland, *Neurophilosophy*, p. 317.

97 See ibid., pp. 317–23.

98 Cf. ibid., p. 317.

99 See Margolis, *Culture and Cultural Entities*, ch. 1.

100 Patricia Churchland, *Neurophilosophy*, pp. 324–5. See also Mario Bunge, "Emergence and the Mind," *Neuroscience*, XI (1977).

101 Patricia Churchland, *Neurophilosophy*, p. 325. See also R. W. Sperry, "Mind–Brain Interaction: Mentalism, Yes; Dualism, No," *Neuroscience*, V (1980).

102 Patricia Churchland, *Neurophilosophy*, p. 340 (italics added). Here she follows Dennett, *Brainstorms*.

103 Patricia Churchland, *Neurophilosophy*, p. 376. Her phrasing agrees with Paul Churchland's; see Paul M. Churchland, "Critical Notice: Joseph Margolis, *Persons and Minds*," *Dialogue*, XIX (1980).

104 Patricia Churchland, *Neurophilosophy*, pp. 412–13, 426. See Margolis, *Culture and Cultural Entities*, ch. 6; and Paul M. Churchland, "Cognitive Neurobiology: A Computational Hypothesis for Laminar Cortex," *Biology and Philosophy*, I.

7

Intentionality, Institutions, and Human Nature

To have provisionally neutralized bottom-up programs in psychology and in the cognitive and human and social sciences is not yet to have provided a positive basis for admitting the distinction or the special rigor of the latter disciplines. The requirements of such an effort are admittedly strenuous. But there is no hope of capturing them – let alone taming them – without a close inspection of the peculiar puzzles of intentionality. There probably is no compendious way to collect the topic, though there are well-known efforts to do so. So the first order of business in this new undertaking will be to provide clues to show

1 that the intentional cannot be easily dismissed;
2 that its usual treatment, where it is favorably affirmed, remains inadequate or too simple; and
3 that its characteristic complexities signify that the human sciences (that accommodate intentionality) are bound to breach the canon of the unity of science and the strong reductive and extensionalist constraints historically linked with it.

So let us make a start.

I

Addressing the question of intentionality, we do well to orient ourselves by provisional extremes. There is a temptation to politicize the discussion by suggesting which are the right-wing and left-wing poles of the debate, but the slightest good sense shows that one would thereby only betray one's own allegiance and one's own assessments. Still, the point of the temptation rests with the remarkable division of interest in the philosophical community.

W. V. Quine, for instance, has never wavered from the early pronouncements of *Word and Object* (and before), down to his review, in 1985, of P. F. Strawson's *Skepticism and Naturalism*, in which he (Quine) effectively reminds us of his complete contempt for the idiom of intentionality. In *Word and Object* he says very plainly,

> In the strictest scientific spirit we can report all the behavior, verbal and otherwise, that may underlie our imputations of propositional attitudes, and we may go on to speculate as we please upon the causes and effects of this behavior; but, so long as we do not switch muses, the essentially dramatic idiom of propositional attitudes will find no place.[1]

He means, of course, to dismiss intentionality as a mere ornament of language. "One may accept," he adds,

> the Brentano thesis either as showing the indispensability of intentional idioms and the importance of an autonomous science of intention, or as showing the baselessness of intentional idioms and the emptiness of a science of intention. My attitude, unlike Brentano's, is the second. To accept intentional usage at face value is . . . to postulate translation relations as somehow objectively valid though indeterminate in principle relative to the totality of speech dispositions. Such postulation promises little gain in scientific insight if there is no better ground for it than that the supposed translation relations are presupposed by the vernacular of semantics and intention.[2]

In reminding us of *his* rejection of the intentional idiom, Quine's object is simply to offset (without argument) the fact that Strawson shows the reasonableness of putting the philosophical burden squarely on the "rejectionist," since there are obviously so many important contexts in which the intentional idiom is natural, where to say so is not merely to favor the vernacular.[3] Quine has plugged the dike for many years in the flatlands of analytic philosophy, but Strawson's small leakages hint at the possibility of a new flood. The point is simply that the very important tradition that, in an altogether fair sense, Quine commands – the tradition that directly descends from the best work of the Vienna Circle and the unity-of-science program *via* the subversive and purifying arguments of "Two Dogmas of Empiricism" (largely directed against Carnap himself)

– never really does demonstrate how, in the real-world terms of actual empirical inquiry, intentionality could ever be eliminated. Quine himself was quite content to indicate *only* what a formal elimination of intentionality would look like *if it were viable in accord with actual cognitive practices*. He himself never addressed the question posed; and no one else has supplied the needed argument, so convinced have his admirers been that there is no need to do so. The formal rigor of Quine's paraphrases, it has been thought, rightly takes precedence over any merely empirical complaint of the sort Strawson may be said to have voiced. This for example is the essential pivot of Donald Davidson's continued exclusion (effectively) of intentional idioms, based (it seems) largely on his having construed the extension of Tarski's semantic conception of truth to the whole field of natural languages (contrary to Tarski's own view) as valid *a priori*, within that extension, without any need to examine empirically whether actual languages do in fact yield to the strict extensional treatment Tarski felt obtained only for selected portions of formal languages.[4]

The theme of this kind of analytic philosophy, then, is the principled elimination of intentionality – without the loss of the phenomena of any domain. There is another, also analytic, tradition, largely focused on the work of Edmund Husserl (and gathering up in a significant way, with inevitable controversy, the original studies of Franz Brentano and Alexius Meinong, with a looser debt to St Thomas Aquinas and Aristotle) that insists on the ineliminability of intentional phenomena. Merely to put matters this way may be taken to threaten to distort the relevant history – for example, by favoring a strongly nonrelational conception of intentionality as opposed to a relational account.[5] But the point here is not to report on the history of the concept or even to do justice to all the viable versions of intentionality that the line of influence barely mentioned may be said to collect. It is only to identify an ample, well-defined program of analysis devoted to an examination of intentional phenomena, one that compares very favorably in rigor with the tradition – largely preferred, in recent decades, in Anglo-American philosophy – that means to eliminate intentionality altogether. And, in fact, in drawing attention to it, not much more is intended than to acknowledge its special labors, in order to move on to forms of intentionality that are not particularly featured (or adequately featured) in that second tradition.[6]

The second tradition is at least primarily centered in the analysis of perception, although intentionality is nowhere meant to be confined to perceptual phenomena. In Brentano, where it is also prominently focused on perception, intentionality is, as is well known, taken to be the mark of the mental.[7] More troublesomely, although Husserl rejected Brentano's relational view of intentionality (without denying intentional objects) as well as Brentano's narrowly psychological restriction of the notion, insofar as Husserl's own phenomenological conception is inseparable from some form of psychological embedding or instancing, it seems impossible to construe Husserl's characteristic account as not primarily centered in mental phenomena (as the prominence of the perceptual issue confirms) – which may (on Husserl's argument) still be distinguished but not separated from the merely psychological. So seen, whatever, on textual grounds, may be the accurate reading of Husserl's notion of the perceptual *noema*, there is a noticeable difficulty in applying intentionality, in the Husserlian sense, to phenomena *not* centered in *mental acts* of some kind.

Thus, one characteristic quarrel among Husserlians asks whether Husserl's *Sinn* (or *noema*) is not the same as Frege's (linguistic) *Sinn*. J. N. Mohanty, redeeming a distinction usually admitted to be muddled by the commentators, says that they are not the same: "The Husserlian *Sinn* of nonlinguistic acts, though expressible (within limitations), is nonconceptual, and when it is 'expressed,' the meaning of the expression and the *Sinn* are 'congruent' but not identical. The *Sinn* or noema is always ideal. 'Ideality' and 'conceptuality' are not the same."[8] On the other hand, Izchak Miller, following Dagfinn Føllesdal's reading of Husserl, maintains that all mental acts or acts of consciousness have *noemata* associated with them – which accounts for their directedness or intentionality (where "'*Sinn*' in the *Logical Investigations* corresponds to 'noema' in *Ideas*"); that "*every noema is a proposition*," although the "noematic *Sinn*" (which is that "part" of a noema "in virtue of which an act relates to its object" – if, indeed, the act has an object, which, against Brentano and Meinong, it need not) may (in effect) be either a proposition or a term (or, "individual concept"); and that therefore the Husserlian *Sinn* is (despite other important differences) very much like the Fregean *Sinn* ("linguistic meaning"), since "noema" designates "a certain kind of meaning, or sense, namely, a meaning of an *act*" – for which reason "noemata are but a sub-class of propositions."[9]

The resolution of this quarrel is not our concern here. Both readings demonstrate that, for Husserl, intentionality is essentially a matter of characterizing mental life, mental acts, even if against earlier tendencies to construe mental life psychologistically (that is, naturalistically). The mere inability of the "rejectionists" to accommodate, without referring to intentionality, the range of phenomena under dispute certainly confirms the need for a measure of care in discarding what we may not be able to do without. But, constrained in the way Husserl explores the notion, and despite the fact that Husserl means to do justice, in intentional terms, to societal, cultural, and historical phenomena, he does not seem to provide anything like the developed use of the notion of intentionality with respect to such phenomena that he offers with respect to individual mental acts.

Labels hardly matter here. We may, if we wish, think of phenomenology as the analysis (along whatever lines) of intentional phenomena. But we need not encumber ourselves, in exploring the intentional, with Husserl's particular theories about his own transcendental method, his "idealism," his notion of a "science," his particular claims about *noemata*, unless, on independent grounds, particular claims seem well worth acknowledging. There is for instance some reason to think that, once one escapes the simplicity of what Husserl pejoratively labels "objectivism" (in both Kant and Brentano, for instance), there may well not be a sufficiently sharp disjunction to be drawn any longer between phenomenological and non-phenomenological (naturalistic) speculation of the sort Husserl at least sometimes (perhaps even often or characteristically) insists on. And, as it happens, the new measure of conceptual freedom is very nearly rampant in contemporary philosophy, on all sides – undoubtedly in part, but only in part, through Husserl's own indefatigable critical skills.

II

The ulterior point being pressed here may be focused in the most telling way by reflecting on language itself – on which the quarrel just rehearsed is itself centered. For, *it is impossible to give a satisfactory account or analysis of language in terms of individual acts, mental or otherwise, even though actual speech is invariably taken to be manifested in individual acts, the acts of competent speakers.* Roughly, the limitation of the Husserlian approach to intentionality (whatever its elastic possibilities may be supposed to be) is fairly captured by reminding ourselves of the puzzle of what Karl Popper has labelled the thesis of "methodological individualism." One cannot identify an individual act of speech except within some kind of linguistic "space"; and that "space" cannot itself be characterized in terms of any aggregate of individual linguistic acts – nor, *a fortiori*, of sentences. The intentionality of linguistic acts must be both (1) conformable with the structure of that linguistic space and (2), as acts, specifically different from it. For, on the argument intended, linguistic acts cannot obtain except in that space and, as acts rather than the space within which they obtain, they have their own forms of directedness. This way of viewing matters strongly favors the thesis that language, the space of linguistic acts, has its own distinctive forms of intentionality; also, that what holds for language holds, *ceteris paribus*, for all cultural and historical phenomena. It is, at the very least, a socially shared space, in which individual agents

  (i)   are aware of sharing it;
  (ii)   are affected by and mutually affect one another's sharing and one another's use of it;
  (iii) are tacitly aware that its properties are not essentially, primarily, totally, or in any similar sense originally generable from their own individual, private, random, independent or arbitrary intentions, however aggregated;
  (iv)   are also aware that their own linguistic behavior does or may affect the properties of their shared linguistic space but does so primarily only by first and continuously conforming to some set of socially shared constraints already in place – however flexible or accommodating they may be; and hence
  (v)   are aware that each is conforming with practices favored within that space only from the vantage of having internalized some partial, idiosyncratic segment of a common space that no one can have actually internalized completely.

There is a metaphor here that needs to be addressed – the notion of linguistic space. Return to Popper for a moment. Popper of course is a staunch champion of what he calls methodological individualism. He means by this to oppose

  (a)   (what he calls) "methodological essentialism" (Aristotelian science, directed to essences);
  (b)   "holism" or "collectivism" (or, by association particularly with Marx and Hegel, "historicism": the thesis that "the social group is *more* than the mere sum total of its members, and is also *more* than the mere sum total of the merely personal relationships existing at any moment between any of its members" – hence, the thesis that the social group has "a *history* of its own"[10]); and
  (c)   (what he calls) "methodological psychologism" ("the reduction of social theories [sociology in particular] to psychology, in the same way as we try to reduce chemistry to physics").[11]

Popper's very nice point is that a (reductive) psychologism "is forced, whether it likes it or not, to operate with the idea of a *beginning of society*, and with the idea of a human nature and a human psychology as they existed prior to society."[12] He opposes to this notion the thesis that "we have every reason to believe that man or rather his ancestor was social prior to being human (considering, for example, that language presupposes society). But this implies that social institutions, and with them, typical social regularities or sociological laws, must have existed prior to what some people are pleased to call 'human nature,' and to human psychology."[13] Nevertheless, although Popper effectively opposes psychologism by taking note of the psychological irreducibiity of institutions, *he nowhere offers an analysis of institutions themselves*. And for that reason he never actually clarifies, fully, methodological individualism: the thesis that "the task of social theory is to construct and to analyze our sociological models carefully in descriptive or nominalist terms, that is to say, *in terms of individuals*, of their attitudes, expectations, relations, etc."[14]

There is a sense in which Husserl's principal project *cannot*, in its own terms, provide an account of the full import of this same irreducibiilty of institutions – although there is no logical reason why an adequate (even a phenomenologically inspired) account or an independent account sympathetic with Husserlian phenomenology could not be reconciled with Husserl's own essential project. The weakness, however, of Husserl's enterprise in *this* respect remains unaffected through his entire career. At the risk of oversimplification, it may be put this way: for Husserl, "intersubjectivity," the presence of the "other" (or, "Other"), or of other egos, other selves, other persons, other humans, is a phenomenological *datum* for his constituting "subject," *not* a condition prior to the cognizing projects of any such reflecting subject. That is, Husserl is only ambivalently attracted to something more than a methodological solipsism, even though the operation of phenomenological reflection yields a sense of the co-presence of other similar beings: our fellow creatures are *encountered* in our experience, but the institutional structure of a human society is not acknowledged as a *precondition* (a precondition *first* fixed in "first philosophy") *for* the very reflection that admits that encounter. The result is that "intersubjectivity" means, for Husserl, primarily, the encounter of an *aggregative* co-presence of similar creatures, not the *structurally constitutive* precondition for every such distinctly human or cultural activity. This is the essential theme, surely, of the famous Fifth Meditation of Husserl's *Cartesian Meditations*. In Husserl's somewhat murky prose,

What concerns us is ... *an essential structure, which is part of the all-embracing constitution* in which the transcendental ego, as constituting an Objective world, lives his life ... *What is specifically peculiar to me as ego, my concrete being, as a monad*, purely in myself and for myself *with an exclusive ownness*, includes my every intentionality and therefore, in particular, the intentionality directed to what is other. ... The second ego, however, is not simply there and strictly presented; rather is he constituted as "alter ego" ["constituted an ego ... as mirrored in my own Ego, in my own monad"].[15]

Taken strictly, this means that Husserl has invented a phenomenological counterpart of what Popper terms methodological psychologism – which fails (despite its avoidance of psychologism itself) by reason of its own (analogous) difficulties. In particular, it means that Husserl's transcendental phenomenology cannot as such illuminate the phenomenon of institutions, and that whatever Husserl may have contributed to an undistorted analysis of the latter must first be disengaged from the idiom accommodating that frame of reference or vision. It also means that Husserl could not have entirely grasped (or was not entirely willing to concede the implications of admitting) that phenomenological reflection itself could not fail to be a moment within the biologically, historically, praxically contingent life of worldly creatures; and that, therefore, such creatures could never genuinely know if or when or how or even to what extent they succeeded in bracketing any and all tacit complexities of their own existence and of whatever within it they might choose to consider in pursuing apodictic knowledge. Generously construed, this historicized or praxical theme is the powerful notion Hegel and Marx jointly support – that subverts Husserl's solipsistic detour in advance of its many trials, and that need not (one supposes) fail to meet Popper's constraint on methodological individualism or, correspondingly, need not fail to offset Popper's equation between historicism and methodological holism. (The fate of historicism itself is not at stake here.) It is at least in this sense (perhaps in no other) that Hegel and Marx share common ground with Merleau-Ponty and Gadamer – against Husserl.

Of course, Husserl himself, particularly in his last efforts, came to insist on much of this, in developing his notion of the life-world (*Lebenswelt*) – which is pre-given for subjects in some sense. But, at the very moment he does so, he also insists that "this insight" – that is, that even "the supposedly completely self-sufficient logic which modern mathematical logicians think they are able to develop, even calling it a truly scientific philosophy ... as the universal, *a priori*, fundamental science for all objective sciences, is nothing but naiveté" – "*surpasses the interest in the life-world which governs us now*" (*a fortiori*, governs the logicians among their delusions). For, at that very moment, he continues to affirm that "Only when this radical, fundamental science exists [which he regards as entirely possible, which is '*required*' by all the sciences, which in particular is required by the mathematical sciences said to be *a priori* 'in the usual sense,' and which in effect is an *a priori* science of 'the pure *life-world*' itself which it thereby '*surpasses*'] can such a logic itself become a science."[16]

Still, to take note of all this is certainly not to deny that, particularly in his criticism of "the regressive method" of Kant's transcendental philosophy, Husserl effectively fixes the profound contingency of the "pre-given" or preformational life-world for all objective knowledge. At a certain level of (phenomenological) reflection, he holds, all human cognition and behavior obtain within and only within the horizon of a cognate life-world – which Kant failed to penetrate, which is present in any Kantian-like achievement "only in concealment,"[17] and which was later to become the master theme for the so-called "existentialized" (the radically contingent) phenomenologies of Heidegger, Merleau-Ponty, Sartre, and (stretching things) Gadamer. Here, Husserl certainly addresses the irreducibility of cultural history, societal structures, human institutions in flux; and these are the principal concerns of the later phenomenologists who eschew Husserl's search for the apodictic. But Husserl himself never abandons his primary conviction and project, despite all variations: "it is not the being of the world as unquestioned, taken for granted, which is primary in itself; and one has not merely to ask what belongs to it objectively; rather, *what is primary in itself is subjectivity, understood as that which naively pregives the being of the world and then rationalizes it (what is the same thing) objectifies it.*"[18]

Recalling that language is our exemplar of institutional phenomena, there can be no question that

1   Husserl's theme of the pre-given structure of a historicized societal world is essential to an analysis of institutions; and therefore
2   his sense of the *irreducibility* of the life-world to the aggregated acts of natural individuals is similarly essential.

But any adequately balanced picture of the relationship bewteen the individual and the societal within the life of apt human agents – that is,

3   the *inseparability* of the individual and the societal; and
4   the symbiosis of the two

is regularly jeopardized by Husserl's ulterior dream of a phenomenological science. The latter two features cannot be drawn from an analysis of mere subjective *acts* (or states) of any kind. So we may acknowledge that, in a manner perhaps more profound than Popper's but not more precise, Husserl identifies much of what is needed in a theory of institutions and of what must be focused in an account of their distinctive intentionality.

Popper of course, as already remarked, merely draws attention to the *need* for an independent notion of societal institutions that – now, returning to the metaphor introduced – provide the palpable reality of the conceptual "space" of language, culture, tradition, practice, human history and the like. He does not supply the notion needed. Put in the most elementary terms, *if* language and other institutions are *real* but *not* analyzable psychologically (as properties or aggregative relations of or among the actual and quite separate agents of the human world), *not* collective entities themselves, *not* the properties of collective entities, then there is only one possible solution to pursue: *institutions must be nominalized abstractions* drawn from the real, actual, institutionalized properties directly ascribable to aggregated individual agents. One may usefully think here of dialects, argots, codes; genres, period styles; epochs, the variable ethos of different ages; rules, etiquette, initiation rites, customs. They are simply not analyzable without remainder in terms of native psychological powers or properties taken as logically prior to such powers. Specific accounts of institutions are idealizations projected from different vantages; but, construed realistically, institutions form the societal dimension of the aggregated habits of thought and action *of* the apt members of a given society. That there is such a dimension is theoretically postulated by an inference to the best explanation; what its properties are at particular times and places is an artifact of alternative idealizations fitted to aggregated behavior reflexively and diachronically characterized as culturally significant. The critical claim, therefore, is that human behavior cannot fail to be ascribed institutional properties embedded in, rather than merely supervenient or heuristically specified with respect to, whatever intentional properties are rightly restricted to the scope of the minimal biological capacities or biologized agency of distributed individuals (possibly extending even to the higher mammals or other animals).

Perhaps it will not be too much to say that the essential clue – but hardly more than a clue – to what is required here is the principal theme of Ludwig Wittgenstein's notion of "forms of life." For, the Wittgensteinian notion is a notion in accord with which societal practices or institutions are instantiated only and always in the actions of individual humans (and what, consequently, they produce or effect), and in accord with which human individuals cannot fail to act in a way that manifests such practices and institutions (no matter how else they may be said to act: improvisationally, inventively, freely or the like). In short, on the Wittgensteinian notion, the individual and the societal, the psychological and the institutional, are symbiotically inseparable in the life and activity of man: the individual cannot be reduced to the societal, and the institutional cannot be reduced to the psychological; and, at the very level of human culture (not perhaps biologically or at some nativist level), human behavior must be characterized jointly in terms of both kinds of features. It must be so characterized in terms of the awareness, on the part of each individual, of sharing practices with others – of being aware to some extent (even in advance) of how one's own thought and action are affected by, and affect, the thought and action of others similarly aware of the open but still orderly social space they share. They share an intentional space that, from their limited vantage they are inclined to idealize variably in intentionally focused terms – which affects their subsequent intentions, the intentions of others, the institutional space in which they act, and aggregated efforts at an interpretive consensus to maintain an efficient and harmonious use of their common space. In short, they share a space that, however real, supports variable and diverging intensional ascriptions at any reflexive moment of review, variable and diverging extensions at any moment of active commitment, and variable and diverging idealizations at any moment of consensual mediation.

Mentioning Wittgenstein risks muddying the conceptual waters, however. For Wittgenstein's own style of writing, as well as the evident range of his speculations, indicates how little he actually has to say about institutions, practices, even "forms of life." He has no account of human history; he has no account of the actual development of human infants growing up among the mature, apt members of particular societies;[19] he has no account of *praxis* or of causal factors that continually transform a society; he has no account of how, diachronically, individual agents affect and alter the structure of the contingent practices of their own societies; he has no account of the interpretive or hermeneutic complexities of a pre-given social world that no one can completely internalize; he has no account of the division, or import of the division, of social or cultural labor. He has no account of intentionality itself. Yet he has some remarkably telling recommendations about intentionality.

In spite of all that, it may be Wittgenstein more than any other recent Western philosopher who has identified the most strategic implications of acknowledging a society's language and institutions as the preformational space within which individual agents are groomed to become the apt agents they are and, by their competence, to be able to affect in variably inventive ways the structure of the continuing tradition within which others, groomed also but later within that altered social world, are subsequently to exercise the same ability. For Wittgenstein makes absolutely central to the *Philosophical Investigations*, *On Certainty*, and related portions of his jottings that (adding to our previous tally)

5   the preformational function of social institutions *precludes the need for* apodictic, foundational, correspondentist, cognitivist grounds, without at all losing the sense both of the necessity and sufficiency of *grounding* cognition and cognitively informed practice in the effective institutions of a given society; and

6   private mental life presupposes the *sharing of public institutions* – in particular, a public language.

One of the marvelous economies the Wittgensteinian thesis yields is the nearly irresistible finding that knowledge itself has a distinctly social dimension – not merely criterially but constitutively; although to say that is not yet to specify its structure. One may acknowledge that only individual agents or aggregates *have* knowledge but insist as well that to have knowledge is to be in some state that cannot be satisfactorily characterized in terms that omit institutional factors.

In a sense, if he is right, Wittgenstein. demolishes at one stroke Husserl's entire presumptive project: this for instance is part of the implied meaning of such a remark as "The *truth* of certain empirical propositions belongs to our frame of reference."[20] The grooming of new members of a society, in virtue of which they become apt agents, includes, essentially, the mastery of a common body of empirical truths – *not* higher-order rules or principles. The entire system, then, rests on a *reliable contingency* and *cannot and need not escape that condition.* "All testing, all confirmation and disconfirmation of a hypothesis takes place already within a system," says Wittgenstein; "The system is not so much the point of departure, as the element in which arguments have their life." It is learned "purely practically, without learning any explicit rules." And to the foundationalist question (posed by G. E. Moore), Wittgenstein answers, "As if giving grounds did not come to an end sometime. But the end is not an ungrounded presupposition; *it is an ungrounded way of acting.*"[21] This may well be the single most important feature of the distinctive intentionality of human institutions:

7   The stability of functioning societies rests on the contingencies of salient empirical truths that reflect and organize the shared

experience of a people; so that whatever the individual members of a society do and think cannot fail to be linked to, informed by, generally congruent with, a diachronically (perhaps slowly) shifting, central network of empirical truths.

Such truths are institutional in a double sense: first, the beliefs (that form the distinctively shared knowledge of a people) themselves exhibit, as *shared* knowledge, an institutional dimension; and, second, the tacit sharing of such contingently collected core knowledge (alterable under the influence of shifting social experience) itself forms an institutional practice. Institutions, then, are embedded in, and embed, such experience.

In effect, Wittgenstein shows how it is possible to provide, within the framework of an empirically contingent societal structure, all the conditions for cognitive and behavioral fluency and the stability of truth-claims, spontaneous interpersonal intelligibility, cultural stability through historical change, individual improvisation and the effectiveness of one's acts, without ever pursuing the various empirical questions we noted just above he has nothing or little to say about. Or, to put matters in a less partisan manner, we have been able to extract at least seven essential conditions bearing on the analysis of institutions, by way of reviewing the clearly incomplete accounts of certain pertinent questions pursued in Husserl, Popper, and Wittgenstein. Together, they show the need for a distinctive kind of intentionality ascribable to institutions, that cannot be extracted in any straightforward way from the tradition that has entrenched the study of intentionality, any more than from the tradition that has sought to eliminate it altogether.

Possibly, Wittgenstein's master stroke, here, is obscure. But one way of isolating it may be particularly instructive. In opposing Bertrand Russell's theory of language and thought, Wittgenstein elliptically remarks, "If you exclude the element of intention from language, its whole function then collapses. What is essential to intention is the picture: the picture of what is intended."[22] Wittgenstein's point is that Russell intrudes a "third event," just the recognition of, say, the match – in contexts of truth – between a thought and a fact, "an external relation"; whereas, for him, "recognition" (as in the "picture conception") is "seeing an internal relation."[23] At one stroke, therefore, Wittgenstein repudiates the correspondentist picture of truth (reducing the apparent contrast between the *Tractatus* and the *Investigations*, and distancing himself from Russell), rejects an externalist account of understanding, cognition, intention and related features of human existence (subverting thereby any simple extensionalist doctrine and any simple disjunction of realist and idealist aspects of knowledge), *and implicitly embeds his own account in a larger notion of institutional practices* (of the stably contingent sort we have already remarked). For the "rules" of a language are not applied externally to the match between thought and fact but obtain in a way internal to the intention with which, *now*, the "picture" is compared with reality.[24] That is, the "picture" (now, not at all confined to the atomism of the *Tractatus*) is in effect whatever can be envisioned in accord with the body of relatively focal, relatively stable empirical truths that, through historical contingencies, have shaped *the forms of life* of one society or another. This, then, is the ingenious and powerful way in which Wittgenstein approaches his well-known insistence that the practice and rules of language (hence, also, of thought and other "lingual" activity) are not to be construed as hierarchically ordered in any sense like that of a formal calculus.[25] Wittgenstein's strategy also puts to rest the pretense of the adequacy of any analysis of intentionality in terms of mere "mental acts," whether in terms of the "thetic" structure of such acts or in terms of their "objects."[26] It is impossible, on the admission of institutional phenomena and on the admission of an institutional encumbrance affecting every cognitive undertaking, to restrict intentionality merely to the heuristic or the fictional or the discursive or the functional or the role-playing or the like.

III

All this may be put in a single line. The polar disputes among the interpreters of Husserl cannot, apart from pursuing interesting comparisons between Husserl and Frege, treat the propositional or proposition-like structure of the *noemata* of mental *acts* as applicable to the intentional structure of the institutional practices of a society – under the condition, pressed in rather different ways by Popper and Wittgenstein, that institutions cannot be reduced to the behavior or acts of any aggregate of individual agents. This is the clear theme of the *sui generis* nature of the intentionality of institutions, about which Husserl, Popper, and Wittgenstein have actually rather little to say. *The intentionality of institutions cannot be structured as propositions or in proposition-like ways.* Furthermore, as the Wittgensteinian criticism of Russell more than implies, the eliminative programs regarding intentionality favored by the Vienna Circle and the Quinean development (which dovetails in a way with Russell's notion of external relations regarding truth and thought) have really denied what they cannot themselves fail to have acknowledged, however covertly, since they distinguish and must distinguish between sentences and the use of sentences.

Intentionality with respect to *acts* involving language is required at two distinct foci: understanding the meaning of utterances, an obvious precondition for assessing truth and falsity (on which the dispute between Wittgenstein and Russell is crucial), and the use of sentences in the performance of particular acts (with respect to which the analysis of sentences cannot be managed independently of their embedding speech acts). On the latter issue, it is primarily P. F. Strawson's very telling criticism of Russell, and J. L. Austin's pioneer study of illocutionary acts, that have, within the Anglo-American tradition, fixed the ineluctable intentionality of language acts.[27] But the extension of these considerations to language as an institutionally structured phenomenon has remained quite primitive within that tradition, as may be readily seen from the efforts of John Searle and H. P. Grice, which extend the work of Strawson and Austin.[28]

The fact is that institutions, the institutional structure of linguistic space – within which particular speech acts and linguistically informed mental states obtain – pretty well drops out of Searle's closer account of intentionality. On Searle's view, it is true that intentionality is not restricted to acts; nevertheless, the directedness of intentional phenomena (states and events included) is effectively restricted to the mental or psychological;[29] and, with respect to the idiom of so-called mental acts, the distinction between acts and states is not at all decisive, since a propositional model of directedness for mental states is normally invoked.[30] (Here the convergence between Husserl and Anglo-American philosophers of language is reasonably clear.)

Nevertheless, there are two fatal weaknesses in Searle's account, which demonstrate at a stroke the ease with which the usual discussions of intentionality fail to grasp the complexity of the relationship between the acts and mental states of individual agents *and* the structured "space" of the institutional order of societal life within which they obtain. Searle's view is compactly summarized in the following dual thesis: "An Intentional state only determines its conditions of satisfaction – and thus only is the state that it is – given its position in a *Network* of other Intentional states and against a *Background* of practices and preintentional assumptions that are neither themselves Intentional states nor are they parts of the conditions of satisfaction of Intentional states."[31] For, first of all, by a "network" of intentional states Searle clearly means an *aggregation* of intentional states – that is, the intentional states of aggregated individuals. As she says, "Intentional states consist of representative contents in the various *psychological* modes ... the Intentional content which determines the conditions of satisfaction is internal to the Intentional state; there is no way the *agent* can have a belief or a desire without it having its conditions of satisfaction."[32] Furthermore, "Intentional states do not neatly individuate."[33] So, Searle's conception of a network of intentional states is meant primarily to offset an infinite homuncular regress by way of agents' intentions regarding the intentional content of beliefs, and to accept a form of holism against the strong individuation of beliefs. The concept of a network, therefore, has nothing to do with the functional role of institutions. Secondly, by a "background" for the effective functioning of mental states (and speech acts, for instance), Searle specifically means to identify "nonrepresent- ational mental capacities" on which the functional success of intentional states depend.[34] That is, "background" capacities are not intentional (on Searle's view) because and only because they are not representational.

So, the representational function of intentional states is said to depend on a *network* of aggregated intentional (or representational) states *and* a *background* of deeper *non*representational (or *non*intentional) capacities of various kinds. The latter may be subdivided as the "deep Background" – namely, those capacities "common to all normal human beings in virtue of their *biological* makeup ... walking, eating, grasping, perceiving, and the preintentional stance that takes account of the solidity of things, and the independent experience of objects and other people"; and as the "local Background" (or "local *cultural* practices"), including "such things as opening doors, drinking beer from bottles, and the preintentional stance that we take toward such things as cars, refrigerators, money and cocktail parties."[35] The important point to emphasize here is simply that, when, as in speaking of networks, Searle admits a larger societal setting implicated in pertinent mental states and speech acts and the like, he treats that setting in aggregative terms only – contrary to what we have already remarked as the irreducibility of the institutional, in considering the issue of methodological individualism; and, when, as in speaking of backgrounds, Searle admits an even wider and deeper setting implicated in such states and acts, he treats that setting in a peculiarly mixed way – as *biological* (hence, on his own view, as nonintentional and noncultural) and, where *cultural*, as specifically *nonintentional*. Needless to say, in the Husserlian tradition what Searle regards as biological and what he regards as cultural but not intentional would be analyzed in an utterly different way.[36] Our concern here is not to redress Searle's account of mental states, cognitively informed acts and capacities, but to redeem (1) the ineliminability of the *cultural* or *institutional* space in which these states, acts, and capacities obtain; (2) its *sui generis* nature, as being not reducible in terms apt for mere individual mental states and acts or aggregative relations among them; and (3) *its distinctive* *intentionality*, as implicated in such states and acts and capacities.

We need, now, to collect the distinctive features of discourse about institutions – leading to the thesis being prepared. The following seems to offer a reasonable tally based on the foregoing:

(i) Discourse about the institutional involves the use of distinctive predicates; institutions are not collective entities but attributes applied to human individuals, aggregates of individuals, and, selectively, what they produce, make, create, invent, effect, operate, engage in, and do (as artworks, wars, modes of production and the like).

(ii) Predicates designating institutional attributes are *sui generis*, and *nonaggregative* (one may say they apply "societally" or "collectively," and only, *via* collective ascription, distributively, as in speaking of traditions, practices, histories, styles, eras and the like).

The model of an institution in the sense intended, in a sense in accord with (i)–(ii), is of course language. It is unique in being not only the most salient and ubiquitous of all institutions but also in being both the precondition and the essential ingredient feature of all cultural life (even of what, as with painting and dancing, we regard as nonlinguistic: such phenomena are really, to borrow a term usefully favored through Gadamer's work, "lingual" but not languaged, in the sense that linguistic aptitude is entailed in these activities but their products are not specifically linguistic).[37] There are of course all sorts of subtle puzzles about how the cultural depends upon linguistic ability and how linguistic ability itself depends on biological, specifically sublinguistic, conditions;[38] but these are not our present concern. The issue is rather one of accommodating the peculiarly "structured openness" of language, its bearing on what amounts to a remarkable division of (linguistic) labor among apt speakers, and its characteristic form of intentionality. To pursue these matters is, effectively, to pursue additions to the tally just offered, features that catch up not merely the formal aspects of discourse about, but also the substantive nature of, institutions. Intentionality, so seen, cannot fail to have *some* ontological import, since language is not merely a formal tool for representing aspects of human life: human existence *is* cultural, lingual, institutional.[39] We are, therefore, attempting to define *human nature*.

In fact, these substantive features of human linguistic practices are really different aspects of one and the same phenomenon. Human persons, agents, selves – on any analysis – develop, appear, emerge, manifest themselves in and only in the acquisition of a natural language, by means of which they become the *lingually* apt creatures that they are (that is, both linguistically apt and apt in nonlinguistic ways not separable from acquiring linguistic aptitude). In this sense, persons are culturally emergent with respect to the members of a certain gifted biological species.[40] In speaking of human nature, we regularly equivocate or at least shift between the biological and the cultural. What is at stake here is the nonreducibility of the cultural to the biological – a matter distinct from the fact that cultural (or institutional) predicates cannot be reduced to mental or psychological predicates even *where the latter are construed in linguistic or similar ways.* That is the reason, of course, why the intentionality of individual acts and individual mental states cannot capture the intentionality of the cultural, the societal, the institutional, the historical, the stylistic or the like. If human nature is lingual (in the sense intended), then (perhaps not trivially) the cultural world is distinguished by its possessing or exhibiting *lingual significance.* All aspects of distinctly human life exhibit or may acquire "significance" in a sense that cannot be restricted to the intentional life of psychologically apt individuals: both because such significance may be ascribed primarily through societal means by reference to institutional structures, and because whatever might justify intentional ascriptions in psychological contexts is itself already so characterized always and only by reference to such structures – to what (at the risk of misunderstanding) Husserl had designated the life-world. Alternatively put, we have here merely reformulated the famous puzzle of philosophical hermeneutics, the symbiosis of psychological and societal attributes.[41]

To ascribe lingual significance to the entire range of cultural phenomena is, effectively, to ascribe intensionality to such phenomena: everything human (in the sense of everything cultural) has assignable linguistic meaning or has some sort of import, meaning, semiotic function or the like that can be aptly indicated, if not actually paraphrased, linguistically (as, a flag "symbolizing" a political state, tears "expressing" one's sorrow, a portrait "representing" Wellington, *Alice in Wonderland*'s "presenting" a certain imaginable world, Lavoisier's experiment with mercury "signifying" the combustible nature of oxygen, rain clouds "meaning" impending rain.)[42]

This is an extraordinarily strategic and pivotal advance. To glimpse – but only to glimpse – its importance, consider the following. To admit intensionality in this sense is not yet to claim that a thoroughgoing extensionalism is untenable; but extensionalism then must face difficulties of two sorts: first, the problem of reducing the lingual or linguistic to physical phenomena that, on familiar assumptions, may be treated in entirely extensional terms; and second, the problem of treating the intensional within the lingual or the linguistic (not reduced in the first sense) extensionally.[43] These are serious difficulties for all forms of the unity-of-science program, since, on the thesis here favored, cultural phenomena and attributes are actual phenomena and attributes.[44] Of course, there are many other important philosophical issues linked to this matter of the intensional that it would be inappropriate to pursue here.

If we grant two further concessions – one, perhaps completely uncontested, the other, admittedly controversial but certainly reasonable – our task will be nearly at an end. The first draws attention to the fact that every apt speaker must have internalized a portion of the admissible practices of a given natural language in such a way as to accommodate improvisation, as well as to the fact that no apt speaker can have internalized the entire array of the admissible practices of any such language. This is to be taken to apply to the use of vocabulary, the use of grammar, the use of sentences and other linguistic utterances in the changing and even novel contexts of human existence and experience. (The very notion of an intelligibly changing, even new, range of experience signifies that the world thus encountered may itself be lingually identified in improvisational ways.) On the argument, what is true of language in this respect is true, *ceteris paribus*, of all other lingual phenomena – that is, of the entire institutional life of man. In effect, then, societal existence involves a seemingly paradoxical division of labor, whether linguistic or lingual. For institutions are collective or societal attributes ascribed to individuals and aggregates of individuals (*methodological individualism*), but the division of labor regarding whatever is psychologically internalized with respect to pertinent practices *cannot by itself account for the smooth functioning of linguistic or lingual aptitudes*. There is only one possible explanation of such functioning under that condition: the members of a linguistic or lingual community must, in acquiring the requisite skills, acquire the skill

(a) to improvise from their own internalized "segment" harmoniously with the improvisations of others;
(b) to interpret the lingually significant acts, states, and products of others in ways coherent with what they understand of the regularities governing or guiding their own segment of what they take to be a common practice; and
(c) to be able, also interpretively, to form a consensus, at least implicitly, by which to acknowledge particular acts or improvisations *as* admissible instances of common practices and as having therefore assignable significance in accord with those practices.

Broadly speaking, to concede (a)–(c) is to construe institutions or institutional life as inherently *hermeneutic*.

How the hermeneutic dimension may be described with respect to logical closure – hence, to the puzzles of extensionalism – will depend in large part on the second concession we must consider. *If* language is not a strongly autonomous phenomenon – that is, if the structural regularities of language are, in principle, a function of man's shifting, variable, improvisationally open experience of the world (in particular, if linguistic – and lingual – structures are a function of *a radically open, multiply and divergently interpretable history*) – then no fixed structure can be assigned to natural languages and no totalized system of possible structures can be formulated for natural languages (as in the manner of the so-called structuralists or of Chomsky[45]). In effect, then, it is just these historically contingent institutional regularities (resting in whatever way they do on deeper biological regularities) that mark the culturally pluralized "nature" (and the projectible range of the "nature") of man himself. We may, then, readily add to our tally regarding the institutional the following:

(iii)   Institutional attributes are essentially intensional.

(iv)   Ascriptions of institutional attributes are inherently hermeneutic and consensual (paradigmatically, reflexive[46]).

We may also add the following final item, which, however, will need to be explained a little more fully.

(v)   Institutions are "habituating."

By "habituating," we may understand an intensional attribute that accommodates the provisional internalizing of lingual regularities, possibly focused on salient but contingent exemplars linked to an individual's distinctive acquisition of particular practices; an improvisational capacity to extend practices along the lines sketched above; a capacity gradually to replace the guiding exemplars of one's own internalized segment of a common practice (for instance, through the force of history and shifting experience); and a capacity fo facilitate the acquisition of such changed practices, consensually altered and confirmed, by the members of the next generation.[47] The idea is that what is internalized by apt individuals is determin*ate* up to a point, in being linked to relatively determinate exemplars; and determin*able* in a certain sense, in being extendable in open, plural, and improvisational ways that affect the regularities and our perception of the regularities thus instantiated. This, of course, is precisely what Wittgenstein stresses – despite the absence in his reflections of a developed account of the historical nature of linguistic and lingual practices – in his insistence that rules and practices are not hierarchically ordered with respect to one another and that the "rules" of a practice are improvisationally projectible only from within a practice mastered: which is itself, as already remarked, "an ungrounded way of acting."

Viewed in a larger way, human animals evolve or emerge as persons or fully formed selves – *culturally* – by internalizing some distributed part of an aggregated aptness for thinking and behaving in accord with the contingently salient truths and practices of a given society. Doubtless, they have, as clever animals, a native capacity for such internalization. So the question of the biological grounding of self and linguistic ability is a most strategic one – as, say, between theorists as opposed as Merleau-Ponty and Lev Vygotsky, or Chomsky and Hilary Putnam. But enriching the intentional complexity and power of the biological poses puzzles primarily for the alleged adequacy of physicalism and extensionalism to account for the cultural; whereas conceding the decisiveness of cultural formation confirms the sense in which the unity of self or person is a societally defined achievement paralleling the achievement of lingual aptitude. For the latter thesis admits an achievement capable of accommodating both transient and profound anomalies of self-awareness and diachronic and synchronic divergences within and between societies.

It is in this sense, in the sense of conditions (iii)–(iv), that the distinctive intentionality of institutions may be said to be *"rule-like"* or *"regulating."* Institutions provide the generic form of all cultural intentionality (leaving animals aside), because all acts, states, products and the like must exhibit "habituating" features insofar as they exhibit whatever forms of intentionality belong to them; and because institutional practices form the nonreducible matrix within which alone pertinent acts and the rest obtain. Human agents intend, we may say, the intensional pertinence of whatever they do, *vis-à-vis* the institutional matrix of their own culture: they cannot do otherwise, although no one can intend merely to act congruently with that matrix. The lingual significance of whatever we do or make is a specification of some intensional attribute accessible within that matrix; the matrix is only the habituating structure (hence, largely determin*able*) within which the other is determinately done or made; and doing and making, as well as acts done and things made, are similarly subject to the interpretive tolerance of that habituating structure. To offer one skewed but particularly compelling illustration, in an interview between Michel Foucault and some "Maoist militants" regarding the possibility of setting up, in France, a "people's court to judge the police" (June 1971), Foucault offered the following: "my hypothesis is not so much that the court is the natural expression of popular justice, but rather that its historical function is to ensnare it, to control it and to strangle it, *by re-inscribing it within institutions which are typical of a state apparatus.*" He adds, "[the] role [of 'proletarianization'] is to force the people to accept their status as proletarians and the conditions for the exploitation of the proletarian."[48] There you have a touchstone for testing the adequacy of any theory of institutions; for it must be able to accommodate, as intelligible, gathering incommensurabilities.

In defining the intentionality of institutions, therefore, we have also sketched a good part of a definition of human nature.

Notes

1   W. V. Quine, *Word and Object* (Cambridge, Mass.: MIT Press, 1960), p. 219.
2   Ibid., p. 221.
3   P. F. Strawson, *Skepticism and Naturalism: Some Varieties* (New York: Columbia University Press, 1985), ch. 4.
4   The sense of Davidson's program is perhaps reasonably explicit, in this regard, in Donald Davidson, "Semantics for Natural Languages," *Inquiries into Truth and Interpretation* (Oxford: Clarendon Press, 1984).
5   See A. N. Prior, "Intentional Attitudes and Relations," in P. T. Geach and Antony Kenny (eds), *Objects of Thought* (Oxford: Clarendon Press, 1971).
6   On the characterization of Husserl's conception of intentionality, I have been particularly helped – with regard to the differences among Brentano, Meinong, and Husserl, as well as between the "traditional" reading of Husserl and "the new Fregean interpretation" of Husserl – by Dagfinn Føllesdal, "Husserl's Notion of Noema," *Journal of Philosophy*, LXVI (1969); J. N. Mohanty, *Husserl and Frege* (Bloomington: Indiana University Press, 1982), especially ch. 3; and Izchak Miller, *Husserl, Perception, and Temporal Awareness* (Cambridge, Mass.: MIT Press, 1984), especially ch. 1. See also, Aron Gurwitsch, *Studies in Phenomenology and Psychology* (Evanston, Ill.: Northwestern University Press, 1966).
7   Franz Brentano, "The Distinction between Mental and Physical Phenomena," *Psychology from an Empirical Standpoint*, ed. Oskar Kraus, English edn trs. Antos C. Rancurello et al., ed. Linda L. McAlister (London: Routledge and Kegan Paul, 1973). See also Roderick M. Chisholm, "Brentano's Descriptive Psychology," in Linda L. McAlister (ed.), *The Philosophy of Brentano* (Atlantic Highlands, NJ: Humanities Press, 1976); and Roderick M. Chisholm, *Perceiving: A Philosophical Study* (Ithaca, NY: Cornell University Press, 1957), ch. 11.

8   Mohanty, *Husserl and Frege*, pp. 75–6.
9   Miller, *Husserl, Perception, and Temporal Awareness*, pp. 23, 29, 31, 39.
10 Karl R. Popper, *The Poverty of Historicism*, 3rd edn (New York: Harper and Row, 1961), p. 17.
11 Ibid., p. 157. See also Karl R. Popper, *The Open Society and its Enemies*, rev. edn (Princeton, NJ: Princeton University Press, 1950), ch. 14.
12 Popper, *The Open Society and its Enemies*, p. 285.
13 Ibid., p. 286.
14 Popper, *The Poverty of Historicism*, p. 136.
15 Edmund Husserl, *Cartesian Meditations*, trs. Dorion Cairns (The Hague: Martinus Nijhoff, 1960), Fifth Meditation, pp. 93–4. See also Edmund Husserl, *Formal and Transcendental Logic*, Dorion Cairns (The Hague: Martinus Nijhoff, 1969), p. 257.
16 Edmund Husserl, *The Crisis of European Sciences and Transcendental Phenomenology*, trs. David Carr (Evanston, Ill.: Northwestern University Press, 1970), pp. 141, 139 (italics added). In a sense, these mixed remarks both confirm and preempt the criticism of Husserl levelled in Jacques Derrida, *Speech and Phenomena and Other Essays on Husserl's Theory of Signs*, trs. David B. Allison (Evanston, Ill.: Northwestern University Press, 1973), ch. 5. They are also of course the key regarding the focus of Merleau-Ponty's phenomenology, and of the importance of the newer Nietzschean-like programs of Foucault and Derrida.
17 Husserl, *The Crisis of European Sciences*, p. 103.
18 Ibid., p. 69.
19 See Joseph Margolis, "Forms of Life: Wittgenstein's Template for Psychological Development," in Michael Chapman and Roger Dixon (eds), *Wittgenstein and Developmental Psychology* (forthcoming).
20 Ludwig Wittgenstein, *On Certainty*, tr. Denis Paul and G. E. M. Anscombe, ed. G. E. M. Anscombe and G. H. von Wright (Oxford: Basil Blackwell, 1969), section 83. See also Ludwig Wittgenstein, *Philosophical Investigations*, tr. G. E. M. Anscombe (New York: Macmillan, 1953).
21 Wittgenstein, *On Certainty*, sections 105, 95, 110 (italics added).
22 Ludwig Wittgenstein, *Philosophical Remarks*, ed. Rush Rhees, tr. Raymond Hargreaves and Roger White (Oxford: Basil Blackwell, 1975), sections 20–1 (p. 63).
23 Ibid.
24 Ibid., section 24 (p. 65).
25 See Wittgenstein, *Philosophical Investigations*, sections 81–6, 186, 198–202, 219, 225, 239–41.
26 See ch. 6, above.
27 P. F. Strawson, "On Referring," *Mind*, LIX (1950); J. L. Austin, *How to Do Things with Words* (Oxford: Clarendon Press, 1962).
28 John R. Searle, *Speech Acts* (Cambridge: Cambridge University Press, 1969), especially ch. 2; H. P. Grice, "Logic and Conversation," in Peter Cole and Jerry L. Morgan (eds), *Syntax and Semantics*, vol. 3 (New York: Academic Press, 1975).

29 John R. Searle, *Intentionality* (Cambridge: Cambridge University Press, 1983), especially ch. 1.
30 See P. T. Geach, *Mental Acts* (London: Routledge and Kegan Paul, n.d.).
31 Searle, *Intentionality*, p. 19.
32 Ibid., pp. 18, 22 (italics added); see also p. 141.
33 Ibid., p. 21.
34 Ibid., p. 21.
35 Ibid., 143–4 (italics added).
36 See for instance Maurice Merleau-Ponty, *The Structure of Behavior*, tr. Alden L. Fisher (Boston, Mass.: Beacon Press, 1963), and *The Visible and the Invisible*, tr. Alfonso Lingis, ed. Claude Lefort (Evanston, Ill.: Northwestern University Press, 1968).
37 See John B. Thompson's "Notes on Editing and Translating," in Paul Ricoeur, *Hermeneutics and the Human Sciences*, tr. and ed. John B. Thompson (Cambridge: Cambridge University Press, 1981), p. 28. I am taking a deliberate liberty, here, with Gadamer's term *Sprachlichkeit* (or the adjective formed from it that Ricoeur translates as *langagier* and Thompson as "lingual"). I use "lingual" to signify cultural phenomena that presuppose linguistic aptitude but that do not necessarily involve actual speech (the dance, for instance); that, because of that dependence, are capable of being construed semiotically, symbolically, representationally, intensionally.
38 A number of the pertinent disputes may be gathered by comparing, for, instance, Noam Chomsky, *Rules and Representations* (New York: Columbia University Press, 1980); Jerome Bruner, *Beyond the Information Given: Studies in the Psychology of Knowing*, ed. Jeremy M. Anglin (New York: W. W. Norton, 1973), and *Child's Talk: Learning to Use Language* (New York: W. W. Norton, 1983); L. S. Vygotsky, *Mind in Society: The Development of Higher Psychological Processes*, ed. Michael Cole et al. (Cambridge, Mass.: Harvard University Press, 1978). See also Joseph Margolis, *Culture and Cultural Entities* (Dordrecht: D. Reidel, 1984).
39 This goes decidedly contrary to Searle's view, for Searle attempts to disengage the logical status of the intentional from questions regarding its ontological status; see for instance *Intentionality*, pp. 14, 16. But, of course, *what* "logical" properties the intentional exhibits is inseparable from *what* are (ontologically) conceded to be intentional phenomena.
40 For the development of this theme, see Margolis, *Culture and Cultural Entities*, ch. 1; also Joseph Margolis, *Art and Philosophy* (Atlantic Highlands, NJ: Humanities Press, 1980), chs 2–3.
41 See Hans-Georg Gadamer, *Truth and Method*, tr. Garrett Barden and John Cumming from 2nd German edn (New York: Seabury Press, 1975), pt II.
42 See Joseph Margolis, "Puzzles of Pictorial Representation," in Margolis (ed.), *Philosophy Looks at the Arts*, 3rd edn (Philadelphia: Temple University Press, 1987). The rain-cloud example is meant to counter H. P. Grice, "Meaning," *Philosophical Review*, LXVI (1957).
43 Donald Davidson is probably the foremost advocate of both forms of

reduction. See Davidson, *Inquiries into Truth and Interpretation*, and *Essays on Actions and Events* (Oxford: Clarendon Press, 1980).

44 See ch. 10, below.

45 See Joseph Margolis, "'The savage mind totalizes,'" *Man and World*, XVII (1984).

46 This is perhaps the central point stressed, and most misunderstood, in Peter Winch, *The Idea of a Social Science* (London: Routledge and Kegan Paul, 1958).

47 I find Pierre Bourdieu's notion of "habitus" to be particularly felicitous in this regard – and to combine (at least implicitly) the best features of Wittgenstein's "forms of life" and Husserl's "life-world." See Pierre Bourdieu, *Outline of a Theory of Practice*, tr. Richard Nice (Cambridge: Cambridge University Press, 1977).

48 Michel Foucault, "On Popular Justice: A Discussion with Maoists," *Power/ Knowledge: Selected Interviews and Other Writings 1972–1977*, tr. Colin Gordon et al., ed. Colin Gordon (New York: Pantheon, 1980), pp. 1, 14 (italics added).

Part Two

Science without Unity

8

Science as a Human Undertaking

We analyze intentionality with a sense that endorsing its puzzles is tantamount to challenging the hegemony of the unity of science; and we analyze the conceptual features of the human sciences with a sense of needing to reconcile the natural and the cultural. Technology straddles both concerns and is plausibly linked to the fortunes of both the unity and bifurcation models of science. So it affords a convenient economy as well as an important clue for any serious effort to picture the relationship between the natural and the human sciences.

I

Insistence on the human standing of technology and the physical sciences seems like a wasteful reminder. There is no technology or science that is not a human undertaking – so the emphasis is entirely superfluous; technologies are essentially characterized in terms of facilitating specifically human interests – so the emphasis is quite redundant; and the validity of the physical sciences is thought to mark the fortunate discovery of the law-like regularities of an independent natural world – so the emphasis is merely intrusive. Insistence on their human aspect would indeed be no more than a conceptual annoyance if these familiar adjustments were conceded to identify the executive themes of technology and science. Of course, we might query why these pursuits are uniquely human. We might ask ourselves how human nature is affected by the technology it spawns. And we might challenge the glib assurance that science simply discovers the structures of an independent world. Here, we would be bound to accommodate adjustments in terms of the symbiosis of self and world, of the indissoluble realist and idealist strands of our accomplishments, and of the incompletely penetrable and shifting preconditions of the life-world within which these other regularities are provisionally identified. Such exercises are hardly pointless. But they can be made to seem noticeably tired and tiresome, if they have no more to say about technology and science than can be drawn out primarily by merely turning to examine the human condition itself. We should then only have tricked ourselves by an inflated rhetoric. Are there, however, some genuinely pointed findings about the principal features of science – about, say, natural laws and causality and explanation – that we fail to grasp when we neglect the implications of insisting on the human face of science? Are there some genuinely compelling distinctions regarding such matters that reflecting on the differences between human technology and the canonical picture of the physical sciences might reveal; or that, in becoming apparent, might encourage or force a sizable, further revision in the canonical picture itself? Here, surely, is the sensible, small locus for any serious insistence on the theme of the human structure of human science and technology.

The concessions so far slimly eked out are already hardly negligible. But they are as much illuminated by the pointed study of the technological and the scientific as ever they illuminate science and technology. At any rate, it is just the reciprocal benefit that is the promised contribution of our present effort. We have in fact managed to insinuate a double constraint on the review intended, one – so we may claim without demonstration – that represents the strongest convergent themes of the entire Western philosophical literature centered on the theory of the sciences: namely, first, that *all* foundational, privileged, cognitively transparent, correspondentist, objectivist, logocentric, essentialist, apodictic, or totalized access to the structures of reality, phenomena, whatever is "given" in epistemically pertinent ways, is taken to be an untenable, theoretically indefensible presumption;[1] second, that, as a result, there are *no* discernibly different levels of cognitive *profondeur*, variably placed approximations to the "originary" sources of certitude or objectivity or necessary structures affecting our best theories of reality. The convergence of these two themes, though it may not be immediately obvious, is (compendiously put) the point of intersection of the most promising forms of naturalism, phenomenology, and deconstruction. Each of these orientations harbors its own familiar, excessive claims; but the potential benefit of their intersection (even of their union) rests in good part with the issues we mean to pursue here. If we adopt these two themes, however, we have effectively refused to admit a theoretical privilege to Kantian-like, Husserlian-like, or Nietzschean-like reflections regarding the legitimacy of contributions motivated chiefly by the concerns of any such sources – supposing, with all candor, that the best philosophical vision possible in our own time requires a reconciliation of the master themes (without privilege to any) favored in Kant, Husserl, and Nietzsche.[2] The resultant form of speculation cannot but be unidimensional (however complex) – horizontal, so to say, in terms of the interconnections among these three perspectives.

The upshot may be characterized in naturalistic terms, if we mean by that that no such speculation could be remotely pertinent if not reflexively addressed to whatever is most salient, least likely to be dismissed or denied or deformed, within the consensual experience of an actual human community. It may be characterized in phenomenological terms, if we mean by that that such reflection disallows any conceptual fixities regarding the putatively factual order of the first and that we should nevertheless attempt to discern – within the historicized conditions of societal life that we cannot hope to penetrate except in accord with (to an unknown extent enabled by) the very aptitudes social history makes possible – whatever conceptual strictures or necessities bear on the referential and the predicative saliences of our factual discourse. And it may be characterized in deconstructive terms, if we mean by that that no conceptual schema (organizing factual claims) can be shown to be timelessly fixed or inviolate or comprehensively adequate for all possible evolving experience and that we have and in principle can have no rational clue by which to assess the sense and extent to which our present schemes are approximations or partial fragments of any ideally totalized conceptual vision. There is no doubt that these rather sanguine abstractions drawn from the entire tradition of Western philosophy may be distributively assigned to such authors as Kant, Husserl, and Foucault (without presuming to rank their importance); but it is equally clear that the standard abuses of the three movements identified are also assignable to these selfsame authors. Our objective, however, is to exploit their better natures (according to our lights) rather than to locate them with perfect justice within the historical record.

If, then, we agree to treat the naturalistic, the phenomenological, and the deconstructive as no more than inherently one-sided concerns (when isolated from their fellows) within any minimally adequate reflection on the condition of man and on his relation to reality, we may well reconsider otherwise partial, potentially distorting views of technology or science generated more or less exclusively from one or another of these three orientations – and gain thereby a surprisingly fresh picture of how an appreciation of what technology confirms about the human condition bears in a decisive way on the adequacy of (what we are here calling) the canonical theory of the sciences. That effort promises in fact a profound recovery of the sense in which science is a human undertaking.

Once again, the point of our intended review is to isolate those respects in which such notions as law-like regularity, causality, and explanation in the sciences – notably, in theorizing about the physical sciences – invite a contest of options usually neglected or dismissed or muffled in the literature favoring the canon, precisely because, there, the full role of the human inquirer is clearly deemed marginal for any explication of the powers of science (though it could not of course be marginal to the exercise of those powers). One might say, in the spirit of that canon: man is the only scientist, but science is a well-formed practice adjusted to the way the world is – a practice that man happens, contingently, not everywhere in the world, to have fortunately hit upon. *If*, however, science were constitutively inseparable from technology, *if* technology were no more than a selectively focused study of human nature and human aptitudes (including, inseparably, the aptitude for science), then the properties of science itself could not be more than astute projections or proposals of reasoned idealizations of how man masters and comes to understand his world (and himself in his shifting life-world). The contrast thus formulated still seems bland enough. But the question arises (now more insistently), is there a fundamental difference between a theory of science essentially linked to the concept of technology (here barely sketched) and the theory of science favored by the received canon?

II

Of course, there is no actual canon. There certainly have been attempts at formulating a canon, which in a reasonably fair sense may be variously characterized as a thesis bounded, for our own century, by such early statements of Rudolf Carnap's as are provided in *The Unity of Science* and in "Psychology in Physical Language," in Carnap's brief autobiographical appraisal of the period of the Vienna Circle and of his work in the United States, in the extremely balanced overview of the unity movement provided by Carl Hempel and variously refined and skewed by different hands, notably by Paul Oppenheim and Hilary Putnam – all clearly counted among the principal contributors to a developing canon.[3] There can be no question that the unity movement accommodates a variety of views that are incompatible with one another on a number of issues. There is no possibility of formulating a central set of doctrines to which all loyal adherents subscribe. So, for example, J. J. C. Smart would deny emergent laws and emergent properties of any kind – since, as Smart explains, "animals and men are [no more than] very complicated mechanisms."[4] Oppenheim and Putnam are, on the other hand, quite open to the prospect of emergent (empirical) laws and emergent properties. Hence, the partisans of unity are divided with regard to reductionisms of a very strong physicalist sort and views that tolerate various forms of emergence. Carnap himself was increasingly inclined to acknowledge emergent laws, though this was not true at first.

In developing his argument, Smart is quite explicit about genuine laws of nature. They have "one very important feature," he maintains: "These laws are universal in that it is supposed that they apply everywhere in space and time, and they can be expressed in perfectly general terms without making use of proper names or of tacit reference to proper names." The trouble is that biology "does not contain any laws in the strict sense," the sense just given.[5] The would-be laws of biology implicitly refer to the earth. Smart's view is, of course, hopelessly sanguine *if*, as it is currently fashionable to hold, science and the history of science are indissoluble. For then, on grounds quite neutral as between naturalism, phenomenology, and deconstruction, it would be impossible to obviate "tacit reference to proper names" (in Smart's sense); or, alternatively put, the use of "perfectly general terms" would be no more than an idealized expectation of the law-like standing of earthbound generalizations – idealizations never actually confirmed as law-like, possibly even incapable in principle of being explicitly formulated (if the Leibnizian theme that indiscernibles are identical is false, if numerically different referents cannot always be distinguished in real-time terms, or if reference to inquiring agents and societies affects in a constitutive way the structure of our would-be laws).[6]

Here we come as close as possible to an essential requirement of the canon: the laws of nature must be *invariant* regularities. This way of putting matters is meant to remain neutral on the question of counterfactual and noncounterfactual readings of laws, on regularity conceptions of laws as opposed to views invoking nomological necessity, and on disputes between deterministic and statistical interpretations of laws – all of which obviously divide the loyalties of unity advocates. Wesley Salmon, for instance, admits counterfactuals in science, but Bas van Fraassen does not.[7] Hempel had originally not accommodated statistical laws but subsequently made provision for them when he was able to reconcile their role with what he took to be the logical form of scientific explanations. Salmon offers what he terms an "ontic conception" of natural laws, which he puts in the following way (reviewing alternative readings of Laplacian explanation, although he himself rejects determinism): "*To explain an event is to exhibit it as occupying its* (nomologically necessary) *place in the discernible patterns of the world.*"[8] Convergence on the theme of invariance is impressively widespread.

One may therefore be reasonably disposed toward glossing the notion of the invariance of laws along the lines recently offered by Adolf Grünbaum – that, for instance, "explanations in physics are generically based on context-free, ahistorical laws" – except that Grünbaum actually offers the formula as an incorrect and utterly misguided view of physical laws fostered by both Jürgen Habermas and Hans-Georg Gadamer, who (on Grünbaum's reading) exaggerate the theme in order to bifurcate the natural and the human sciences.[9] The quarrel involving Grünbaum, Habermas, and Gadamer is worth tarrying over, because there is a great muddle involving all three and because its resolution actually bears in a decisive way on the contested views of laws and causality that were to be drawn (as promised) from contrasting the import of technology and the canonical view of the sciences. Let it be said at once that Carl Hempel had clearly espoused the doctrine that "there is no difference [in attempting the impossible task of giving a '*complete explanation* of an individual event'] between history and the natural sciences: both can give an account of their subject-matter only in terms of general concepts, and history can 'grasp the unique individuality' of its objects of study no more and no less than can physics or chemistry." When he wrote this, Hempel still held to the view that a "general law" is "a statement of universal conditional form which is capable of being confirmed or disconfirmed by suitable empirical findings."[10] But, in conceding statistical laws, Hempel intended no concessions at all regarding a possible disjunction between historical and physical laws and explanations – which of course would have been inimical to all versions of the unity program.

In that sense, Grünbaum's insistence is a little surprising: not because Grünbaum tolerates the bifurcation of the sciences – his own close study of the methodology of psychoanalysis is obviously designed to strengthen just the opposite view; but rather because he apparently believes he can reconcile the *invariance* or *nomological necessity* of scientific laws *with* the denial that they are "context-free" and "ahistorical."[11] But of course those claims, applied to the human sciences, actually formed the very basis on which Habermas and Gadamer had expected to disjoin the natural and the human sciences. It may be fair to say that Habermas and Gadamer are not altogether clear about the nature of the physical sciences; but it is equally fair to say that Grünbaum has somehow lost the thread of the pertinently challenging issue they raise (however weakly) and has, in addition, conflated two entirely different senses of "history" and "context." On just such disputes as these depends, we may say, the entire contrast intended between the import of technology and the canonical view of science.

Our question has suddenly become very complicated. For one thing, *if*, following Grünbaum, physical laws are not context-free or ahistorical, then it is difficult (if not impossible) to see how they can be genuinely universal or invariant. *If*, for another, following Smart, genuine laws make no use of proper names (or surrogate indexicals), then it is impossible to reconcile Smart's and Grünbaum's views of the logical properties of laws, and an intuitively necessary condition for the invariance of laws is placed at risk. *If* the "universal" scope of genuine laws does not require a universal form, then, on our two assumptions (the denial of cognitive transparency and the denial of a graded cognitive privilege favoring either naturalism, phenomenology, or deconstruction), it is difficult (if not impossible) to provide a theoretical basis for Salmon's and Grünbaum's insistence on nomological invariance. *If*, pursuing an additional complication, the invariance of laws is itself a posit or projection, within the historical context of an actual, practicing science, then the difference between the "canonical" view of science (which reduces technology to an application, in some sense, of the independent achievements of science) and a technologized view *of* science (which construes science as itself a disciplined idealization of man's interventions in and within nature) cannot fail to be radically diminished if not altogether erased. Finally, *if* that difference is erased, then the physical (or "natural") sciences cannot serve as an independent paradigm for the human (or "cultural") sciences, and their own rigor would be judged to be a function of whatever rigor could rightly be assigned the latter sciences. Differences among the different disciplines would remain of course, and they might reasonably support the "bifurcation" of the sciences. But the decisive claim of nomological invariance would clearly have been systematically weakened (or altogether lost). In fact, it would then prove difficult (if not impossible) to deny that the human sciences (perhaps even the natural sciences) were under no *conceptual* obligation to admit that the phenomena they studied fell under invariant laws. They might even consider the option of treating the "canonical" explanation of such phenomena under invariant laws as entailing the use of a (useful) fiction.

To return: Grünbaum is admirably explicit in his objections against Habermas and Gadamer. He introduces an example from classical electrodynamics, which yields the finding: "at ANY ONE INSTANT *t*, the electric and magnetic fields produced throughout infinite space by a charge moving with arbitrary acceleration depend on its own PARTICULAR ENTIRE INFINITE PAST KINEMATIC HISTORY!"[12] Grünbaum draws the general conclusion,

Though the individual histories of each of two or more charged particles can be very different indeed, the electrodynamic laws accommodate these differences while remaining general. The generality derives from the *form* of the lawlike functional dependencies of the electric and magnetic field intensities on the earlier accelerations, velocities, and positions of the field-producing charge. But the latter's individual history consists of the infinite temporal series of the particular values of these kinematic attributes (variables)

and then applies it specifically against Habermas and Gadamer:

As against Habermas, I submit that these electrodynamic laws exhibit context-dependence with a vengeance by making the field produced by a charge for any one time dependent on the particular infinite past history of the charge. And to the detriment of Gadamer, these laws are based on replicable experiments but resoundingly belie his thesis that "no place can be left for the historicality of experience in science."[13]

The point of misunderstanding is simplicity itself. Habermas's and Gadamer's notions of "history" and "context-dependence" entail, whatever else may be said of their views, the denial of the kind of invariance *strict (physical) laws exhibit* (on the canonical view) – *when applied* to the regularities of human history. They do not deny the invariance of physical laws. Grünbaum is quite misleading, therefore, when he insists that, *if* Habermas's argument "were legitimate, it could *likewise* serve to establish the following absurdity: The elementary law of thermal elongation in physics [say, regarding the expansion and contraction of metals under changing temperatures] does not exhibit the nomic invariance of the causality of nature after all. ..."[14] Habermas had intended (indeed, explicitly so stated) that there is a fundamental difference (following Hegel) between the "causality of fate" and "the causality of nature."[15] His argument may or may not work. But it is clear that Habermas intends to bifurcate the notion of causality (*a fortiori*, the sciences themselves) – dividing causality between the natural and the human sciences. It is a further irony that Grünbaum himself insists *on* the invariance of the laws of nature, just when he also insists on the historical and context-dependent features of physical phenomena. The fact is that, on Grünbaum's view (which of course endorses the unity canon), the historical and context-dependent features in question *instantiate the invariance of physical laws* – which is just the reverse of Habermas's express claim. They must also instantiate therefore – on his argument regarding the would-be science of psychoanalysis – the *invariance* of any genuine psychoanalytic laws (if there be any). So it is clear that there is a profound misreading of the bifurcationist's thesis (however difficult it may have been to avoid) in Grünbaum's treatment of the temporal ordering of physical events (and the dependence of certain causal effects on such ordering) *as* conceptually equivalent to the notions of history and context-dependence intended by Habermas and Gadamer. To have allowed the difference would have been to acknowledge the irrelevance of any of his analogies between cases in the physical sciences and cases in the human sciences (in particular, cases in Freudian psychoanalysis). So the champions and antagonists of the unity model are not really brought together in Grünbaum's account, though that is the apparent purpose of Grünbaum's extended review of hermeneutic conceptions of the human sciences.

Now, it must be admitted that, for his part, Habermas bungles the required contrast – inviting the attack he receives – by attempting to distinguish between "invariance of nature" and "invariance of life history." Here is what Habermas says, interpreting Freud's remarks regarding the *causally* pertinent pathology of language and behavior:

Following Hegel we can call this the causality of fate, in contrast to the causality of nature. For the causal connection between the original scene, defense, and symptom is not anchored in the invariance of nature according to natural laws but only in the *spontaneously generated invariance of life history, represented by the repetition compulsion, which can nevertheless be dissolved by the power of reflection.*[16]

It must be borne in mind that Habermas has remarked just a moment before that psychoanalysis "achieves more than a mere treatment of symptoms, because it certainly does grasp causal connections, although not at the level of physical events."[17] And it must be conceded that Grünbaum quite rightly perceives the *radical* nature of this reading of the methodological import of Freud's work – whether or not it correctly represents Freud (who after all subscribed to the main tenets of nineteenth-century positivism[18]), and whether or not the repetition compulsion may rightly be said to function in the manner assigned. Grünbaum offers an instructive (but argumentatively inappropriate) analogy between Habermas's reasoning and a similar form of reasoning that might be applied to the law of thermal elongation; and he concludes, "In neither case can there be any question at all of 'dissolving' or 'overcoming' a causal connection between an initial condition *I* and an

effect E on the strength of terminating E by a suitable alteration of I."
His point is that the effect actually instantiates and must instantiate the
law in question, and does not and cannot "dissolve" it *by* instantiating
it.[19]

The "causality of fate," it should be said, is rather a nice notion. What
it suggests is that there are contingent, identifiable causal complexes that
function in a sufficiently regular way such that we may cast them in a
form at least suggestive of genuinely law-like regularities – so that we
come to regard them as analogous in behavior and explanatory power
when compared with genuine law-like regularities. There is no logical
or conceptual *need* to regard them *as* actually law-like; there is not even
a need (*pace* Habermas) to regard them as invariant. They merely need
to be regular, salient, reasonably congruent with familiar and effective
sorts of explanation in the human sciences. "Fate" may be a trifle
melodramatic; but we are usually not misled when, say, we speak of
one's being "fated" by his own gullibility – regularly disappointed (in
causally relevant ways) by virtue of the trust he places in others. There
is also no reason whatever for thinking that one could not "dissolve" or
overcome such a tendency by exercising other pertinent aptitudes. There
may be no sense in which thermal elongation exhibits a parallel case;
but that is hardly a reason for thinking that a compulsion neurosis may
not be rather like gullibility (whatever Freud may have thought).
Grünbaum's analogy is hardly Habermas's.

The trouble is that Grünbaum takes proper laws to be *invariant* –
very much in the sense in which Salmon speaks of nomological necessity:
they signify not merely the human formulation of some presumed
regularity in nature, but also that there actually are such regularities in
nature, and that, for *any* domain that presumes to be scientifically
accessible (in context: psychoanalysis), invariant laws (nomological
necessities) will there obtain. But this is precisely what Habermas denies
– and what Grünbaum never actually shows to be true or convincing.

Habermas, however, commits the fatal blunder that *he* introduces
"invariances of life history" meant to parallel "invariances of nature";
so *he* cannot really dispute Grünbaum's objection. The only way – the
only plausible way – to resist Grünbaum's argument is to *deny that
causality must be invariant*: to disjoin, conceptually, *causality* and
nomological necessity. Once that is done, then it becomes entirely
possible to argue (regardless of whether it is true or not) that "dissolving"
that regularity (not a strict invariance) in the life history of a patient –
marked, say, as a repetition compulsion – *could* in principle effect a
psychoanalytic cure. The compulsion, on that reading, would yield a
sense both of the causal syndrome affirmed (separated from the issue of

invariance) *and* of the "replicability" of the causal phenomena adduced (within whatever limits are imposed in the absence of strict invariance).

The remarkable consequence is that there would then be absolutely no principled objection to raise against Habermas's notion that the compulsion neurosis – construed now as a relatively isolable causal syndrome that exhibits a certain sufficient regularity or uniformity or loose or provisional "invariance" (justifying something like psychoanalytic "laws" or "fateful" regularities) – *could* be "dissolved." Of course it could. One has only to think of it as a *habit* or a *character trait* or the like that is not unalterable (analogous, it may be usefully remarked, to what is involved in speaking of social *institutions, traditions, practices, styles* and the like). It is not merely that the compulsion neurosis is a causal factor instantiating a "law-like" regularity (as Grünbaum not unfairly insists); it is also that it embodies "law-like" regularities (if we may favor the term thus) *that are not invariant* in the canonical sense. The proper quarrel concerns the question of whether or not there can be causality without nomologicality, or causality without nomological necessity, or causality without strict invariance; or whether such causality may be found in the human sciences independently of whether it also obtains in the physical sciences; or whether the meaning of "causal law" and of cognate notions is simply equivocal as between the natural and the human sciences.[20] Grünbaum does not address the question at all. For our part, particularly given the notorious difficulty of characterizing a genuine law of nature and the associated difficulty of providing a decisive sense in which would-be laws may be said to be confirmed or *pertinently* strengthened on empirical grounds, we may mark the fact that *there is no known argument that conclusively demonstrates that it would be contradictory (or even without merit) to affirm causal processes and causal effects while denying strict invariance, nomologicality, or nomological necessity.*[21]

There is, therefore, a theoretical option regarding causality and laws that is usually muffled, as already remarked, in the large literature addressed to the nature of science motivated in the spirit of the unity canon. Furthermore, given the suggested convergence between naturalistic, phenomenological, and deconstructive orientations already remarked, it is very reasonable to explore the bifurcation of the sciences – it may even be a logically inescapable finding – if physicalist and similar reductive programs regarding the analysis of the self (or eliminative programs for that matter) collapse or fail to be suitably vindicated. In a word, it may well be that if one accepts the symbiosis of self and world, under historicized conditions that disallow any and all forms of cognitive privilege, one cannot then avoid admitting at the very least that the human sciences cannot be brought to heel in accord with the canonical view. Admitting its constructive effect on the physical sciences as well, it may be that the canon will prove to be profoundly misguided even with regard to the physical and life sciences.

III

The essential pivot of competing intuitions about causality and laws that are drawn from the canonical picture of science and from reflecting on what may be called the "technic" (or technological) experience of man, the immemorial experience and activity of domesticating natural processes for human purposes – largely tacit, largely spontaneously apt, pretheoretical or only partially "entheorized,"[22] distributively successful – rests squarely with the issue of invariance. The Humean conception of regularity or constant conjunction cannot in principle be assigned any discernible or determinate invariance or necessity, except on the basis of prejudice, habit, nonrational disposition or the like. Theorists of science, particularly those committed to the canon (since Hume did not assume its need), have tried one way or another to invest the *phenomenological* regularities of the various sciences (law-like regularities, if you please) with the full import of the *theoretical* (or fundamental or explanatory) laws of those sciences: which is to say, they have tried to invest them with a strict invariance, an invariance applying to all possible instances of the empirically open-ended set of phenomena said to be "governed" by the laws in question, an invariance that just *is* nomological (or natural) necessity.

The difference between the two kinds of laws cannot be formulated in terms of a perceptual/nonperceptual disjunction or a nontheoretical/theoretical disjunction, but depends rather on competing norms of methodological rigor and economy regarding what, in the indissoluble mixing of theory and perception (*and* experiment and technological intervention) would or would not justify pronouncements of the theoretically *implicated* strict invariance of given phenomenological regularities. Those favoring phenomenological laws over theoretical laws, at least as far as inductive, experimental, and observational work is concerned, short of canonical explanations (though not necessarily incompatibly with the requirements of theoretical *explanation*), are not in any way logically obliged to treat "phenomenological laws" *as* nomologically necessary or strictly invariant, or (failing to meet that constraint) as laws only in a Pickwickian sense.[23]

The more-or-less standard view holds that the very validity of treating phenomenological laws as laws depends on *their* derivability *from* fundamental laws – which, then, are taken to have a very strong, antecedently assignable realist character.[24] But it is entirely possible to defend the realism of the "content" of phenomenological laws independently of the presumptive realism of explanatory laws and, therefore, to defend the standing of such laws without a commitment to nomological or natural necessity (the nomological necessity of the fundamental laws). On Nancy Cartwright's view, for instance, "The great explanatory and predictive powers of our theories lies in their fundamental laws. Nevertheless, the *content* of our scientific knowledge is expressed in the phenomenological laws." On the "ultra-realist" view of fundamental laws, the realism of phenomenological laws derives – causally or logically – from the realism of the explanatory laws. This is the usual reason (though it is not logically entailed) why a realism about fundamental laws is often closely linked with a reductionism regarding the world *of* our phenomenological laws (which includes all macroscopic objects, not the least of which are the very scientists who pursue these matters).[25] From the perspective of favoring fundamental laws, on the most generous view, phenomenological laws are said to have realist import because "the fundamental laws say the same things as the phenomenological laws which are explained" – only they say them better, more abstractly and more generally, and they say them in a way that permits the phenomenological laws to serve as "evidence" for the truth and realism of the fundamental ones.[26]

If, however, we disengaged the grounds for the realism *and* nomic status of phenomenological laws from the standing of so-called fundamental laws, then

1   we would not need to presuppose that the two sorts of law say "the same things" (make the same affirmations of fact, have the same realist import);
2   we would secure a natural basis for holding that the validity and realist import of would-be fundamental laws depend instead on the prior validity and realist import of phenomenological laws; *and*
3   we would not need to concede that phenomenological laws either are (on inductive or experimental grounds), or need be, or even can be shown to be, "nomologically necessary."

Once these themes are in place, it becomes reasonably clear that, because of the very process of inductive, experimental, and technologically inventive procedures for intervening in nature, phenomenological regularities depend essentially *on generalizing* (*via* whatever theories we employ) *from discrete, individual cases that are observationally anchored and judged to have the realist import they do* (*on grounds not logically dependent on first being able to establish laws of either the phenomenological or fundamental kind*). It is more than reasonable to suppose that regularities among the candidates for phenomenological laws are confirmable without invoking arguments that justify construing such regularities *as* such laws – which is not to say that they can be confirmed on grounds that are theory-neutral or empirical· in any sense close to would-be empiricist economies. What the argument shows, then, in a particularly perspicuous way is

(i)   that causality and nomologicality are logically and conceptually quite distinct notions; and

(ii) that their theoretical linkage (or the denial that they are so linked) is an artifact of competing visions of realism, of scientific realism in particular, and of the realist import of theoretical explanation.

But to concede these findings is in itself to be hospitable at once to

(a)   notions of *causality not* restricted to the Humean notion of constant conjunction or to the canonical notion of strict invariance or nomological necessity; and

(b)   analyses of *empirical regularity*, underlying any would-be laws, that heavily depend on notions of causality in accord with (a).

To focus the argument in a very spare line: only a grasp of the technic or technological experience of man could possibly supply an alternative account of causality opposed to the Humean and canonical theories.

There is only one promising line of reasoning linking causality in accord with (a), empirical regularities in accord with (b), and a theoretical reliance on the technic experience of man: the exemplars of causality must be or must be drawn from (effective) *human agency*. Agency may well be attenuated in being extended from human exemplars to animals to inanimate forces; and in that declension, the theme of agency may become vestigial and may be replaced (as the history of the theory of science confirms) by stronger and stronger notions of either Humean conjunction or nomological necessity.[27] Also, in admitting agency as a paradigm of causality, it is hardly necessary to disallow competing exemplars – that is, either the Humean or the strict invariance (the natural-necessity) view. The essential point is that there is a natural and entirely viable model of causality that

1 precludes or at least does not require nomological necessity;
2 has strong realist credentials, or has them to the extent that human agents (selves or persons) cannot be ontologically eliminated; and
3 is capable of accommodating, on an adjusted view of the realism of science, whatever viable explanatory practices are assigned science on non-agental readings of causality.

So seen, the presumptive, *realist* import of experimental intervention and technological invention *involving unobservable theoretical entities that are assigned causal efficacy* is conceptually dependent on the interpretation of the very meaning of the deliberate and controlled exercise of human agency itself. On this view, it is a straightforward move (in an evidential respect) to assign a causal role to unobserved theoretical entities when, on a theory, a particular, deliberate human intervention produces changes in *them* which thereupon produce observable changes of a technologically significant or successful sort. To produce for the first time a light bulb that actually works, for example, may be reasonably taken to involve the causal "agency" of pertinent theoretical entities operative in, or as a direct result of, the exercise of human agency. To turn on a light bulb in one's home is compellingly causal in nature – *not* because of any assurance that there is a law-like invariance or nomological necessity covering the phenomenon (though there may be one), but because the very achievement of that technological marvel provides a natural basis for extending our familiar sense of ordinary human agency in routinized particular cases. Agency does not rule out nomologicality, but it does not entail it either. The point is captured in an admirably compelling way in Ian Hacking's formulation:

The vast majority of experimental physicists are realists about some theoretical entities, namely the ones they use. I claim that they cannot help being so. ... Experimenters are often realists about the entities that they *investigate*, but they do not have to be so. Millikan probably had few qualms about the reality of electrons when he set out to measure their charge. But he could have been sceptical about what he would find until he found it. He could even have remained sceptical. Perhaps there is a least unit of electric charge, but there is no particle or object with exactly that unit of charge. Experimenting on an entity does not commit you to believing that it exists. Only *manipulating* an entity, in order to experiment on something else, need do that.[28]

In effect, Hacking offers us a criterion of "ontic commitment" (rather than of reality *sans phrase*) in precisely the same spirit (but he hardly adheres to the same letter) as Quine famously favored.[29] *That* criterion (Quine's as well, if the truth be known) pretty well commits us to the reality of human agents (with whatever properties they exhibit – the power of causal agency in particular) even if they cannot be reductively streamlined in accord with the spirit of the unity canon. (Hacking does not explicitly address the question of entailments between causality and nomologicality; but he is distinctly sympathetic to Cartwright's account of intervening explanatory models and even indicates a reason for tolerating inconsistent laws.[30])

The important point about agency is a dual one: in the first place, persons are persons only in virtue of being effective agents (this is a paradigm notion of what it is to be a person – a notion entirely capable, incidentally, of accommodating difficult moral cases that do not happen to bear on our present issue), and apt persons are normally cognizant of being causally effective *as* agents through their deliberate, particular intervention in the world;[31] secondly, effective agency provides one of the principal grounds, possibly the most important, by which we fix what we take to be salient and real. This, of course, is just the point of Hacking's insistence on "manipulation" – including the manipulation of unperceived theoretical processes and entities. The upshot is that to endorse the agency conception of causality is inevitably to admit a basis for reasonably detecting causal processes *in particular cases*, without having to invoke causal laws at all, *and* without being committed to any entailment between instances of actual causation and their subsumability under laws of any kind – fundamental laws, phenomenological laws because derivable from fundamental laws, or phenomenological laws *tout court*. Donald Davidson, it may be remarked, admits the reasonableness of ascribing causality in individual cases (even when agency, in the human sense, does not obtain and even when no pertinent would-be covering laws are known to obtain); he nevertheless insists – without supporting arguments – that causality *is* nomological, is necessarily nomological (which of course is not equivalent to insisting that natural laws entail a certain natural necessity).[32]

The point of appealing to the effective intervention of human technology in clarifying the import of causality lies precisely here. For effective human agency or human technology – in whatever reasonable sense we concede a pertinent continuity between the most ancient human technology (largely tacit, hardly recognized, essentially pretheoretical) and the most up-to-date inventions and experiments of science – *proceeds largely by perceived similarities between particular cases, or perceived analogies ranging over a small number of cases.* Effective technology and effective agency yield, on reflection, a remarkably strong sense in which realism, real similarity, and causality are irresistibly linked as provisional saliences without ever presupposing or entailing invariance, universal scope, nomologicality, or nomological necessity.

It is the natural accessibility of this account, its coherence and economy, and the difficulty of discounting it, even the difficulty of demonstrating that it cannot accommodate the most sophisticated work of the sciences, that recommends it. It shows quite clearly that, once we prefer the phenomenological to the explanatory, once we key the phenomenological to an agental conception of casuality, we mark the very notion of *uniform, law-like regularities* as a problematic one. We cannot simply convert the informal, analogically stipulated, case-by-case uniformities of technological intervention into nomological (or, even more strenuously, naturally necessary nomological) regularities without a further argument. Such an argument may be ready at hand, but it is not logically entailed or presupposed in any way by (affirming) the reality of causally effective human agency *or* of what, in nature, is pertinently implicated in man's successful technological interventions. There is no clear reason to suppose, *if* the phenomena of human existence cannot be eliminated or reduced to the phenomena of basic physics, that the paradigms of causality suited to the explanatory objectives of a realist physics could, in principle, obviate the agental conception of causality *or* the logical disconnection between causality and nomologicality. Furthermore, once we see matters in this way – first, acknowledging agency as causality, and, second, disconnecting causality and nomologicality – we cannot fail to see that the very legitimacy of the Humean and the natural-necessity readings of causal invariance presupposes an operative procedure for actually detailing the strict, *uniform properties* of things that mere informal similarities (moving from case to case) cannot directly vouchsafe. The upshot is that, *assuming* the reductive programs of all those theorists who would eliminate the "human" or "folk" phenomena from the range of legitimate science (for instance, Sellars or Dennett or Stich or the Churchlands), it begins to *seem* possible to insist on the realism of the invariance reading of causal law. But the admission of the "human" *and* the failure of reductionism (as opposed to eliminationism) places that exclusionary conception of causality and causal law in essential doubt – where it should be anyway. The point of the challenge is not satisfactorily perceived by theorists who regularly appeal to the invariance conception, for instance Davidson and Grünbaum.

Causality, in the agental sense or in a technological sense bound to the agental (rather than keyed at once to the canonical nomological view[33]), has absolutely no need to insist on the strict detection of formulably uniform properties that strict nomologicality requires. A strong *realist* reading of natural laws cannot fail to be *essentialist*, in the sense Karl Popper has forcefully identified (perhaps too passionately in his vendetta against Aristotle):

> By choosing explanations in terms of universal laws of nature, we ... conceive all individual things, and all singular facts, to be subject to these laws. The laws (which in their turn *are* in need of further explanation) thus explain regularities or similarities of individual things or singular facts or events. And these laws are not inherent in the singular things (Nor are they Platonic ideas outside the world.) Laws of nature are conceived, rather, as (conjectural) descriptions of the structural properties of nature – of our world itself.[34]

What Popper's argument shows, however, is that a realist reading of laws (but not of causality) cannot be managed nominalistically;[35] that the conceptual linkage between causality and nomologicality (if preserved) need not be construed in realist or essentialist terms; that strict nomologicality cannot be demonstrated on inductivist grounds; and that there need be no conceptual entailment between causality and nomologicality.[36] The very enterprise of science, read in terms of the canonical view of causality – which (1) construes causality in realist terms, (2) construes causality in terms of nomological invariance, and (3) validates assignments of causality in particular cases on the grounds of their subsumption under such regularities – is completely stalemated *without an independent provision for confirming the actual, strict uniformities of the properties in question.* It is clear, therefore, that the canonical view gains an enormous advantage by stipulating that the uniformities of phenomenological laws, however informal and untidy they may be, can be satisfactorily corrected, regimented, brought into strict accord with the requirements of nomologicality just because the fundamental laws of any domain *fix – in theory –* the very uniformities that (would-be) phenomenological laws exhibit (otherwise, indiscernibly or untidily). Hence, *once* the bond between phenomenological laws and fundamental laws is broken (with regard at least to denying the logical priority of the latter), once the bond between causality and nomologicality is broken, we are pretty well forced to retreat (or rather, to turn) to the paradigm of agency *and*, in doing that, we are relieved at once of all essentialist pretensions regarding causal regularities *and* of the need to defend a realist account of causal counterfactuals on either nomological grounds or on the grounds of essentialist uniformities or both.[37] (There is an instructive analogy, we may observe, between the alleged connection between phenomenological and explanatory laws in science and the connection between the grammatical structure of the uttered fragments of actual speech and the deep structure of the idealized sentences assigned such fragments in Noam Chomsky's theory of language.[38])

Furthermore, *if* the canonical linkage between the phenomenological and the theoretical is broken, *if* the linkage between causality and nomologicality is broken, *if* the agental conception of causality is favored, then, *if* these notions are reconciled with our opening postulates (the rejection of transparency and the rejection of any hierarchized ordering of the privilege of scientific inquiry), it is but a step to concede that man's technological intervention extends to his linguistically formulated theories, to the competing conceptual networks he entertains, to his very thinking – to the "artifactual" (though not idealist) structure of the "real world" he effectively explores, changes and manipulates. In a word, *what* counts as technology is radically different on the canonical and technologized views of science; and *what* science can accomplish is seen to be radically different on those opposed views.

IV

It is important to grasp that in contesting the various loose and strict versions of the unity-of-science account of causality and nomologicality we are breaching one of the most dominant *naturalistic* accounts of how man discerns the properties of the real world. It is noticeably difficult to make the counterargument plausible – frankly, because of the longstanding professional hegemony of the canonical view. But it is very clear, on conceptual grounds, that such a countermove would be a reasonable one, once, as proposed earlier on, the usual epistemic privilege of naturalistic science is discounted and naturalism itself is reconciled with moderate forms of phenomenology and deconstruction similarly shorn of their own characteristic presumptions. Such a policy (which would make no sense *if* realist forms of reducing or eliminating persons, human experience and agency, were actually open to straightforward validation) clearly encourages the view – within the terms of salience but not of privilege – that the technic or technological interventions of man count as the ineluctable channel by which man minimally discerns, *distributively*, the real structures of the world. It is the effectiveness of deliberate human agency – of successful invention and intervention affecting the processes of nature, sensed and preserved pretheoretically (as in the manufacture of fire, doubtless) or improvised for reasons inseparable from theorizing about relations between unseen entities and processes and those we do perceive (as in producing the atom bomb) – that yields the most compelling and enduring stratum of what we take to be real. Notoriously, in all of his speculations about testing the sense of reality among alien speakers, Quine never once considers the role of species-wide technic experience in testing the so-called radical indeterminacy of translation.[39] If he had, he would perhaps not have given up ontological relativity; but he could not have secured it in "radical" terms unless (contrary to his own severe extensionalism) he subscribed to a comparably radical form of incommensurabilism.

Once we concede this, however, we grasp as well the sense in which a strong disjunction between naturalistic and phenomenological views of reality (or a disjunction between ontic and pre-ontic reflection) is merely contrived and unconvincing. Technology or technic experience (agency, intervention, experiment, invention, even observation therefore) signifies the intersection of *all* humanly pertinent modes of discerning, contacting, affecting and being affected by *whatever is real*. The technic, in short, is the principal way in which we move from a merely *holistic* sense of reality (since, after all, the race survives as a result of being "somehow" in touch with the effective structures and processes of things) to a *distributive* sense of those structures. The only warning required is that that transition must be confined to what is salient rather than privileged – hence, subject to phenomenological, critical, deconstructive (and further naturalistic) constraints. But that, precisely, is the essential theme of Heidegger's dualized account of technology, of *Zuhandenheit* and *Vorhandenheit* – which is to say (relative to *Zuhandenheit*: "readiness-to-hand") that the reality of worldly things is already centered in a global and nameless effectiveness of the worldly career of what Heidegger calls *Dasein*, and (relative to *Vorhandenheit*: "presence-at-hand") that the distributed real objects of the world, including explicit tools and more (man for instance), are thus articulated (and rightly thus articulated) within the global effectiveness of the other.[40] *Dasein*, for Heidegger, is never, of course, confined to any worldly articulation of "man"; but there may be many ways of expressing the required intersection of the naturalistic and the phenomenological. At any rate, from both naturalistic and phenomenological perspectives, it is reasonable to insist on the peculiarly supple sense in which the technological provides a quite different vision of science, in particular of causality and laws, from that afforded by the canonical picture.

The argument also shows that it is very unlikely that that thesis can be given a fair inning without going very far afield of the plainly narrow concerns of analytic conceptions of causality and laws. With an eye to argumentative strategies, the perceived weakness of the canonical picture depends on the difficulty at least of resolving the famous problem of universals.[41] Here, we need take note only of the fact that the importance of the problem of universals in resolving the further problem of natural laws, together with the importance of the agency conception of causality, confirms the profound sense in which the sciences are distinctly *human* undertakings, even when (as in the physical sciences) their domain is defined without reference to the distinctly human. The natural way in which this theme presents itself, consistently with a sanguine respect for the realism of science (on any of a wide variety of views) features – perhaps must feature – the salience of man's technic experience. It is a sense, needless to say, that cannot be easily discounted as merely picturesque or as the vestigial persistence of a mere "folk" conception that is better exorcised than welcomed.

Perhaps a final word may be said about the problem of universals, not so much with the intention of settling that issue quickly as with the hope of affording a clearer sense of the relevance of technology and related ways of intervening in the world. We may remind ourselves very briefly, first, of one of Freud's characteristic claims – which Grünbaum (not at all unreasonably) takes Freud to have construed in terms of his own penchant for viewing psychoanalysis as a natural science. Grünbaum himself regards the following as a perfectly eligible scientific thesis the empirical evidence for which (on Grünbaum's account) Freud failed to supply – indeed, Freud apparently never got beyond a *post hoc proper hoc* argument: the thesis "that all parapraxes whose causes are unknown to the subject are the result of repressions."[42] The point of importance is that the *detection* of a string of instances that could rightly be characterized as parapraxes, in virtue of which a law-like regularity is adduced (or a regularity suitably linked to a covering psychoanalytic law, if it cannot be taken to be a law itself), requires an observational procedure by means of which to characterize a given string of instances *as* parapraxes.[43] It is important to notice that the vexed issue of whether psychoanalysis *is* a natural science (even if it is one without confirmed causal laws) depends on whether the conditions under which such distinctions as parapraxes and repressions *are empirically detectable are suitably congruent with the conditions under which empirically detectable regularities would support a natural science* – a natural science addressed to strong invariance or nomological necessity, in the sense already supplied. (Grünbaum does not discuss the matter, which, it may be argued, *is* one of the decisive sources of contention regarding the disputed status of psychoanalysis.) For our present purpose, it is perhaps enough to indicate that it may be claimed that parapraxes and repressions are essentially intentional in a way that cannot be straightforwardly matched in the physical sciences *and* (therefore) do not lend themselves to the enunciation of pertinent uniformities as the simplest sensory similarities are alleged to do. This is not to deny, of course, that we may rightly speak of psychoanalytic uniformities, or that such uniformities may support the thesis that psychoanalysis is a natural science, or that intentional complications affect the uniformities of sensory perception as well.

The scientific standing of psychoanalysis rests, nevertheless, on being able to show that what is entailed in collecting a set of personal histories and anecdotes *as* a string of pertinent uniformities could or could not serve, in the psychoanalytic setting, as evidence for a causal law *in just the sense in which the cognate connection is taken to obtain in the natural sciences.* Furthermore, it is a reasonable conjecture that the intentional complexities of psychoanalytically relevant attributions are bound to be very closely akin to the intentional complexities of the general agental life of humans. Hence, if the idiom of technological intervention both encourages a picture of science at odds with the canonical account and is essentially of a piece with the intentional idiom of psychoanalysis, Freud's (or Grünbaum's) sanguine views about the natural-science status of the discipline may well be misplaced (the matter need not be decided here); and how it should be decided is bound to be affected, one way or another, by the methodological implications of first fixing psychoanalytically pertinent uniformities.[44]

Uniformities of sensory discrimination of the most elementary sorts indicate even more compellingly the pertinence of methodological links to agental and technological considerations. This may be shown most suggestively by considering Niko Tinbergen's well-known experiments with stickleback. Tinbergen has shown of course that the fighting and courtship behavior of the stickleback is decisively triggered by visual stimuli falling within almost calibrated limits of variation (what he calls the "dependence of innate behavior on sign stimuli") – as of shape, posture, abdominal swelling, proximity, "red belly," and "zigzag dance."[45] First of all, what, *behaviorally*, counts as uniformity with respect to pertinent visual discrimination cannot be decided in terms of uniformities adduced on the basis of purely physical laws (as of light frequencies) but depends instead on an "intentional" model of courtship and aggression – even though Tinbergen is expressly opposed to intruding, in a "scientific" study, any reference to "subjective" phenomena in animals, any "phenomena that can be known only by introspection."[46] So the pertinent sensory uniformities are conceptually dependent on analogies between the agental behavior of stickleback and that of humans.

What is particularly pretty about Tinbergen's work is that it shows us how a science such as ethology can function very much as a natural science despite such conceptual features, and at the same time helps us understand why success in at least this range of ethology cannot ensure, in principle, a similar success in psychoanalysis. Furthermore, the very fixing *of* the "intentional" behavioral model for stickleback – in particular, for the "Innate Releasing Mechanism" that Tinbergen has refined as the key to instinctual functioning – depends in a crucial way on the technological manipulation *of* the releasing mechanism (theoretically postulated), by means of experiments keyed to observing the behavior of the fish in the presence of a variety of ingenious dummies Tinbergen has fashioned. Here, we cannot escape the ubiquity of technic experience with respect both to the canonical conception of a natural science and to the possible bifurcation of the sciences.

Finally, it is clear that, on the theory of the releasing mechanism itself, creatures exhibiting innate responses must be taken to have discriminated a relatively invariant, "abstract" uniformity, as of shape or color or behavior, ranging over a finite spectrum of graduated changes in particular instances – when, that is, an animal responds uniformly in those particular instances.[47] So the perceptual uniformities pertinent to Tinbergen's ethological studies are complex universals of some sort that *cannot* be specified (not merely, not detected) except in a way that conceptually depends on the intervention and technic experience of the human investigator. If this is true in the ethological case, it is bound to be true with a vengeance in the psychoanalytic (a matter Grünbaum never addresses); and there (and elsewhere) it may force us to concede that the relevant universals will not support a unity-of-science model.

In general, one may argue that a conceptual linkage obtains between the methodology of a science and the nature of the properties of the phenomena the domain collects. It is a notorious fact, for instance, that, in strengthening his account of linguistic universals in natural language. Noam Chomsky retreats increasingly from his original close-grained empirical study of actual linguistic *performance*, to postulating underlying universals of *competence* relatively indifferent to the conditions of contingent behavior. Roughly, this means that Chomsky's account shifts from the description and interpretation of sentences actually produced in the natural contexts of human discourse to the context-free generation of sentences in accord with the underlying theory. Terry Winograd, for instance, has rather tactfully insisted on the point in his own attempt at a computer model of English, observing, "A program for parsing language is as much a 'generative' description of the language as is a set of rules for producing sentences. The meaning of 'generative' in Chomsky's original sense ... is that the grammar should associate a structural description to each permissible sentence in the language."[48] The upshot is that Chomsky is able to *present* his linguistic theory in a form intended to accord closely with the model of causal laws and causal explanation favored in the canon, without really attending in a close way to the "surface" conditions – the conditions of cognitive pickup – under which *linguistic invariances in the contexts of actual use* are detected and tested. It is one thing, however, to model language in accord with the canon; it is quite another to confirm that natural languages actually exhibit structures – in contexts of use – suited to canonical explanation or to the unity of science. (We shall return to the issue in a later chapter.)

In general, we may claim that, insofar as the uniformities essential to this or that would-be science critically depend on intentional considerations that *do not yield in a law-like way* to physical reduction or elimination or even to approximative mapping, the technic experience involved in fixing such uniformities tends to favor a bifurcation of the sciences rather than a unity of science. The technological straddles both options. But to grasp that very fact is to appreciate why it is that the broad similarities of agental intervention, invention, manipulation, experiment, and observation do not automatically support the methodological uniformity of science; *and* why it is that, by exaggerating the privilege of theory and explanatory rigor, one can easily mask – in the direction of favoring the physical sciences (as in a way Chomsky's speculations also do) – decisive discrepancies among the methods suited to the different sciences themselves. The fact remains that the agency conception of causality provides (in different ways) for both the bifurcation and the unity thesis, for both the denial and the affirmation (attenuated along the lines of sub-agental frequencies) of a strong connection between causality and nomologicality; and this dualized articulation of agency signifies the very role of technic experience in fixing our sense of what is distributively real – in a way in which the naturalistic, the phenomenological, and the deconstructive cannot be disjunctively sorted in terms of any executive priority or privilege.

V

One immense, enormously important final topic needs to be broached here, a topic that is quite naturally linked to the agental conception of causality – that is in fact quite naturally linked to what is shared between the agental notion and the notion of universals introduced in discussing Freud and Tinbergen. This is the right place to mention it; but we can do little more than make room for it, without at all supposing we are doing justice to the topic. If we acknowledge the agental model as one of at least two basic models of causality – in particular, the model naturally favored in the human sciences – we are bound to apply it in cultural contexts in intensionally encumbered ways. The most unguarded sort of case in which this way of speaking is invoked occurs when we explain (and when we suppose that what we thus mention does actually correspond with the events) that we wanted this or that and, wanting it, acted to acquire or achieve it. The sense intended is that wanting plays a causal role in our acting as we do. This much at least accords completely with Donald Davidson's well-known and influential example of alternative descriptions of what is putatively one action:

> I flip the switch, turn on the light, and illuminate the room. Unbeknownst to me I also alert a prowler to the fact that I am home. Here I need not have done four things, but only one, of which four descriptions have been given. I flipped the switch because I wanted to turn on the light and by saying I wanted to turn on the light I explain (give my reason for, rationalize) the flipping. But I do not, by giving this reason, rationalize my alerting of the prowler nor my illuminating of the room.[49]

Davidson characterizes such descriptions within rationalizations as "quasi-intentional" and marks the feature in a formal way thus:

> C1 R is a primary reason why an agent performed the action A under the description d only if R consists of a pro attitude of the agent towards actions with a certain property, and a belief of the agent that A, under the description d, has that property.[50]

But why, we may ask, is the characterization "quasi-intentional," why not just intentional – or better, *intensional*? Davidson compounds the mystery by acknowledging

C2   A primary reason for an action is its cause.[51]

In fact, we may reasonably claim that, *if* I flip the light switch (which, as an action, is identical with illuminating the room), and *if* the event of the prowler's being alerted is not identical with the action I perform, then my flipping the light switch *causes* the prowler *to be alerted to the fact* that I am home. On any plausible reading, *that* causal connection would have to be intensionally specified – for it is quite clear that my action could not be supposed to cause the prowler to be alerted to each and every extensionally equivalent or logically entailed "fact," could not extensionally isolate the key fact among other potentially pertinent facts, and could not (solely on the account given) reduce the state of "alert" to some nonintentional state. Davidson insists, of course, that "my nonintentional alerting of the prowler" is not different from my flipping the switch, is not "just its consequence."[52] But he nowhere explains why this is so or why we should take it to be so. Furthermore, it would be very easy to adjust the example to make it out that alerting the prowler to the fact in question *was* a consequence of what I did: for instance, my flipping the switch (and illuminating the room) might well have caused the prowler (under various scenarios) to infer that I was home. The intensional feature of the causal sequence would not then be so easily tamed. But in any case, on the face of it, if my primary reason for flipping the switch was to turn on the light, then my *reason* (the reason I had) for acting as I did was the *cause* of what I did, *whether* described as turning on the light or as alerting the prowler – even though it could not rightly be specified as the reason "I had," *within* my rationalization.

What it is important to appreciate here are the alternative strategies that recommend themselves, not the final solution of the puzzle. Davidson makes one additional suggestion regarding these cases: he insists, as already noted, that causal sequences fall under covering laws even if those laws are unknown; now he adds,

> The laws whose existence is required if reasons are causes of actions do not, we may be sure, deal in the concepts in which rationalizations must deal. If the causes of a class of events (actions) fall in a certain class (reasons) and there is a law to back each singular causal statement, it does not follow that there is any law connecting events classified as reasons with events classified as actions – the classifications may even be neurological, chemical, or physical.[53]

Davidson's solution clearly depends on his independent advocacy of a form of token identity (or, by analogy with provisional functionalist strategies) on narrow-gauge type identities facilitated by distributed token identities.[54] We need take notice here only of the fact that, *if* type and token identities fail (if, in particular, Davidson's anomalous monism forms an inconsistent triad[55]) and if construing actions as no more than "primitive actions" token-identical with simple "bodily movements" (such as moving one's finger[56]) falls with the general failure of token identity, *then we should have to admit that there was no effective way of barring intensionally complex causes from obtaining within the space of the human sciences.* But the protases mentioned are more than plausible.

The agental model of causality confirms the reasonableness of admitting that "we are usually far more certain of a singular causal connection than we are of any causal law governing the case."[57] This is Davidson's own wording – against a possible reading of Hume's view. But Davidson comes to Hume's rescue in maintaining that, even if it is wrong to hold that "'*A* caused *B*' entails some particular law involving the predicates used in the descriptions '*A*' and '*B*,'" it is nevertheless right to hold (and that alone would vindicate Hume) *that* "'*A* caused *B*' entails that there exists a causal law instantiated by some true descriptions of *A* and *B*."[58] But the agental model also confirms the reasonableness of holding that, in some contexts, causality does not entail nomologicality at all; at any rate, there is no conceptual incoherence in so maintaining. Furthermore, the option would be strengthened *if* causes in agental contexts were prominently or characteristically intensionally complex. So the viability of Davidson's claim (clearly, the canonical and conventional wisdom), that causality entails nomologicality, depends on independent evidence that intensionally specified causes – notably, reasons "had" or "primary reasons" – *can* be reduced token-wise in physical terms, so that "true descriptions of [some relevant] *A* and *B*" can replace the unwelcome descriptions of the *admitted* causes mentioned in our rationalizations. But, as we have already argued, there is no satisfactory demonstration that token identity obtains (let alone type identity): on the thesis that causality is (ultimately) invariably extensional, identifying reasons as singular causes is identifying what, *qua* reasons, can be specified extensionally – which is incompatible with Davidson's view of the holism of the mental; and, on the thesis that causality entails nomologicality, what are identified as singular causes can, as such, enter into (type-wise) law-like generalizations – which is incompatible with Davidson's anomalous monism.[59] If intensionally complex causes prove irreducible, then it is a foregone conclusion that causality does not entail nomological-

ity. So, at the very least, Davidson has steadfastly turned his back on a clear (if unpleasant) option that his own cases force us to consider.

In fact, Elizabeth Anscombe was very frank to admit, reviewing Davidson's paper "Causal Relations" shortly after its original appearance, that "I have no sure insight into the sources of the conviction that causal statements are extensional."[60] She offered an interesting and helpful example of causality of the agental sort – although, quite instructively, she herself pursued the extensional question only in terms of causal explanations and in terms of the extensionality of reference in explanatory contexts. *If* it were the case that contexts of causality and contexts of causal explanation could not be segregated, then of course (as Davidson himself implicitly acknowledges), since explanatory contexts are nonextensional, causal contexts could not but be intensional as well. But there seems to be no reason to deny that causal contexts are separable from explanatory contexts – particularly if we may know that "*A* caused *B*" without the least inkling of its would-be covering law.

Anscombe's example involves three causal statements differing only in the substitution of co-designative expressions identifying the putative causal agent:

There is an international crisis because "moi, de Gaulle" made a speech.

There is an international crisis because the President of the French Republic made a speech.

There is an international crisis because the man with the biggest nose in France made a speech.[61]

Her reasonable point is that the explanatory use of "because" is affected by the intensional differences among the descriptions (including the epithet) offered. True enough. But what Anscombe fails to mention is that the referential use of the expressions in question does not clarify and does not transparently address the nature of the causes in question (no such uses ever would); hence, that the solution of the extensional puzzle of fixing causes within explanatory contexts does not directly bear on the solution of the extensional puzzle regarding the individuation of particular causes. It is most difficult – in fact, it seems impossible – to bring Anscombe's example into line with Davidson's view that causality entails nomologicality *and* with Davidson's further view that, in accord at least with the strictures of token identity (and of the narrow-gauge type identities that might be discovered thereupon), the relevant covering laws "may even be neurological, chemical, or physical."

This is simply to say that Davidson is here adhering (in however ingenious a way) to Hempel's conception of covering laws.[62] To see this is to see as well how the admission of the agental model of causality threatens the unity conception of science. To admit that singular causes can be discerned without reference to covering laws favors the agental model. To admit that "primary reasons" are actual causes risks the irreducibility of intensionally complex causes. To confirm the irreducibility of primary reasons token-wise is to subvert the claim that causality entails nomologicality. And to reject the entailment and to affirm the irreducibility in question is to concede the bifurcation of the sciences.

It may be difficult to appreciate the enormous power of this conclusion. But perhaps one way of focusing its force is to remind ourselves of the misunderstanding of the puzzle of intensionality encouraged by Daniel Dennett's much repeated question, "What sort of thing is a different thing under different descriptions? . . . Intentional sentences are *intensional* (non-extensional) sentences."[63] The point is that causes are not "different" under different descriptions. It's only that intensionally complex causes cannot be reliably reidentified extensionally under different descriptions. It is for that reason that they cannot be collected under covering laws. But, of course (since the matter must be put vacuously), a cause remains self-identical under whatever descriptions single it out – or, what serves as a cause in a causal or explanatory context remains self-identical under whatever descriptions single it out (whether as a cause or not). To know that "every thing is self-identical" is to know nothing as yet about how to reidentify "any thing." So the admission of intensionally complex causes does not lead us to support the absurd thesis Dennett rightly rejects; it also makes one wonder why Dennett offers it as a likely interpretation of the intensionality thesis. In any case, the admission of intensionally complex causes need not be taken to preclude causal explanations in the human sciences, or the scientific standing of particular human studies, or a plausible sense of objectivity for those disciplines. It does mean, however, that, whatever we make out to be the objectivity of causal accounts in the human sciences, such objectivity will have to be construed in terms of consensually developed interpretations. This surely is the minimal consequence of admitting intensionally qualified causes within a space of inquiry in which the subjects observed and those who observe them are one and the same – that is, members of one and the same society (by way of bilingualism and biculturalism) of an extended and enlarged society that could acknowledge diverging, even alien, forms of life within its compass.[64] The objectivity of the human and social sciences presupposes a sufficient range of such common practices: a Quinean field linguist, for example, *simply has no basis at all* for claiming a pertinent or "basic" form of objectivity.[65] The issue also bears in a decisive way on the pretended ease with which culturally pertinent universals can be regularized for the purposes of inductions within the human sciences – for instance as in Grünbaum's assessment of the validity *and* testability of Freud's hypotheses. In any case, it makes more sense to adjust our theory of causality to what appear to be the saliencies of the human sciences than it is to tailor our theory to antecedent prejudices drawn from entirely different sources (the physical sciences) that have not yet vindicated (and appear unable to vindicate) their exclusive rights to the field at stake.

Finally and very briefly, *if* phenomenological laws or lawlike regularities are disjoined from and take precedence over explanatory laws, *if* an agental conception of causality is disjoined from an invariance conception and given empirical primacy over the other with respect to both the physical and human sciences, *if* the agental conception with respect to both the physical and human sciences cannot be freed of intentional and intensional complications affecting the very formulation of similarities, regularities, uniformities, invariances, universals, then we *have* all but installed the general thesis of the *praxical* reading of science itself.

By the "praxical," one means here, at least minimally, the insistence: (i) that man's cognitive aptitudes and achievements are continuous with and incarnate in his technic or technological interaction with the environing world; (ii) that his interventions in this regard incorporate to a significant extent activity that is tacit, pretheoretic, keyed to quotidian survival;[66] (iii) that conceptualizing the regular features of the world is an artifact of man's technic experience; (iv) that such conceptualizing is itself tacitly preformed in a way that cannot be entirely or undistortedly exposed by any critical reflection *via* the praxically grounded beliefs and formative structures of the environing culture in which, in an historically contingent way, each generation masters the language and societal practices of preceding generations; (v) that the abstractions and generalizations of any viable science or rational inquiry or explanatory theory or descriptive classification are projections incorporating (always tacitly) the tacit dispositions of the species (and, within the species, of particular, divergent, sub-species-wide cultures) to detect similarities, uniformities, causal mechanisms and the like; (vi) that, therefore, the effectiveness of social (possibly but not necessarily revolutionary) action depends on the extent to which idealized, ideologized, programmatic undertakings are actually in touch in a causally effective way with the effective technic forces that produce and sustain the (perceived) social structures viewed as needing to be altered; and

(vii) that the very condition of human existence commits man to an active, practical, interventionist orientation with respect to favorably altering or sustaining the conditions of social life, regardless of how such activity is reflexively or ideologically represented.

It is very much in this spirit – in an explicitly Marxist idiom – that Roy Bhaskar rather prettily remarks that "philosophy [and of course if philosophy, then science as well] is ... soiled in life."[67] In fact, it may fairly be claimed that (i)–(vii) afford an orderly articulation of Marx's *Theses on Feuerbach* shorn of any doctrinal privilege dividing "substratum" from "superstructure" or favoring the revolutionary role of the proletariat or anything of the sort (also, of course, not skewed in such a way as to preclude the defense of any such thesis). Marx is sufficiently Hegelian to oppose the eighteenth-century – particularly the Enlightenment – conception of the faculty of reason. He is sufficiently anti-Hegelian to oppose (the abstractionism of) Hegel's locating the effectiveness of rational analysis and rational action in the idealized *Geist* of an age. In the first regard, he treats reason and its conceptual proclivities as an historical formation; and in the second regard, he treats a historicized reason as capable of much more (both in analysis and action) than an abstract or utopian exercise. He opposes, for instance, whatever may reasonably be collected in Edward Gans's Hegelian pronouncement: "Whatever is produced by a people at a determinate epoch is produced by its force *and by its reason*."[68] But in opposing (this version of) the Hegelian formula, Marx means to preserve the sense in which thinking is fundamentally practical, that is, directed to fulfilling projects, to altering one's life and the conditions of one's life, to being or becoming what one is or becomes through practical activity, and to being committed to and formed for such activity as the "essential" but structurally or historically entailed purposiveness of human existence itself.[69]

Nevertheless, although there can be no doubt that the modern conception of *praxis* has been most sustainedly explored in the Marxist tradition, the key notion itself is by no means absent in various partial forms from antiquity to Hegel. In our own time, it may fairly be ascribed, without prejudice, to such widely different thinkers as Dewey, Heidegger, and Foucault – which is to say, to the historicized currents of naturalism, phenomenology, and (Nietzscheanized) post-structuralism. (The radical difference between Foucault and Derrida must not be ignored, of course, but one can make out a reasonable case that something like a praxical orientation – admittedly a thin one – can be found in Derrida, despite or perhaps because of his disagreement with the French Marxists.[70]) In any case, the point of pressing the praxical in the present context is primarily to afford a sense of just how radical is the growing repudiation of the unity of science model and its characteristic abstractionism (notably regarding the supposed invariances or nomological necessity of causal processes). It affords, therefore, a sense of the profound import of insisting that the sciences are human undertakings and that the physical sciences themselves depend in an ineluctable way on the discipline of the human sciences.

Notes

1 The full argument is given in Joseph Margolis, *Pragmatism without Foundations: Reconciling Realism and Relativism* (Oxford: Basil Blackwell, 1986).
2 See ch. 2, above.
3 See Rudolf Carnap: *The Unity of Science*, tr. Max Black (London: Kegan Paul, Trench, Trubner, 1934); "Psychology in Physical Language," tr. George Schick, in A. J. Ayer (ed.), *Logical Positivism* (Glencoe, Ill.: Free Press, 1959); "Intellectual Autobiography," in Paul Arthur Schilpp (ed.), *The Philosophy of Rudolf Carnap* (LaSalle, Ill.: Open Court, 1963), particularly pp. 50–3; and "The Philosopher Replies," in the Schilpp volume, *passim*. See also Carl G. Hempel, *Aspects of Scientific Explanation* (New York: Free Press, 1965); Paul Oppenheim and Hilary Putnam, "Unity of Science as a Working Hypothesis," in Herbert Feigl et al. (eds), *Minnesota Studies in the Philosophy of Science*, vol. 2 (Minneapolis: University of Minnesota Press, 1958); and Robert L. Causey, *Unity of Science* (Dordrecht: D. Reidel, 1977).
4 J. J. C. Smart, *Philosophy and Scientific Realism* (London: Routledge and Kegan Paul, 1963), p. 50.
5 Ibid., p. 53.
6 This is the immensely important issue that hangs on Quine's account of "pegasizing." See W. V. Quine, *Word and Object*, (Cambridge, Mass.: MIT Press, 1960), section 37.
7 Wesley C. Salmon, *Scientific Explanation and the Causal Structure of the World* (Princeton, NJ: Princeton University Press, 1984), pp. 148–9; see Bas C. van Fraassen, *The Scientific Image* (Oxford: Clarendon Press, 1980), p. 118, cited by Salmon.
8 Salmon, *Scientific Explanation*, p. 18; cf. p. 190.
9 Adolf Grünbaum, *The Foundations of Psychoanalysis* (Berkeley, Calif.: University of California, 1984), p. 16. See also Jürgen Habermas, *Knowledge and Human Interests*, tr. J. J. Shapiro (Boston, Mass.: Beacon Press, 1971), pp. 272–3 (cited by Grünbaum); and Hans-Georg Gadamer, *Truth and Method*, tr. Garrett Barden and John Cumming from 2nd German edn (New York: Seabury Press, 1975), p. 311 (also cited by Grünbaum).

10 Hempel, "The Function of General Laws in Histroy," *Aspects of Scientific Explanation*, pp. 233, 231.

11 It may be remarked that Popper, whom Grünbaum generally opposes, effectively agrees with at least this much of Grünbaum's thesis; see Karl R. Popper, *The Poverty of Historicism*, 3rd edn (London: Routledge and Kegan Paul, 1961). Popper, of course, is also a partisan of the unity program.

12 Grünbaum, *The Foundations of Psychoanalysis*, p. 17.

13 Ibid., p. 18.

14 Ibid., p. 14 (italics added).

15 Habermas, *Knowledge and Human Interests*, p. 271.

16 Ibid. (italics added).

17 Ibid.

18 See Joseph Margolis, "Reconciling Freud's *Scientific Project* and Psychoanalysis," in H. Tristram Engelhardt, Jr, and Daniel Callahan (eds), *Morals Science and Sociality*, vol. III: *The Foundations of Ethics and its Relationship to Science* (Hastings-on-Hudson, NY, 1978).

19 Grünbaum, *The Foundations of Psychoanalysis*, p. 14.

20 See Joseph Margolis, *Culture and Cultural Entities* (Dordrecht: D. Reidel, 1984), chs 4, 5.

21 For mutually opposing views on this matter, compare D. M. Armstrong, *What is a Law of Nature?* (Cambridge: Cambridge University Press, 1983); and Norman Swartz, *The Concept of Physical Law* (Cambridge: Cambridge University Press, 1985).

22 I borrow the term from Arthur Fine, "Einstein's Realism," in James T. Cushing et al (eds), *Science and Reality: Recent Work in the Philosophy of Science* (Notre Dame, Ind.: Notre Dame University Press, 1984).

23 See Nancy Cartwright, *How the Laws of Physics Lie* (Oxford: Clarendon Press, 1983), Essay 5: "When Explanation Leads to Inference," and Essay 6: "For Phenomenological Laws." See also Ian Hacking, *Representing and Intervening: Introductory Topics in the Philosophy of Natural Science* (Cambridge: Cambridge University Press, 1983), ch. 16; Pierre Duhem, *The Aim and Structure of Physical Theory*, tr. Philip P. Wiener (New York: Atheneum, 1962); van Fraassen, *The Scientific Image*.

24 Cartwright, *How the Laws of Physics Lie*, p. 127.

25 Ibid., pp. 100–3. One of the boldest specimen views of this sort is certainly that espoused by Wilfrid Sellars, "The Language of Theories," *Science, Perception and Reality* (London: Routledge and Kegan Paul, 1963).

26 Cartwright, *How the Laws of Physics Lie*, p. 126.

27 See Tom L. Beauchamp and Alexander Rosenberg, *Hume and the Problem of Causation* (New York: Oxford University Press, 1981), ch. 4. See also J. L. Mackie, *The Cement of the Universe* (Oxford: Clarendon Press, 1974); William Kneale, "Universality and Necessity," *British Journal for the Philosophy of Science*, XII (1961); Karl R. Popper, "A Revised Definition of Natural Necessity," *British Journal for the Philosophy of Science*, XVIII (1967).

28 Hacking, *Representing and Intervening*, pp. 262–3.

29 See W. V. Quine, "On What There Is," *From a Logical Point of View* (Cambridge, Mass.: Harvard University Press, 1953).

30 Hacking, *Representing and Intervening*, ch. 12, particularly pp. 218–19.

31 See Joseph Margolis, *Persons and Minds* (Dordrecht: D. Reidel, 1978).

32 See Donald Davidson, "Causal Relations," "Actions, Reasons, and Causes" and "Mental Events," *Essays on Actions and Events* (Oxford: Clarendon Press, 1980).

33 See for instance Mario Bunge, "Toward a Philosophy of Technology," in Carl Mitcham and Robert Mackey (eds), *Philosophy and Technology*, 2nd edn (New York: Free Press, 1983); also Joseph Margolis, "Three Conceptions of Technology: Satanic, Titanic, Human," in Paul T. Durbin (ed.) *Philosophy and Technology*, vol. 7 (Greenwich, Conn.: JAI Press, 1984). See also Albert Borgmann, *Technology and the Character of Contemporary Life* (Chicago: University of Chicago Press, 1984), particularly pt I; and Jon Elster, *Explaining Technical Change* (Cambridge: Cambridge University Press, 1983), particularly pt I.

34 Karl R. Popper, "The Aim of Science," *Objective Knowledge* (Oxford: Clarendon Press, 1972), p. 190.

35 This is what is so puzzling about Nelson Goodman's account of induction *in* the context of a practicing science. It provides the key to grasping the peculiar vacuity of Goodman's notion of "entrenchment" – correct in a purely formal sense but utterly inoperable *in* quotidian scientific induction *when construed nominalistically*. This issue does not at all depend on Goodman's careful restrictions on the use of the terms "law" and "causality." See for instance Nelson Goodman, *Fact, Fiction, and Forecast*, 2nd edn (Indianapolis: Bobbs Merrill, 1965), pp. 20–2 – in the context of Goodman's "Seven Structures on Similarity," in Lawrence Foster and J. W. Swanson (eds), *Experience and Theory* (Amherst: University of Massachusetts Press, 1970), and his *Of Mind and Other Matters* (Cambridge, Mass.: Harvard University Press, 1984), pt II.

36 It also shows, inadvertently, that Popper himself wavers considerably in his rejection of a realist reading of *nomologicality*– by way of his doctrine of verisimilitude: "although I do not think that we can ever describe, by our universal laws, an *ultimate* essence of the world, I do not doubt that we may seek to probe deeper and deeper into the structure of our world or, as we might say, into properties of the world that are more and more essential or of greater depth" ("The Aims of Science," *Objective Knowledge*, p. 196).

37 A similar theme is developed, in terms of possible-worlds semantics, however, in David Lewis, "Causation," *Journal of Philosophy*, LXX (1970). See also Nicholas Rescher, "Lawfulness as Mind-Dependence," in Nicholas Rescher et al. (eds) *Essays in Honor of Carl G. Hempel* (Dordrecht: D. Reidel, 1969).

38 See ch. 12, below.

39 See Quine, *Word and Object*, ch. 2.

40 See Martin Heidegger, *The Question Concerning Technology and Other Essays*, tr. William Lovitt (New York: Harper and Row, 1977), and *Being and Time*, tr. John Macquarrie and Edward Robinson from 7th German edn (New York: Harper and Row, 1962), sections 15–16.

41 The connection is explicitly pursued in D. M. Armstrong, *What is a Law of Nature?* (Cambridge: Cambridge University Press, 1983). There is an extremely perceptive account of causal laws, that construes natural necessity more in terms of pragmatic "resiliency" than of "real" necessity, offered in Brian Skyrms, *Causal Necessity* (New Haven, Conn.: Yale University Press, 1980), pt I. By "resiliency" Skyrms means (roughly) the degree of invariance of some statistical probability; hence, "High resiliency can be thought of as a statistical notion of *necessity*" (pp. 11–12). But Skyrms does not here consider the question of the operative relationship between detecting relevant invariances and the solution to the problem of universals.

42 Grünbaum, *The Foundations of Psychoanalysis*, p. 197; see also pp. 1–9.

43 Ibid., p. 198; see also pp. 10–15.

44 This, it should be noted, is (at least to the extent suggested) sympathetic with Grünbaum's own criticism of the "hermeneutic" confusion regarding the priorities of methodological and ontological speculation (and commitment) in Freud; see for instance ibid., pp. 5–9.

45 See N. Tinbergen, *The Study of Instinct* (New York: Oxford University Press, 1969), pp. 37–43.

46 Ibid., pp. 3–5.

47 Ibid., ch. 2.

48 Terry Winograd, *Understanding Natural Language* (New York: Academic Press, 1976), p. 22; see also p. 42. See also Noam Chomsky, *Syntactic Structures* (The Hague: Mouton, 1957) and *Aspects of the Theory of Syntax* (Cambridge, Mass.: MIT Press, 1965).

49 Davidson, "Actions, Reasons, and Causes," *Essays on Actions and Events*, pp. 4–5.

50 Ibid., p. 5

51 Ibid., p. 12.

52 Ibid., p. 4, n. 2.

53 Ibid., p. 17.

54 See Davidson, "Mental Events" and "Causal Relations," *Essays on Actions and Events*.

55 See Margolis, *Culture and Cultural Entities*, chs 4–5; also Joseph Margolis, "Prospects for an Extensionalist Psychology of Action," *Journal for the Theory of Social Behavior*, XI (1981).

56 See Davidson, "Agency," *Essays on Actions and Events*. See also Arthur C. Danto, *Analytical Philosophy of Action* (Cambridge: Cambridge University Press, 1973); Alvin I. Goldman, *A Theory of Action* (Englewood Cliffs, NJ: Prentice-Hall, 1970); and Joseph Margolis, review of Goldman's *A Theory of Action*, *Metaphilosophy*, V (1974).

57 Davidson, "Actions, Reasons, and Causes," *Essays on Actions and Events*, p. 16.

58 Ibid.

59 Cf. Davidson, "Mental Events," *Essays on Actions and Events*.

60 G. E. M. Anscombe, "Causality and Extensonality," *Journal of Philosophy*, LXVI (1969), 155.

61 Ibid. Anscombe adds rather prettily, "I owe this pleasing example, as well as the thought about temporal connectives, to P. Geach."

62 See Hempel, *Aspects of Scientific Explanation*, pt IV. An extremely careful development of a theory of action that departs from Davidson's account is offered in Myles Brand, *Intending and Acting* (Cambridge, Mass.: MIT Press, 1984). Unaccountably, however, it simply does not address the issue of intensionality that Davidson's model forces us to consider. Brand takes "intention" or "intending" rather in a psychologically restricted sense – which of course is a perfectly pertinent ingredient of the phenomena in question.

63 Daniel C. Dennett, *Content and Consciousness* (London: Routledge and Kegan Paul, 1969), pp. 28–9.

64 This, I believe, is the essential and basically correct (Wittgensteinian) insight of that much-maligned book, Peter Winch's *The Idea of a Social Science* (London: Routledge and Kegan Paul, 1958).

65 Quine, *Word and Object*, ch. 2.

66 See Michael Polanyi, *Personal Knowledge*, corr. ed. (Chicago: University of Chicago Press, 1962).

67 Roy Bhaskar, *Scientific Realism and Human Emancipation* (London: Verso, 1986), p. 16; see, further, the whole of ch. 1.

68 Cited from Gans's *Das Erbrecht in Weltgeschichtlichen Entwicklung* (1826) in Louis Dupré, *Marx's Social Critique of Culture* (New Haven: Yale University Press, 1983), p. 67. See, further, the whole of ch. 2.

69 See Richard J. Bernstein, *Praxis and Action* (Philadelphia: University of Pennsylvania Press, 1971), part 1; Richard Kilminster, "Theory and Practice in Marx and Marxism," in G.H.R. Parkinson (ed.), *Marx and Marxisms* (Cambridge: Cambridge University Press, 1982); Nicholas Lobkowicz, *Theory and Practice* (Notre Dame: University of Notre Dame Press, 1967); Adolf Sanchez Vazquez, *The Philosophy of Praxis*, tr. Mike Gonzalez (London: Merlin Press, 1977); Joseph Margolis, "Pragmatism, Transcendental Arguments, and the Technological," in Paul T. Durbin and Friedrich Rapp (eds), *Philosophy and Technology* (Dordrecht: D. Reidel, 1983).

70 See Jacques Derrida, *Positions*, tr. Alan Bass (Chicago: University of Chicago Press, 1981).

9

The Structures of Functional Properties

In contesting the canonical picture of science, generally a relatively tidy affair, one is always aware of certain sprawling questions that, however essential, are rarely directly canvassed. They may be of decisive importance but, because they cannot be trimly managed, because they have been systematically neglected and even denied legitimacy outright, they are hardly ever collected in those orderly, if provisional, ways that, pressed with a little more precision and stamina, might bring us to an unexpected crux. The analysis of functional and intentional properties is very much that sort of question. It may not be amiss, therefore, to seize a convenient opportunity here to focus the promised benefit of scanning the matter before actually doing so. It may help to give a clearer sense of the larger *agon* within the running argument.

Two very brief remarks drawn from Paul Churchland's convenient handbook of the mind/body problem will serve our purpose. There is no particular reason to single Churchland out – except for the economy. Churchland's view is of course a slightly adjusted version of Sellars's; this is the point of the near-equivalence of Churchland's popular use of the expression "folk-," as in "folk-physics" and "folk-psychology," and Sellars's term "manifest image".[1] Sellars's doctrine is, currently, the single most influential version of eliminationism we have – is very nearly the godfather of the most strenuous recent efforts of American analytic philosophy (those of Dennett, Stich, Paul and Patricia Churchland) to eliminate from the human sciences all intentionality, all functional attributes. In truth, therefore, Churchland's remarks are a bit more than merely convenient.

Here they are:

the standard evolutionary story is that the human species and all of its features are the wholly physical outcome of a purely physical process. ... If this is the correct account of our origins, then there seems neither need, nor room, to fit any nonphysical substances or properties into our theoretical account of ourselves. We are creatures of matter. And we should learn to live with that fact.

... [where "properties" consist] only in the subject item's being *recognized, perceived*, or *known as* something-or-other ... such apprehension is not a genuine property of the item itself, fit for divining identities, since one and the same subject may be successfully recognized under one name or description, and yet fail to be recognized under another (accurate, coreferential) description. Bluntly, Leibniz' Law is not valid for these bogus "properties."[2]

These are statements of two important prejudices widespread among all those who favor the canonical picture of science – especially eliminationists.

The first prejudice is an extremely strategic one, also extraordinarily easy to topple if one is willing to pay the conceptual price. More important: to disarm the first prejudice is instantly to install a vigorous alternative to the usual reductive and eliminative accounts more or less tethered to the canonical unity-of-science program. What we need here is no more than a sketch of the required counterargument, a sense of the linkage between what we shall pursue (in a moment) with respect to functional and intentional properties and the larger question of the very nature of science. We may admit, as entirely uncontroversial, the truism that the methodology appropriate to any science must be adjusted to the nature and structure of the domain that that science addresses – even if it is the case that what the determinate structure of its assigned domain actually is is a disciplined finding of that science itself. Now, given this truism, if we but add the generally acknowledged truth (hardly a truism) that science can count on no form of cognitive transparency, the first prejudice collapses at once.

All one need consider is that, *if* transparency fails, then every scientific inquiry cannot but be encumbered by the preformational conceptual limitations (however contingent or essential) under which actual scientists labor. Consequently, it cannot but be an egregious *non sequitur* to argue, as Churchland does, that, since (admittedly) our physics and biology confirm that the human species must have evolved from some primordial physical soup, the creatures we are (having thus emerged) must be such that the description and explanation of all our processes *must* be manageable in terms of "purely physical processes" – say, in terms of a homonomic physical vocabulary.[3] That would hold only if, in addition to the evolutionary fact, *transparency obtained*. But, since it does not (as Churchland is entirely willing to admit – insisting on the "theory-ladenness of perception"[4]), *it cannot be shown that the description and explanation of* anything, *at any emergent level of reality*, must *be manageable in the vocabulary of physics*. Inquiry, as a structured human capacity, has and must have endogenous limitations as a direct entailment of its own powers.[5]

This is a breathtakingly simple, altogether irresistible argument. But what is even more important is that, grasping its meaning, one is obliged to admit that the scientific description and explanation of phenomena – *both* at the human level and at any level between the human and the putatively most basic level of the original primordial soup (itself posited by human science) – are conceptually (or artifactually) encumbered by the categories reflexively employed by human scientists in characterizing their own inquiries into any matter of any kind: as a result, we are not able to penetrate the structures of any given phenomena except in the sense in which *we* actually do so. If true, then, contrary to Churchland's easy claim, it is not at all necessary, in any logical or conceptual sense, that all empirical description and explanation be able to satisfy the extreme reductive and eliminative requirements Churchland presses (in order to qualify *as* the work of a *bona fide* science) and it need not be possible, in however tolerant a spirit we construe what would fall within a "purely physical" vocabulary, that the *scientific* description and explanation of any given domain accord with the constraints of *that* vocabulary. In short, science is not obliged to proceed bottom-up, *from* distinctions restricted to the putative features of the original primordial soup (or restricted to whatever would explain its processes adequately) *to* distinctions addressed to the phenomena of any emergent order only insofar as their would-be properties are suitably linked (by way of homonomic constraints, say) to those ground-level distinctions.[6] No; science may proceed top-down, at the level of the emergence of complex human phenomena – *a fortiori*, at any intervening level posited by human inquirers – without fear of violating any normative condition of what it is to be a science. Proceeding that way does not entail any evolutionary discontinuity – that is, any evolutionary discontinuity *in nature*: it does entail discontinuities *in the explanation of nature*. But that is just what we should expect *if* science is subject to the intransparency of nature, to the discontinuities of the paradigms and research programs of particular investigators, to the praxical concerns of human societies tacitly influencing the choice and pursuit of scientific projects, to open-ended historical changes in the conditions of human existence, to the incapacity of reflecting inquirers to understand their own labor except in terms of what they grasp of their own culture, language, history, and psychology. There are, one must admit, threateningly heterodox implications embedded in this concession (for instance, regarding the laws of nature), but that is not our present concern. Our concern is served by conceding that eliminationism (and related programs) may well be internally coherent, initially eligible visions of the canon of science, even if, on the empirical evidence, they prove improbable (perhaps even impossible) within real-time terms and under the encumbering conditions of actual inquiry.

Once this conclusion is in place, it is plainly preposterous to insist without further reflection that the properties, "being recognized," "being perceived," "being known as" and the like, are "bogus 'properties'" (the second prejudice). It is entirely possible to hold that Leibniz's Law *is to be applied* to an interpreted vocabulary – quite apart form its purely formal import (which, admittedly, is entirely compelling) – *only if* that vocabulary is first suitably prepared.[7] But *to* prepare it thus is *not*, indisputably, to prepare those vocabularies that alone can service a "*bona fide* science." It may be that intentional properties are both real and conceptually irreconcilable with an unrestricted use of Leibniz's Law. So much the worse for the range of application of the Law (not the Law itself).

We may avail ourselves at this point of another large gain. *If* eliminationism and reductionism are untenable, and *if* Cartesian dualism (so-called) is to be avoided, then, on the argument, functional and intentional properties must be *incarnate* properties. That is, if (to take the paradigm case) "mental" properties are (1) real and (2) causally efficacious, then they must also be (3) indissolubly complex and (4) such that we may "abstract" with respect to them physical and functional "features" though they are not in any sense composed of separable physical and functional properties. (We may deny that they form separate properties but admit that they justify the use of distinct predicates.) Any alternative would commit us to reductionism or eliminationism, on the one side, or to dualism, on the other. If, for instance, functional properties were abstract *and* efficacious (as Hilary Putnam seems once to have held[8]), then we should be committed to an analogue of Cartesian dualism. (Functional predicates cannot designate separable functional properties – except dualistically.)

Now, Churchland emphasizes (and challenges) the familiar thesis that "mental states and properties are *irreducible*, in the sense that they are not just organizational features of physical matter."[9] He means this in a familiar, quite tendentious, even strongly dualistic sense – in which it is clearly unacceptable: namely, the sense in which the "mental" is "novel" and utterly unlike the physical. Hence, even though he insists on "the *neural dependence* of all known mental phenomena," Churchland specifically supports the claim in a way that would utterly preclude the mental.[10] This is enough to show that he never considers – in fact, one may claim in all truth that no prominent contemporary reductionist or eliminationist ever directly considered – the conceptual possibility that the functional and the intentional (prominently, the mental or psychological) is complex in the sense of being incarnate. By an *incarnate* property, again, is meant a property satisfying conditions (1)–(4) above.

There is an important equivocation involved. For to say that the mental is "irreducible" to the physical may mean either that it is utterly unlike the physical (in the sense classically formulated by Descartes – or in some similar way) or that, however it may involve the physical, it is not merely reducible to the physical or not reducible to the *merely* physical.[11] Thought, for instance, may be such that, reflecting, we are able only to recover the informational or significative or propositional "features" or "content" of our thoughts, even though thought itself (in a sense more profound than Churchland admits) may entail "neural dependence": thinking may be an incarnate phenomenon inseparable (token-wise) from the neural or physiological, but it may not be characterizable solely in terms of the physiological; it may, in short, have functionally, informationally, "mentally," semiotically, significatively, intentionally ineliminable "features." For instance, a fearful thought may not cause (Cartesian-wise) sweating and a quickening of the pulse: it may be, or rather it may include indissolubly, such phenomena within its own complex structure; and when it does enter into causal relations, it may do so only as such a complex phenomenon.[12] (The individuation of intentional phenomena is, therefore, clearly problematic.) This single adjustment could reconcile at one stroke the top-down model of science alluded to, the continuity of the natural evolution of mind, and the irreducibility of mind and the human sciences.

These, then, are the governing considerations in the inevitably sprawling account that follows.

I

Now for the sprawl.

There is a deliberately careless instruction that recent literary criticism (theorizing about itself) has fixed upon – in penetrating the common ground between literature and science. We inhabit language, the lesson reads, and to do that is to live in fiction. The extravagance is recognizably Nietzschean, and is nowhere more seriously pursued than in the paradoxical play of Paul de Man's literary speculations. If we think of language as referring to an independent world – and, therefore, as demystifying the world encountered, whether by science or literature – we fail (de Man maintains) to perceive the "presence of a nothingness" in our zeal to claim a privileged access to the "something" that our science and literature aspire to disclose. Hence, we victimize ourselves in "the confusion between the imaginary loci of the physicist and the *fictional* entities that occur in literary language."[13] The intended *aporiai* deserve a certain debunking and have received their due.[14] The doctrine itself is a most sly anticipation of the deconstructive mode and theme drawn together from Husserl and Nietzsche by way of Rousseau. But it has a certain perverse merit – which lies in its indirection, in exaggerating the encumbrance of language in order to preclude utterly the option of recovering the realism of science without first acknowledging that physics must address the world under the same conditions as literature does.

Our problem is to understand what those conditions are. In pursuing it, we shall be threading our way through the very dimly charted sea of *functional* distinctions – in particular, *intentional* distinctions as a subset of the other. De Man's deliberate paradox is really an expression of bafflement. The canonical theory of the physical sciences usually discounts a realism of intentional properties. If the charge holds, de Man (rightly) insinuates, then physics is as much concerned with the unreal as is fiction. But he himself is unable to offer an adequate corrective. What, we ask ourselves, are the prospects and conceptual costs of reconciling a realism of functional and intentional properties with a realism normal to the physical sciences?

Strange as it may seem, a very similar finding may be drawn from the most extreme forms of scientific realism – that favor the privilege of naturalistic science over phenomenological and deconstructive objections, but that hold that mere elimination, *now*, of the apparent space of persons, of their language and culture, of their current grasp of science (and morality) – would be premature, would fail, in Wilfrid Sellars's terms, "to transcend the dualism of the manifest and scientific images of man-of-the-world."[15] Sellars rejects (1) dualism, (2) the elimination (at present) of persons and "manifest physical objects in favor of the exclusive reality of scientific objects," and (3) a merely "'calculational'" interpretation of scientific theories serving the realist "affirmation of the primacy of the manifest image."[16] But his ulterior intention *is* to confirm that, in an adequate scientific realism, "the objects of the observational framework [the objects of the 'manifest image'] *do not really exist – there really are no such things*. [Such a realism] envisage[s] the *abandonment* of a sense and its denotation," but also the recovery (at the level of microtheory) of the qualitative, relational, practical, and emotional complexities of whatever *is* manifest at the manifest level.[17] It should be remarked, that, as in criticizing Rudolf Carnap's thesis to the effect that "the framework of observable physical things and their properties" to which our language must be disciplined in the interests of a meaningful science "has an absolute reality" of its own,[18] Sellars recovers, within a naturalistic idiom, what de Man more recklessly formulates in a mixture of phenomenological and deconstructive terms: namely, the denial of epistemic privilege, of the cognitive transparency of the world, of an ahistorical stance with respect to science, of an exit from language, of the segregation of the realist and idealist ingredients of any body of would-be knowledge and understanding.

Here, then, by a slightly eccentric route, we fix the common ground of recent speculations as diverse as those of naturalism, phenomenology, and deconstruction. For the extreme doctrines – conveniently, those of de Man and Sellars – obscure by exaggerated denial their own reasonable themes (that literature addresses the reality of man's world, that the rigor of theoretical physics cannot be undermined by the vagaries of man's transient interests; that there is a physical world that we inquire into, that there is no objective science that is not actually manned and managed by men).

So it is that functional properties and predicates are ubiquitous; for, for every specified would-be attribute, there corresponds another, that of being so designated (by humans); and, whether or not the members of the "first" set of predicates are functional in nature (perhaps "camel" is not, perhaps "blue" is not, perhaps "to the left of" is not), and whether or not they are fully separable in principle from those of the "second" (parasitic) set, the second are certainly ineliminably functional even if we suppose that the realities the first set name or designate or describe are entirely transparent to or unaffected by our cognitive powers and interests. On de Man's and Sellars's respective views (irreconcilable though they are), the first and second sets cannot be straightforwardly arrayed as first-order and second-order in the disjunctive sense suited to the hierarchical structure of a formal calculus. De Man matches Sellars's criticism of Carnap by his own hint of the fairness of turning Husserl's attack on cognitive privilege against Husserl's own presumption.[19] In fact, this is the convergent point of their being jointly made to carry the master theme of contemporary Western theorizing about the very role of concepts. The upshot, in a fair sense, is that all names, predicates, indices and the like *are functional* – are, minimally, assigned, named, designated, used *to* designate what they purportedly do designate.

Sellars is more difficult to fathom in this regard: "the irreducibility of the personal," he claims, "is the irreducibility of the 'ought' to the 'is'." To think of persons, therefore, is to think of "the most general common [normative] intentions of [a] community [comprised of persons] with respect to the behavior of members of the group" – to think of what is "correct" or "incorrect," "right" or "wrong," "done" or "not done" and the like. Sellars adds,

> A person can almost be defined as a being that has intentions. Thus the conceptual framework of persons is not something that needs to be *reconciled with* the scientific image [it is in any case *"in principle* impossible" to reconcile the manifest and scientific images, to bring man into the scientific image], but rather something to be *joined* to it. Thus, to complete the scientific image we need to enrich it *not* with more ways of saying what is the case, but with the language of community and individual intentions, so that by construing the actions we intend to do and the circumstances in which we intend to do them in scientific terms, we *directly* relate the world as conceived by scientific theory to our purposes, and make it *our* world and no longer an alien appendage to the world to which we do our living. We can, of course, as matters now stand, realize this direct incorporation of the scientific image into our way of life only in imagination. But to do so is, if only in imagination, to transcend the dualism of the manifest and scientific images of man-of-the-world.[20]

This extraordinary passage, at once one of the most philosophically optimistic, self-defeating, and influential conceptual recommendations that can be imagined, helps to clarify the sense in which the strong tendencies at present to eliminate altogether the mental, the psychological, the personal, the linguistic, the historical, and the cultural – in the name of a rigorous scientific realism – move to theorize that *all the phenomena within the space of personal life* (the space of the manifest image) *are functionally and only functionally specified and therefore are not real*.

The *aporia* of Sellars's account rests with the intended truth that, if, within the real-time concerns of persons, we adjust our science to our changing interests and intentions and horizons, *we* must understand (in doing that) that, ultimately, in the "completed scientific image," we will be conceptually discharged.

The program is not quite as clearcut as our summary suggests. But it does serve to sketch the sense in which

1   functional distinctions are to be treated as essentially anthropocentric or, in addition, anthropomorphic;

2   functional distinctions are merely heuristic, conventional, interest-centered, provisional, or fictive;

3   observational distinctions are irretrievably anthropocentric; and

4   functional distinctions are, if conditions 1 and 2 obtain, precluded from naming or designating, as such, real things or real properties, in accord with the ultimately reductive requirements of scientific realism.

Conditions 1–4 seal the fate of any realist presumptions favoring the "manifest image."

Certainly, at the level of the "second" (parasitic) set of functional attributes – those attributes that belong to names and terms organizing the "scientific image" favored by persons who identify themselves within the space of the "manifest image" – the very intelligibility of science is risked by the apparently indissoluble linkage between the "first" and "second" sets of distinctions, whether in science or in literature. Furthermore, it is a foregone conclusion that the functional distinctions of the "second" set could not possibly lead as such to universal law-like regularities: on Sellars's arguments, they answer only to the heterogeneous, unsystematic, shifting interests of persons-in-the-manifest-image – itself a thesis irreconcilable in principle with the ultimate scientific image. Consequently, it would be most problematic to expect that distinctions in accord with the "first" set – infected as they cannot but be by the intentions and interests governing the "second" (by which they are generated) – could reliably isolate empirical regularities of a law-like nature.

But that of course goes entirely contrary to the very reason adduced for preferring the scientific image to the manifest image. On Sellars's view, the microtheories of the natural sciences "explain empirical laws and explain observational matters of fact only in the derivative sense that they explain [the empirical] explainers of the latter"; they "*explain empirical laws by explaining why observable things obey to the extent that they do, these empirical laws*"; and they explain not only "why observable things obey certain laws, they also explain why in certain respects their behavior obeys no inductively confirmable generalization in the observation framework."[21] So the superiority of microtheoretical *unobservables* is said to be that they explain the inevitably deformed, approximative function of the observable order: they do that by positing entities and processes that, by definition, instantiate nomic universals in a strict way. Scientific rigor is expected to ensure a testing of the fit; but the entire procedure, after all, rests (and cannot but rest) in the hands of creatures (persons) who can hardly exempt themselves from the intrusive prejudices of their own praxical, historical, observational, interest-driven, "manifest" orientation. In fact, *they* must assess the *fit* of the observable and the theoretical. There's the *aporia* once again – the revenge of Husserl and Nietzsche and Heidegger and Derrida, we may say.

If the space of the personal cannot be eliminated for either or both of two reasons,

1   that the use of conceptual categories would be incoherent without reference to the personal (under the condition of denying the cognitive transparency of nature), and

2   that the elimination of the categories of the specifically personal (the linguistic, intentional, semiotic, agental, psychological, mental, experiential, societal, interpretive, historical, cultural) – servicing the processes of reason 1 – cannot be shown to yield in a promising way to any known strategy,

then functional (and intentional) categories cannot be eliminated from a realist science. There's the pivot of the argument.

This concession commands our attention in two important respects: first, because *all* the categories and distinctions of science would *then* have to be admitted to reflect the world discriminated *at* the level of the "manifest-image" interests of human scientists; and, secondly, because *all* the activities of persons – naming, describing, intending, desiring, believing, thinking, acting, explaining, altering, making, creating, facilitating, reasoning, choosing, deciding, inventing, intervening, adjusting, correcting, and interpreting and the like – could function distributively *only if* they also co-function holistically (relationally rather than atomically). The first would entail that the realism of any science would have to concede the realist import of the functional properties of human inquiries; and the second would oblige us to adjust our conception of the competence of the relevant sciences with regard to the logical peculiarities of functional and intentional properties.

There are, indeed, other categories of a functional sort that would need to be sorted – famously, the functional properties of the sub-human biological order (for instance, the properties of the DNA code and of the selective mechanisms of evolutionary change). But the dual concession remarked is decidedly more important than the latter. First of all, it would signify that there are no distinctions pertinent to the development of a rigorous science that are not hostage (because of the relationship between the "first" and "second" [parasitic] set of predicates) to the irreducible, radically functional interests, observation, activity, inquiry and categories of creatures who do not (and in a sense cannot) understand themselves in terms that subvert their own emergent level of activity and reflection. In effect, this would signify that *all* explanation – even the kind of eliminative, bottom-up explanation Sellars favors in microtheoretical terms – would, necessarily, be an artifact of top-down reflection. Secondly, it would signify that all sub-personal distinctions that might serve to explain the functioning of molar persons themselves could, logically, be introduced only relationally or factorially as the sub-functions of the holistic functioning of such persons (however modified the categories of person-level life might come to be).

This is a most interesting consequence. For it is often supposed that an ultimately reductive or eliminative account of the anthropocentric, functional categories of "folk" theories, or "manifest-image" theories, cannot be resisted in a principled manner *if*, as seems entirely reasonable, man evolved in some causally continuous way from a lifeless order of physical materials and processes – the contingent events of which are explicable in exclusively physical terms. The point is that, *if* what is identified and explained (including the capacity to identify and explain) is conceptually hostage to the emergent talent of humans to reflect on their world and their relationship to it, then even the theory of their evolution from an order of nature described and explained adequately in physicalist terms *is a posit controlled at a level of cognitive entry that may itself not yet have yielded satisfactorily to a physicalist reduction or to a more extreme eliminative option.* It is entirely coherent to entertain the possibility that the very conditions of a *human* science impose threshold conceptual constraints below which we cannot intelligibly explore, or below which we cannot intelligibly explore without providing a companion reduction or elimination of the categories descriptive of human inquiry itself, or below which we cannot presume to explore intelligibly without acknowledging that the would-be replacing categories are themselves the posits of real human inquirers.

Once we grasp the peculiar persistence of what Sellars calls the manifest image, we cannot ignore the patent paradox of his own careful strategy – which would have us not hurry to embrace the options he himself (rightly) deplores but which would also have us believe that an extreme eliminative materialism *will* ultimately be vindicated, *is* already vindicated at least as an ontically sound guide for the genuine sciences.

II

It is in this latter regard that Sellars is surely the inspiration behind the extreme eliminative strategies of Daniel Dennett, Stephen Stich, and Patricia and Paul Churchland. Thus, for instance, Dennett closes his *Content and Consciousness* in the following way (though Sellars is not mentioned in the account):

> The story we tell when we tell the ordinary story of a person's mental activities cannot be mapped with precision on to the extensional story of events in the person's body, nor has the ordinary story any real precision of its own. It has no precision, for when we say a person knows or believes this or that, for example, we ascribe to him no determinable, circumscribed, invariant, generalizable states, capacities or dispositions. The personal story, moreover, has a relatively vulnerable and impermanent place in our conceptual scheme, and could in principle be rendered 'obsolete' if some day *we* ceased to *treat* anything (any mobile body or system or device) as an Intentional system – by reasoning with it, communicating with it, etc.[22]

What is particularly instructive is that the ontic elimination of persons is favored by Dennett because the functional attributes that would have to be conceded *if* persons were irreducibly real – particularly the attributes of intentional life – would not be compatible (on Dennett's view) with a conception of science putatively fitted (already) to the exemplary study of physical events and, *therefore*, normatively implicated in any would-be scientific psychology or social science or the like. Quite apart from the issue of the internal coherence of his thesis about the actual eliminability of "the personal story" (homuncularism), the fact is that Dennett never addresses the question of how in general to test the reasonableness of adhering to a model of science faced with the recalcitrant reality of entities whose (functional) properties *are not* explicable in terms merely of the laws of physical nature.

Dennett actually offers a sketch of what we should regard as a functional property:

Let us mean by a *functional structure* any bit of matter (e.g., wiring, plumbing, ropes and pulleys) that can be counted on – because of the laws of nature – to operate in a certain way when operated upon in a certain way. Obviously just about anything can be viewed as a functional structure from one point of view or another. A functional structure can break down – not by breaking laws of nature but by obeying them – or operate normally. A nail is a functional structure; so is a gall bladder, and an open telephone line between Washington and Moscow.[23]

For Dennett, functional discourse is a *façon de parler* for phenomena described and explained adequately and entirely in physical terms and in accord with physical laws. "Non-intelligent information storage [another functional distinction] is nothing more," Dennett explains, "than reliable plasticity of whatever lies between input and output"; hence, such storage forms the basis for intelligent storage. The evolution of the brain and of higher intelligence simply needs "some intra-cerebral function to take over the evolutionary role played by the exigencies of nature in species evolution; i.e., some force to extinguish the inappropriate."[24]

Dennett's solution is admirably straightforward. "Content" (intentional content) must somehow be ascribed to particular bodily or neural events; but "An event, state or structure can be considered to have content only within a system as a whole" – in effect, only with respect to "what the organism [in some 'expressively' pertinent sense: by verbal expression or behavior or the like] 'takes the signal [to which it is said to respond] to mean.'"[25] So Dennett concludes that "The ideal picture . . . is of content being ascribed to structures, events and states in the brain on the basis of a determination of origins in stimulation and eventual appropriate behavioral effects, such ascriptions being essentially a heuristic overlay on the extensional theory rather than intervening variables of the theory."[26] A so-called "centralist theory" (Dennett's own) will afford "an extensional account of the interaction of functional structures" and "an intentional characterization of [those] structures" as well as of the events and states within them and resulting from them.[27]

So much is good and finely drawn. But it only proceeds in one direction – the wrong one, as it happens, for our needs. Dennett shows us how *physical systems* may be described functionally, informationally, as intentional but not intelligent (possibly even as intelligent). He nowhere shows us how a system – in particular, how the functional life of real human persons – *can be analyzed in terms of the model he provides*. There you have the Sellarsian difficulty again. It's no good saying that the eliminative program will work because the heuristic program works. That is simply not yet a settled finding we have at our disposal.

The Churchlands and Stich tend to favor both Sellars and Dennett. Stich usefully notes the unnecessary restrictedness of Sellars's account: his attack on "folk" psychology goes beyond Sellars's attack on the "manifest image," in the specific sense that "for Sellars the manifest image is restricted [unnecessarily] to the observable, while folk psychology is up to its ears in unobservable 'theoretical' states."[28] Otherwise, the point of these later strategies is very much the same. What is more instructive is their amplification of the notion of functional properties. In fact, there is not likely to be a more revealing inadvertence than the one Paul Churchland conveniently betrays in his otherwise knowledgeable summary of the current literature.

On Churchland's view, "According to *functionalism*, the essential or defining feature of any type of mental state is the set of causal relations it bears to (1) environmental effects on the body, (2) other types of mental states, and (3) bodily behavior."[29] Like Dennett, Churchland treats pain as the natural specimen of a mental state by reference to which the adequacy and force of the thesis may be assessed. The point of functionalism, as opposed, say, to that of behaviorism and reductive identity theories, is that, for the functionalist, "the adequate characterization of almost any mental state involves an *ineliminable* reference to a variety of other mental states with which it is causally connected."[30] Functional states are identified as and only as causally linked to other mental states (or at least to some subset of bodily and environmental phenomena to which mental states are essentially linked token-wise). By virtue of this logical feature, a given mental state may, on further evidence, be *causally correlated* with other physical or mental phenomena. That is, on the functionalist thesis, mental states and phenomena cannot be *first* identified on independent grounds and then, contingently, correlated with other phenomena (mental or physical); they are first introduced as necessarily linked in causal ways with other phenomena that can be independently identified. The great trouble with functionalism (thus construed), is that (ultimately) it cannot countenance what Jerry Fodor has usefully dubbed "mentalism": the position of one who denies "necessarily *P*," where *P* holds (in effect, the behaviorist's thesis, on Fodor's reading though not on Churchland's) that "For each mental predicate that can be employed in a psychological explanation, there must be at least one description of behavior to which it bears a logical connection."[31] The behaviorist reading (on Fodor's characterization) is a sub-species of the functionalist's (on Churchland's), if only because (on Churchland's view) the logical connection need not be restricted to behavior – it may include any pertinent bodily or environmental element. ("Logical," of course, on Fodor's usage, may be as strict or as loose as we please, ranging from strict entailment to Wittgensteinian criterial connections. But Wittgenstein himself need not, for that reason, be taken to be a behaviorist, in the sense that we need not conflate the conditions for the public intelligibility of discourse about mental states with the defining conditions of such states themselves.) The initial (and essential) trouble with Churchland's view of functionalism, then, is that it rather carelessly conflates the issue of the public accessibility of mental states (mental states require cognitive criteria of some sort) with the issue of the defining nature of such states (causal connection cannot be a defining feature of mental states *if* mental states are real). (It should also be noted that, although he professes to be a "mentalist," Fodor is also a functionalist – which is to say, he ultimately fails to discharge his functionalism in favor of his mentalism. The only way to solve the puzzle, *if* one is to avoid eliminationism, reductionism, and dualism, is to reject the thesis that functional properties are first introduced causally and to affirm the thesis that functional properties are, where real, incarnate.)

Now, since he defines mental states as central states apt (in some sense) for producing certain effects,[32] the functionalist is able

1   to deny strong, so-called "type"-identity theories, to the effect that "there is [a] single type of physical state to which a given type of mental state must always correspond" (which, in any case, entails Fodor's mentalism);[33]

2   to affirm the putatively "weaker" thesis of "token"-identity, to the effect that "each *instance* of a given type of mental state is numerically identical with some specific physical state in some physical system of other" (which Churchland assures us "virtually all [functionalists] remain committed to"[34]);

3   to accommodate the possibility that a given *type* of mental state (pain, say) may be manifested in an immense variety of open-ended, unforeseeably novel physical states ("there are *too many* different kinds of physical systems," Churchland advises us, "that can realize the functional economy characteristic of conscious intelligence," pain, or any other mental states[35]); and

4   to support the thesis that "the science of psychology is or should be *methodologically autonomous* from the various physical sciences

such as physics, biology, and even neurophysiology," that it "has its own irreducible laws [functional laws] and its own abstract subject matter [functional isomorphisms]."[36]

Churchland, however, is more sanguine than the functionalist. He defines the functionalist's position but he rejects it. He rightly pursues the import of the functional analogue of properties such as that of *temperature* in physics, of the analogue of the fact that "the physical property of temperature enjoys 'multiple instantiations' no less than do psychological properties" – observing for instance that the intertheoretic identity "temperature = mean kinetic energy of constituent molecules" holds strictly only for the temperature of a gas, that temperature "is realized differently" in solid plasmas and vacuums, and hence that "*reductions* [in the physical sciences] *are domain-specific.*" The upshot, he concludes, is that "we may expect some type/type reductions of mental states to physical states after all, though they will be much narrower" than type-identity theories had supposed, and (as a result) "the radical autonomy of psychology cannot be sustained."[37]

This is a very pretty picture, but it won't do. For, in the first place, even where token-identity is affirmed, the possibility of type–type reductions *of functionally law-like regularities* involving mental states requires some methodologically rigorous procedure for *first* fixing mental *types and their corresponding functional laws.* This clearly shows the inadequacy of Dennett's treating the functional (in particular, the mental) as if it were always introduced by reference to reasonably settled physical laws. But Churchland offers no reason for supposing (short of the dismissive option of the eliminationist, who holds that folk psychology, and functionalism within it, "is an outright *mis*representation of our internal states and activities"[38]) that *we can first fix functional (abstract) isomorphisms in a suitably law-like way* – isomorphisms that may thereupon be viewed as fair ·candidates for the narrower type–type reductions of the physical sort Churchland envisages. *If* the mental is *typed* in ways that are intensionally complex, context-sensitive, possibly interpretively freighted in historically and culturally diverse ways, subject to all the vagaries of manifest-image and folk-theoretic constructions (that, on Sellars's view at least, cannot independently lead us to recognizably law-like regularities), then the very prospect of psychological laws *of* the functional sort cannot but be conceptually and methodologically suspect. Churchland is an eliminationist, so the problem does not bother him. But, if we are obliged to fall back from eliminationism to something like the functionalist's initial position, then the difficulty we are here considering radically threatens the alleged closure and homonomic nature of the physical sciences themselves.[39]

Secondly, if mental ascriptions are introduced holistically– that is, in radically relational, context-dependent, and intensionally open-ended ways,[40] then, if, as Churchland himself affirms, mental states are *defined* in terms of the causal relations they bear to other mental states, the possibility of psychophysical type–type reductions of the narrower sort would require that the *mental effects* of mental states could themselves be suitably regimented (independently) in terms of functional equivalences of a suitably law-like kind. But that would mean that functional (psychological) laws *could* be affirmed on grounds methodologically quite different from those obtaining in the physical sciences before being reduced. It would then be an empirically open question whether such laws could actually *be* reduced in (physicalist) type–type terms. So functionalism, contrary to its avowed purpose, may be taken to challenge in a serious way even relatively loose versions of the unity-of-science program.

Thirdly, perhaps most importantly, *if*, along the lines Sellars and Stephen Stich have most strenuously pursued (against folk psychology and the manifest image), *the domains or sectors of inquiry to which functional* (in particular, psychological) *claims are fitted are themselves essentially described in "folk" or "manifest" terms* – which is to say, also *functionally* – then it is by no means clear that whatever psychological functions are to be secured in a suitably reductive science *can* be meaningfully and methodologically so secured without a prior reduction of "folk" physics and "folk" biology. This is a corollary of the thesis, broached earlier, to the effect that the distinctions favored in human inquiry may well be a function of the cognitively native level at which we investigate the natural phenomena we do. Very possibly, the functional distinctions of psychologically significant life, or of biological evolution, would make no sense and find no place among the distinctions of a ramified and comprehensive microphysics. The difficulty is very clearly remarked by Jerry Fodor:

> it looks, at first blush, as though we might find a path from talk about functional equivalences to the comforting causal talk characteristic of the hard sciences if we take functionally equivalent systems to be those whose effects are indistinguishable. But the appearance is misleading unless we have some reason to believe that the states of affairs upon which functionally equivalent systems converge must not themselves be functionally defined. As things now stand, there is no particular reason to believe this.[41]

Churchland does not consider the threat, which, in a profound sense, is just the consequence of the aporetic elimination of personal-level discourse Sellars and Dennett have favored. But the threat is hardly restricted to the generic consequences of rejecting cognitive transparency or of affirming the conceptual indissolubility of the realist–idealist symbiosis of our cognitive inquiries. It bears quite directly as well on the so-called "many–many" problem (Feigl),[42] on the heteronomy of the phenomena of functional regularities, and on the apparent ineliminability of interlevel theories in the sciences. It is also of course the implicit threat imposed by any merely general insistence on the empirical nature of the sciences themselves – that is, on their dependence on the disciplined experience of professional investigators.

III

Functional predicates are clearly quite heterogeneous on any account sensitive to questions of causality and nomological connection. They are hardly restricted to psychological contexts, in a descriptive sense, although no descriptive contexts of any kind can, *qua* descriptive, be entirely disengaged from psychologically rich considerations. This is simply once again a tribute to the persistence of folk-theoretic or manifest-image notions. In fact, the most strategic first question about functional predicates asks whether they are being introduced heuristically or realistically – or whether, if introduced heuristically, they are to be taken as heuristic renderings of real properties or as ultimately eliminable altogether. For, when we speak of plants "searching for" nutrients, we suppose that our heuristic idiom conveniently enables us to theorize about real structures and processes that are not purposive, intentional, psychological, or linguistic; but, when we speak of human memory and choice, even if our folk-theoretic idiom is hopelessly informal at the present moment, we cannot suppose (unless we favor the eliminationist option) that a corrected or rigorous description of *such* (psychological) phenomena will fail to remain functional in personal-level terms. When we describe the mental states and behavior of animals, for instance – say, lions perceiving, hunting, stalking, attacking, and killing eland – we may well be obliged to fall back to an admittedly heuristic rendering of the *real* psychological life of those animals. But what would be the consequence of holding – in the full, reflexive space in which we examine ourselves psychologically, sociologically, historically, linguistically and in related ways – the realism of the functionally apt descriptions of ourselves we seem unable to discharge or dismiss or subvert or replace or reduce or discipline along the lines of the nonfunctional or benignly (or extensionally) functional vocabulary of the physical sciences together with the realism of that physical vocabulary?

The remarkable implications of these somewhat casual options are easily overlooked – as they have been in much of the central literature on functional predicates and properties, which views them chiefly in terms of the local puzzles of fairly narrow sectors of inquiry. We have admitted three options:

  (i)   a purely heuristic use of functional predicates;
  (ii)   a robustly realist use of functional predicates; and
  (iii) a heuristic use of functional predicates in order to designate and represent *real* functional properties otherwise difficult or impossible to designate more perspicuously.

Option (iii) is the most interesting and the most strategic. (i) is unproblematic where deliberately so intended (as in speaking of plants "searching for" nutrients), even where what is thus designated may, on an argument, still need to be recovered in terms of (ii) or (iii). Naturally, in reductionist and eliminative accounts, (i) is more controversial. On Dennett's view, for instance, regardless of a speaker's intention, functional predicates are invariably heuristic, in the sense (already remarked) that they are *introduced* to designate the behavior or operative change of things *due to* causes entirely subsumable under given physical laws. It is certainly possible that the "searching" of a plant is capable of being redescribed in accord with what, effectively, is Dennett's reading of (i). Dennett, however, offers nothing to establish that his reading should be preferred – whether for plants or for human selves; and, in general, the informational modeling of biological phenomena, particularly with regard to the developmental differentiation of multi-celled organisms from zygotes, the evolution of distinct species, and the functioning of individual organisms in multilevel ways that require joint reference to genetic coding and normal survival behavior may well oblige us to favor options (ii) or (iii).[43]

Furthermore, the treatment of psychological, social, linguistic, and cultural phenomena in terms of a reductive or eliminative reading of (i) does not follow from a successful (such) treatment of biological phenomena of the sorts just noted, and cannot possibly be achieved where we have not yet been successful at the strictly biological level. The interesting question to pose is, What might failure to achieve a universal reductive or eliminative application of (i) signify? The answer returns us to option (iii). In any case, Dennett's account of functional predicates is altogether too sanguine, unguarded, tendentious, and utterly lacking in empirical support. Furthermore, Churchland's emendation –

which provisionally acknowledges functional laws and does not suppose functional predicates to be first introduced on the basis of known physical laws – cannot, by dint of its own commitments to token identity, consistently do more than refine the intent of Dennett's original option. Both Churchland and Dennett are, effectively, eliminationists of that stripe Sellars best exemplifies; although, like Stich, they are not bound to restrict the "folk"-theoretic to the perceptual order of things – in Sellars's way.

The resources and resilience of option (iii), however, are considerable. For (iii) permits us to segregate, for purposes of argument, the grounds for a realist reading of given phenomena from the grounds for favoring particular predicative strategies for describing them within some pertinent science. These are matters not normally distinguished, because it is ordinarily supposed (in "objectivist," "naturalist," "inductivist," or similar terms) that arguments concerning the one are indistinguishable from those concerning the other. But, if we reflect on option (iii), other conceptual possibilities suggest themselves. For one thing, if (as already argued) the categories of *every* science are ineluctably dependent on conceptual constraints due to the conditions under which human inquirers explore any and all sectors of nature and reflect on their own aptitude and behavior in doing so, then all scientific discourse is *functional* in the sense in which (implicitly) it is at least heuristically adjusted to capture how what we take to be real presents itself to such investigators so constrained. Even the nonfunctional vocabulary of the physical sciences (on, say, something like Dennett's view) *is* functional in this sense: the categories of science answer to the tacit interests of human investigators even where the functional objectives of those interests do not explicitly color the distributed categories of a particular science.

This may be the essential point of Thomas Kuhn's notion of the history of the shifting influence of shifting scientific paradigms. Even when registering his own reluctance to repudiate altogether the "epistemological viewpoint that has most often guided Western philosophy for three centuries," the view that "sensory experience [is actually] fixed and neutral" – the view, in effect, that it is true that "Galileo interpreted observations on the pendulum" even though the pendulum did not "really" exist before the fourteenth-century paradigm shift originated by Jean Buridan and Nicole Oresme[44] – Kuhn nevertheless candidly admits that the notion "of a neutral language of observations now seems to me hopeless": "What occurs during a scientific revolution is not fully reducible to a reinterpretation of individual and stable data. ... Rather than being an interpreter, the scientist who embraces a new paradigm is like the man wearing inverting lenses."[45] Kuhn goes even further of course, in the Postscript to *The Structure of Scientific Revolutions*: "There is, I think, no theory-independent way to reconstruct phrases like 'really there'; the notion of a match between the ontology of a theory and its 'real' counterpart in nature now seems to me illusive in principle."[46]

We may understand Kuhn's mixture of a bold rejection of cognitive transparency and the advocacy of a radically historicized inquiry, on the one hand, and his own fiddling (as in the Galileo case) with formulas that do not altogether repudiate a conserving continuity with the transparency canon, on the other, as a series of reflections on option (iii). Narrowly construed, his speculation means to steer a middle course between the extravagances of Feyerabend (who in a most provocative way advocates a "science without experience"[47]) and the inconsistencies of Popper (who, while rejecting essentialism and transparency, holds to a version of the canonical fixity of observation in virtue of which verisimilitudinous "approximations" are still taken to make scientific sense[48]). More broadly construed, Kuhn is insisting on the inseparability of the prevailing network of scientific categories that jointly form and affect observation and theory from the history of paradigm shifts within the range of actual science. In this respect, Kuhn offers a moderate, rather narrow-gauge, almost minimal rendering of the following thesis:

(a)   scientific concepts are functional inasmuch as they presuppose and entail the rejection of cognitive transparency, the concession of a realist–idealist symbiosis, and the reflection, within those terms, of the tacit interests of paradigm-centered communities of scientific investigators.

(We shall pick up the rest of the tally shortly.)

That thesis, even within the narrow gauge of Kuhn's concern, is deepened along sociological lines in Ludwik Fleck's neglected study of the complexities of the linkage between the phenomenon of syphilis and the "Wassermann reaction." Fleck demonstrates that "from the very beginning the rise of the Wassermann reaction was not based upon purely scientific factors alone." But Fleck goes on to support his charge in the context of modeling the social processing of scientific work itself ("journal science" and "vademecum science"), in terms (*contra* Kuhn) of all stages of science (without regard to the distinction between "normal" and "revolutionary" science), and, most important, in terms of treating a scientific *fact* as (as the very title of Fleck's book confirms) an artifactual outcome of perceptual minima and the social processing of such minima.[49] It is true, ironically (perhaps even pathetically), that

Fleck should have cast his finding (in 1934–5) in terms of the "thought style" (*Denkstil*) of a form of collective thinking ("thought collective," *Denkkollektiv*) along Durkheimian lines.[50] Nevertheless, Fleck remains more explicit than Kuhn on the interpenetration of putatively scientific and nonscientific (praxical) orientations. Fleck characterizes "thinking as a supremely social activity which cannot by any means be completely localized within the confines of the individual"; and, thus construed (with due allowance for his attraction to the notion of the collective mind), he provisionally defines a "scientific fact" as "*the signal of resistance* opposing free, arbitrary thinking" – or, more compactly, as "*a thought-stylized conceptual relation which can be investigated from the point of view of history and from that of psychology, both individual and collective, but which cannot be substantively reconstructed* in toto *simply from these points of view.*"[51] The "relation" is not altogether satisfactorily developed in terms of the so-called "active and passive elements of knowledge" (perception, particularly) that are processed in the collective life of a scientific community. But what Fleck brings home very clearly is the potentially executive role, in science itself, of initially unrelated intellectual interests within a given community; its conceptual orientation and "style"; and its capacity to resist, within its own membership, the ideation of actual investigators.[52]

In pursuing these matters, Fleck moves us inexorably in more generalized Marxist, Hegelian, and Freudian directions –

(b) scientific concepts are (or may be said to be) *functional*, in manifesting or reflecting the class- or *Geist*-like or unconscious or related interests of individual investigators and of communities of investigators.

The difference between (a) and (b) corresponds, if we may risk the comparison, to the largest alternative themes of Husserl's notion of the *Lebenswelt*: the indissolubility of the realist/idealist features of an operative science (a) and the ideological, historically skewed, fragmentary, plural, tacitly conative, interested, praxical exploitation of that symbiosis (b). In fact, what Fleck compellingly demonstrates (more clearly than Kuhn) is just how the entire range of praxical concerns – influential within the life of an historical society (hardly scientific in themselves) – color and are inseparable from the governing concepts of scientific explanation that emerge and dominate (for a time) the science of that society. This, of course, is an aspect of theorizing about scientific theorizing that is very nearly completely ignored by strongly elimination-ist programs of analysis.[53]

Step back from the argument for a moment. (a) fixes the functional role of concepts, all concepts, insofar as they manifest the structuring role of a community of investigative selves. That minimal constraint entails, as such, no determinate doctrine of the nature of world or self beyond their symbiosis and beyond the irrecoverability (within the space of that symbiosis) of any and every form of cognitive transparency. From the point of view of functional properties, (a) signifies *the incapacity of human investigators to penetrate (cognitively) the world or self or the society of selves below that level at which all posited conceptual distinctions are, at least tacitly, artifactually affected by the symbiosis of self and world.*[54] It follows at once that option (ii), a robustly realist use of functional predicates (the analogue of a robustly realist use of *physical* predicates), is, in an obvious sense, no more than a posit internal to option (iii). Option (iii), particularly on interpretation (a), effectively absorbs *all* conceptual speculation deprived of cognitive transparency – what, alternatively approached, amounts to *pragmatism* (or, the union of naturalistic, phenomenological, and deconstructive orientations, each serially monitored by the discipline of the others). But what (a) decisively establishes is that, even assuming the evolutionary continuity of human selves from an animate sub-human world and, ultimately, from a lifeless planetary physical soup, the facts of that continuity are themselves posited (in accord with (ii)) within and only within our option (iii) – in which the salient structures of cognizing selves cannot be dismissed, ignored, or guaranteed to be eliminable.

The upshot is that (a) affects in the most fundamental way possible the unity-of-science program and every conception of what a viable science must be like. In confirming the ubiquity of functional concepts, the adoption of (a) also confirms that the natural evolution of the human world from pre-biological, pre-human, pre-cultural, pre-linguistic, pre-historical, pre-intentional sources *does not impose a physicalist or otherwise reductive or eliminative resolution on all coherent and self-consistent efforts to pursue option (ii) – whether for the human or for the physical sciences.* To insist otherwise is to commit an enormous conceptual blunder – to effect a complete *non sequitur* at the very least.

The reason is the dual one we have been pressing in a variety of ways: first, because, *if* science is a real undertaking, then the reality of human investigators is ensured; second, because, *if* every science is a human science in the sense that its achievements are essentially constrained by the intentional life, history, and conceptual horizons of contingently placed investigators, then the methodology of any and every science must accommodate the reflexive limits of human inquiry in general as well as the provisionally agreed-upon salient features of any particular

domain of inquiry. It may help, to fix our ideas, to recall that Nancy
Cartwright had found a plausible way of interpreting relativity and
quantum physics as the orderly achievement of human inquirers – a way
profoundly opposed to Sellars's vision of a reductive microphysics in
which persons and macroscopic objects are simply eliminated; and to
recall that Sellars himself (as well as his better followers) had invited us
not to dismiss *as yet* the intentional complexities of human existence –
that is, before we were in a position to achieve the final "scientific
image." But there cannot be any principled disjunction between a
"correct" methodological preference for ensuring the scientific status of
the human sciences (or of the physical sciences, for that matter) and an
"ontological" commitment or interpretation of such a preference. For,
for one thing, every methodology must be fitted to a supposed domain
– which, at least holistically, entails a realist presumption; and, for
another, distributed ontic neutrality or fictional reference or the like,
within the scope of a given methodology, cannot (quite) be ensured on
the mere say-so of its advocates.[55] When, therefore, Dennett charges,
"psychology *without* homunculi is impossible. But psychology *with*
homunculi is doomed to circularity or infinite regress, so psychology is
impossible ... all homunculi are ultimately discharged,"[56] he fails to
consider that the "circularity or infinite regress" may initially lie not so
much in analyses *internal* to an (optional) intentional or linguistic or
molar or "folk" or "manifest" account of psychological phenomena (in
contest with reductive or eliminative alternatives) as in analyses *reflexively
conducted by* "selves" *upon* "such" phenomena (that is, phenomena
involving the very selves Dennett means to eliminate) and *therefore* to
afford (apparently ineliminably) a "psychology *with* homunculi."
Dennett wishes to discharge all psychologies that aim at self-understanding while they preserve self-understanding selves. *He* assumes that he
can discharge homunculi: "One *discharges* fancy homunculi from one's
[psychological] schema by organizing armies of [more and more stupid
homuncular] idiots to do the work."[57] But Dennett never explains how
that can be actually accomplished empirically; and he never considers
that, if it is impossible, he himself will have to revise his assessment of
psychology quite radically. The conceptual blunder or *non sequitur*
concerns only the latter issue. The former invites an assessment of
another kind. Without the eliminationist's strong reading of option (i),
there is absolutely no warrant for Dennett's rejection of the conceptual
viability and pertinence of person-level or "folk" psychology: there
would simply be no other way to proceed. But there is also no known
knockdown argument favoring the eliminationist's reading – or favoring
even its coherence as a fully developed empirical program. The lesson

to be drawn is a straightforward one: the methodology of a science cannot be reasonably formulated without accommodating the nature of the phenomena of the domain it would investigate – even if our picture of the nature of that domain is (in some sense) an artifact of the contingent discoveries of an ongoing and disciplined science.

IV

We are pursuing, here, a natural declension of option (iii). (a) posits in the least encumbered, least determinate, least tendentious sense possible the respect in which all inquiry manifests, in its very success, an implied limitation upon conceptual invention. It sets no explicit condition favoring one schema over another (not even regarding schemata of the most radically eliminationist sort); it advises us rather of the need to acknowledge, and to remain consistent with acknowledging, the fact that all conceptual networks presuppose and entail the existence and the work of entities (selves) capable of forming and using such networks. *If*, then, the cognizing agents of science cannot be eliminated or conceptually reduced in terms of the categories adequate to the fundamental order of physical nature from which (quite reasonably) we suppose them (ourselves) to have somehow evolved, the realist claims of all the sciences cannot but be functionally generated within the space of option (iii) – that is, in terms of whatever may be the salient constraints under which given human communities are able to understand themselves and their world. (a), therefore, naturally suggests (b), the entire possible array of reflexively *interpretive* schemata, spontaneously, critically, or "suspiciously" applied to our understanding of what we take our understanding of our real world to be.[58] The categories of (b) are explicitly functional (as opposed to those of (a)) in being specifically linguistic, representational, interpretive, intentional, purposive, institutional, praxical, significative, or the like.

(b) captures an enormous array of categories ranged along two interlocking dimensions: first, regarding the historically contingent, largely tacit, enormously varied, incompletely fathomable complexities of all the different ages or epochs or societal islands of self-understanding that fall within the scope of (a); second, regarding the dialectical complexities within the ethos of any particular such island or society affecting the division of the labor in it (of every sort), the incapacity of its aggregated members to internalize completely and distributively its operative distinctions, and the untotalizable inventiveness of its own salient practices. The first dimension is most distinctly associated (again at the risk of misunderstanding) with the work of Heidegger, with the doctrine of *Dasein* – which links (a) and (b) in a compelling reversal of the import of Husserl's notion of the *Lebenswelt* and which also emphasizes the radical provisionality of alternative forms of (b) *as* heuristic renderings (at least) of what, in terms of obvious salience, we cannot fail to treat as real. Admitting Nietzsche's influence on Heidegger and the Nietzscheanized extension of Heidegger's theme, (b) may also be traced in the work of Michel Foucault.[59] (But these are mere hints of possibly useful lines of connection.) The second dimension has in various ways been most notably developed along Hegelian, Marxist, and Freudian lines; also, in somewhat paler but hardly negligible ways, in Wittgenstein, Dewey, the Frankfurt Critical thinkers, and the structuralists.

The relevance of the first dimension is always implied in the second, and the two dimensions together (what we are calling the (b) reading of (iii)) always presupposes (a). Furthermore, once we grant (b), it is impossible to avoid a further declension of (iii), namely;

> (c)   all (necessarily reflexive) conceptions of persons or selves are *functional* insofar as such entities are ascribed linguistic, intentional, deliberative, purposive, reflective, informational, propositional, interpretive, semiotic, creative, productive, praxical, technological, actional and cognitive aptitudes.

(c), therefore, clearly represents the persistence of so-called top–down psychological strategies,[60] now integrated *via* (a) *and* (b) within a large grasp of the import of rejecting cognitive transparency.

Given the immense vigor and proliferation of bottom–up conceptions of the human and psychological sciences, we may draw an extremely provocative – but ineluctable – finding here: *the repudiation of transparency entails the defeat of all reductive, eliminative, bottom–up conceptions of the human sciences.* It entails a robustly realist reading of (c) on an equal footing with any robustly realist reading of physical nature – jointly within the terms of reference of (iii). It explicates the futility and lack of force of Sellars's (and with Sellars, the army of eliminationists') treatment of persons and their characteristic aptitudes as mere fictional or conventional or heuristic or unactual *roles* ascribed to whatever entities (microphysical, on Sellars's view) ultimately prove to be real. By the same token, the argument explicates the sense in which the coherence of Quine's repudiation of intentionality, joined to his charge that the analytic/synthetic distinction is a mere dogma, may be effectively challenged: for the first rejects outright what appears to be finally presupposed by the second. To theorize that we cannot, in correcting for the preformational bias with which we pursue our inquiries except by way of further subterranean shifts in our tacit conceptual orientation, somehow effected through the dynamic conditions under which such inquiries are pursued – a concession fashionably tagged Nietzschean or Foucauldian these days but just as reasonably labeled Husserlian or Heideggerian, or Hegelian or Marxist – *is* to theorize that we cannot completely penetrate the constitutive role we ourselves play in the formation of our own science. There you have the fatal weakness of the eliminationists – fixed with equal facility among naturalists, phenomenologists, and deconstructionists or post-structuralists.

None of this, of course, commits us to any privileged account of the real nature of human selves. There's the beauty of (iii). For (iii) permits us *both* to posit the real functional powers of selves (along the lines of (ii), congruent with whatever realism we may assign the work of the physical sciences) *and* to concede the provisionality of any such posit and the potentially heuristic nature of any particular representation we care to make of our world, ourselves, and our efforts to understand our world and ourselves. This is the upshot of the bearing of (a) and (b), taken together, on (c): we penetrate our own nature under the constraint of an absence of transparency (a) and within the scope of the preformational powers of our *Lebenswelt* (b). The only important finding (c) ensures – it is of course a most important finding – is that the distinctive and salient powers of human selves (roughly, what we collect as the mind) are, paradigmatically, functional and, *qua* functional, real or actual.

(b) and (c), therefore, are linked in rather a different way from (a) and (b). (b) is the subaltern of (a); or, (a) is determinately embedded in (b); or, (a) is transcendentally extracted from (b) – where (b) itself provides our reflexive sense of the reality of actual societal life. (b) and (c), however, are conceptually symbiotic in just the sense in which the societal and the psychological are inseparable at the level of human culture – what has regularly been understood as the problem of methodological individualism.

We may then, for the sake of an orderly account, extend our declension another step or two; thus,

(d) the mental life of animals is *functional*, wherever it is judged to be suitably developed (as among the higher mammals), because it is *modeled* by (and apparently cannot but be modeled by) whatever model we reflexively use in characterizing the human mind.

Here is the clearest instance of a heuristic use of a functional model to represent real (imputed) *functional* properties that we cannot represent in any more perspicuous way. This explains, for instance, what is so misleading in Thomas Nagel's well-known question about the interior life of bats.[61] *Only* human beings can answer Nagel's question. To attribute an intentional life to the higher mammals, for instance, may be to waver unavoidably regarding the full realism of our *description* of animal life but it need not be to waver about the realism of the actual (functional) properties we thereby identify. Thus, if we attribute cognitive states to lions (stalking eland with considerable skill), then, on a familiar argument,[62] if a lion sees an eland (in a cognitively pertinent sense), it must also see *that* an eland is present *or* that something suitably and similarly pertinent obtains. We may well question how perspicuous any particular propositional ascription may be, but we cannot wonder whether animals entertain propositions. They clearly do not in any linguistically pertinent sense: the propositional content imputed to perceptual states in creatures lacking language cannot but be heuristically modeled on (and distributively ascribed by suitable analogy with) the perceptual states of linguistically apt creatures (ourselves). In fact, even in the human case, propositional ascriptions of the "content" of perceptual states are similarly heuristically modeled on what a linguistically apt speaker might reasonably report as what he sees. The argument here affirms only that, conceding the irreducibility of perceptual states, the ascription of such states to animals entails a heuristic use of the notion of propositional content in order to represent the otherwise inaccessible (but real) *intentional* states of animals.[63] (On this reading of course the intentional cannot be confined to the linguistic.)

Animal psychology is in this sense inherently anthropomorphized, however we may dispute the "concepts" possessed by, or the conceptual capacities of, particular species, or of what, consistent with such ascriptions, may be fitted in intentionally perspicuous ways (without attributing any intensionally pertinent aptitudes to animals whatsoever) to the life and behavior of particular members of a given species. The important issue here concerns only the provisionally parasitic standing of animal psychology *and* its coherence (and utility) as such, within a systematic declension of functional properties.

By a similar but different extension, we also arrive at

(e) the properties of machines, computers, robots, and related artifacts – indeed, the properties of artworks, technologies, and all cultural phenomena – are *functional*, simply because they *are nothing but* the work of humans or the products of human work and interchange.[64]

In an obvious sense, the realism of the informational properties of man-made machines, the representational properties of paintings, the semantic properties of linguistic utterances and the like cannot be less robust than the realism of the functional properties self-ascribed by the human agents responsible for conceiving and executing them. (The matter is entirely neutral, of course, regarding disputes about how, methodologically, to ensure correct or accurate ascriptions to those agents or to the phenomena of their artifactual world.)

We have hardly paused here to supply a ramified account of the work required at every stage in the declension of (iii), particularly with regard to whatever, within the scope of (iii), may reasonably be posited as the confirmed content of (ii). But too many large issues have been released together, all at once, in affirming that *the functional properties of (b) and (c) are quite real, every bit as real as the properties of physical nature.* Effectively, to admit this much is to admit a realist reading of (e), since (e) simply collects, even if redundantly, the artifactual or cultural phenomena of the symbiotic space of (b) and (c); and to admit this much as real signifies that it cannot be conceptually crucial, even if it is empirically crucial, to affirm or deny that the phenomena of (d) are real. The principal appeal of the declension of (iii) is just that it provides an extremely compendious, systematic account of the largest conceptual issues affecting the relationship between the human and natural sciences – and it does this by featuring the distinctive properties saliently ascribed to human selves, *as the paradigm instantiations of functional properties.* Understandably, this issue takes precedence over a full account of the issues raised by each of the stages of that declension.

We have, in effect, answered many of the most powerful objections the partisans of reduction and elimination could possibly raise. We have accounted for admitting the reality of selves, by reference to (a): in virtue of which the very reality of the determinate structures of physical nature is fated to be linked to the reality of selves. We have shown that it is entirely coherent to concede the evolution and emergence of human selves from a natural world entirely devoid of any and all properties that putatively distinguish selves, without needing (or being able) to reduce or eliminate such properties and without being obliged to deny their actuality if not reduced to the terms of the physical sciences. We have shown that, if (a) holds, then the work of every realist science cannot but be constrained by whatever may be taken, within the symbiosis of (b) and (c), to form a fair account of the real properties of (cognizing) selves; and that, as a result, every viable science and every viable conception of a viable science must be dialectically reconciled with that constraint. And we have shown that this entire argument is effectively entailed by the generally acknowledged indefensibility of any and all forms of cognitive transparency. What we have not yet explored is what we should understand by the reality of the functional properties that fall within the scope of (b), (c), (d), and (e) and what the import is of admitting such properties, with regard to the methodology and interconnection among the sciences.

V

Many of the pertinent issues have of course occupied us continually, under other guises, through the developing argument. But there is at least one essential question that we need to take up here. That has to do with understanding *what* we are committed to, regarding the sciences, *in* actually admitting the reality of functional properties. The answer is remarkably simple and straightforward. We must recall that, on Sellars's view, functional properties – just those in fact that we have collected as the saliencies of (c) – are mere *façons de parler*. We must recall that, on Dennett's view, functional properties (essentially the same ones, whether applied to humans or machines) are first introduced (for whatever descriptive or explanatory or practical purpose) as designating what, on independent grounds, are known to fall under covering physical laws. We must recall that, on Churchland's view, functional properties (again, the same ones), if they are known or may be shown to be subsumable under functional laws, may be counted on to yield as well to physical laws or law-like regularities of a considerably narrower gauge than (type-) identity theories usually pursue. Finally, we must recall that a large number of functional predicates may be only heuristically or fictively employed anyway, not intended at all to designate any actual properties directly.

Grant only that

1   functional properties are real;
2   functional properties are causally efficacious; and
3   functional properties are not reducible in physicalist terms.

Then, if we are to avoid Cartesian dualism (or an analogous dualism, where functional properties are not restricted to the mental – for instance, as in evolutionary biology), there is only one possible account to give:

4 *functional properties of whatever kind, satisfying theses 1–3, must be indissolubly complex, incarnate, and emergent with regard to some physical or biological sector.*

It is just the difficulty of supporting thesis 4 that encourages the eliminationist's charge – that, ultimately, functional predicates are only and always heuristically employed. This is certainly the common thread running through the accounts of Sellars, Dennett, and Churchland. The alternative option sketched is, however, the only alternative to the dualist's *and* the eliminationist's (or reductionist's) interpretations of functional properties – *consistent with theses 1–3 and (a)*. If we remind ourselves that the emergence of the paradigmatically functional properties of (b), (c), and (e) need not be explained bottom–up (to be rigorously explained), we may quite easily admit the force of the finding: *functional = incarnate*. To deny that functional properties are essentially, indissolubly encumbered by physical or biological constraints is, effectively, to justify the eliminationist's account of such properties. There would otherwise be no viable alternative to pursue.

So it is that Sellars's, Dennett's, Churchland's projects fail by a hair; and so it is that the entire vision of the cognitive sciences, construed as autonomous disciplines addressed to purely abstract functional laws, cannot but be misconceived. The thesis also fixes the failure of Zenon Pylyshyn's purist vision of the cognitive sciences – quite uncontaminated by neurophysiological constraints; it also (and for essentially the same reasons) fixes the matching failure on the part of the eliminationists (Patricia and Paul Churchland, for instance) to differentiate the conceptual prospects of (what we may call) incarnatist and eliminationist strategies.[65]

Roughly, *incarnatism* holds that, wherever would-be functional properties satisfy theses 1–2 of the tally given just above, they also satisfy thesis 3 – and in satisfying 3 must also satisfy 4, must also be indivisibly complex though emergent with respect to some physical order. Any would-be functional property that does not satisfy 3 may be treated as the misleading "shadow" cast by a purely heuristic predicate: it cannot satisfy 1 or 2 either. This means that functional analysis, however successful when *abstracted* from physical, biological, or electronic media, is ultimately and always addressed (in an empirically responsible sense) to *the instantiating incarnate properties* of a given entity. It *is* possible to construct an analysis of an abstract (or abstracted) capacity or disposition in terms of abstractly conceived componential sub-capacities or sub-dispositions without direct attention to the *incarnate components* of any actual instantiating entity or system; and, in doing that, it is entirely possible to hit on regularities binding or fitting a significant variety of incarnate systems (Pylyshyn's vision, say). But to insist on the autonomy of such a functional analysis is both to miss the conceptual conditions *on which* it empirically depends and to fail to come to terms with the full array of alternative theories regarding the very properties and would-be laws it analyzes.

The "components" of entirely abstracted capacities or dispositions need not correspond with the functioning components *of* actual incarnate systems. Robert Cummins has put the point effectively:

> The biologically significant capacities of an entire organism are explained by analyzing them into a number of "systems" – the circulatory system, the digestive system, the endocrine system – each of which is defined by its characteristic capacities. These capacities are in turn analyzed into capacities of component organs and structures. We can easily imagine biologists expressing their analyses in a form analogous to the schematic diagrams of electronics, with special symbols for pumps, conduits, filters, and so on. Indeed, if transplants and implants ever become commonplace, this is the only sort of description that would achieve real generality.[66]

On Cummins's view, any property construed in terms of an abstract capacity or disposition (the capacity or disposition of a supposed system) may be said to be a *functional property*; and its *analysis* in terms of what it is "to be" such a property, what it is for any system to instantiate that property, what the *sub*-properties are whose co-functioning constitutes the property in question, yields a "functional analysis" or "functional explanation" of that property. Hence, as Cummins rightly observes,

> Since we do this sort of analysis without reference to an instantiating system [such as an actual biological system with *its* component sub-systems], the analysis is evidently not an analysis of an instantiating system. The analyzing capacities are conceived as capcities of the whole system. Thus functional analysis puts *very indirect constraints on componential analysis*. My capacity to multiply 27 times 32 analyzes into the capacity to multiply 2 times 7, to add 5 and 1, etc. These capacities are not (so far as is known) capacities of my components: indeed, the analysis seems to put no constraints at all on my componential analysis.[67]

(We may accept Cummins's account without following his lead regarding the ubiquity of true systems.)

Once we see matters in this way, the challenge of incarnatism is clear. Advocates of a purist cognitive science (Pylyshyn) are convinced that their abstract machine programs not only will yield to a complete extensional treatment (they are after all introduced to do that) but also *fit* adequately and perspicuously – nomically, in fact – every physical, biological, electronic or similar system to which functional or informational properties are empirically ascribed. They have missed two essential constraints: first, that the "law-like" regularities of purely abstract capacities may not (probably could not) be the same as the law-like regularities of actual systems, unless first so introduced; and, second, that the functional properties ascribed actual systems – in particular, those involving human selves and human cultures – may not, even in principle, yield to a satisfactorily extensional treatment. (That, after all, was just Quine's complaint.) By parity of reasoning, the eliminationist, whether favoring computer simulations of the mental (Dennett) or the reductive neurophysiological capture of selected mental functions provisionally needed for the explanation of biological systems (the Churchlands), similarly miss two essential constraints: first, that the functional properties empirically first ascribed to molar or emergent selves may be pertinently altered or displaced only on the basis of what we determine to be the *sub*-molar contribution of neurophysiological *components* (of that entity or system) *to that entity or system*; and, second, that token-identity claims at the sub-molar level are themselves conceptually dependent on first demonstrating that the functional properties of the molar entity or system behave logically in just the same way as do the functional capacities ascribed purely physical, biological (sub-human or sub-psychological) or electronic systems. But this presumption actually fixes the essential *petitio* of all current reductive and eliminationist programs.

It also fixes the point of objection to the excessively weak (functionalist) resistance to reductionism offered in Jerry Fodor's account of human thinking. Fodor is primarily concerned to *preserve* a resilient version of the unity-of-science program without reductionism – hence, in effect, to argue that some irreducible functional laws may not be laws at all:

> there is an open empirical possibility that what corresponds to the kind predicates of a reduced science may be a heterogeneous and unsystematic disjunction of predicates in the reducing science, we do not want the unity of science to be prejudiced by this possibility. ... I take it that this is tantamount to allowing that at least some "bridge laws" may, in fact, not turn out to be laws, since I take it that a necessary condition on a universal generalization being

lawlike is that the predicates which constitute its antecedent and consequent should be kind predicates. I am thus assuming that it is enough, for purposes of the unity of science, that every law of the special sciences [a functional psychology in particular] *should be reducible to physics by bridge statements which express true empirical generalizations* [where the "bridge statements are to be construed as species of identity statements"].[68]

The fact remains that Fodor holds that "the laws of basic science are strictly exceptionless." Hence, for Fodor, "token physicalism" (which he apparently accepts – must accept, on his own thesis) is not tantamount to "reductionism." He means this in the same sense as that in which Davidson denies being a reductionist.[69] But the commitment favoring "true generalizations" in the absence of "true law-like generalizations" *still* presupposes a thoroughly extensional treatment of pertinent functional properties (linguistic, cultural, psychological) as well as a "token physicalism" – neither of which Fodor actually vindicates. The maneuver depends, of course, on Fodor's too-easy assimilation of Davidson's extended use of Tarski's semantic conception of truth.[70] The decisive objection remains unchanged: token physicalism may not be a reductionism in the sense in which "type physicalism" is, but it is as much a reductionism as the other in the sense that (for Fodor) functional properties are, distributively, identical with physical properties. Token physicalism still *refers* all functional properties *and* "true [functional] generalizations" to a purely physical domain. Hence, functional and intentional distinctions remain entirely heuristic for Fodor, even if they prove contingently indispensable for an ongoing science.

We have, through these considerations, tracked down every distinctive way in which functional properties are construed in the cognitive disciplines whenever they are not (as normally they are not) construed as incarnate; *when they are not treated dualistically, they are and must be either heuristically introduced and entirely eliminable in principle or else identical with physical properties taken token-wise.* On the latter strategy, functional "laws" must be "danglers" (in Feigl's sense), "true [functional] generalizations" must be rule-of-thumb conveniences or place-holders for narrow-gauge physical laws (in Paul Churchland's sense). In either case, strict functionalists fail to justify the conceptual relationship they assign to physical and functional properties. The best they can say is that they treat psychological properties functionally (by way of a mere *façon de parler*) whenever those properties cannot yet be brought under covering physical laws or linked to such laws perspicuously.[71] These maneuvers, therefore, never seriously entertain the possibility that some significant sub-set of the properties usually termed functional are actually neither eliminable nor identical (token-wise) with physical properties. Surely, the option deserves a hearing.

VI

We are now in a position to collect systematically the various senses of "functional property" – and, in doing that, to move on to a more promising line of analysis. Minimally, any predicate is a functional predicate if it specifies (in Cummins's sense) a capacity or disposition of a putative system or (in Sellars's sense) a role or conventionally assigned function to such a system. But it would be a mistake to suppose that all functional predicates were only heuristically thus employed, never designated real properties. Further, it is entirely reasonable to suppose (in the various ways favored by Fodor, the Churchlands, Dennett, and Cummins) that many *real properties designated by functional predicates* may be nonfunctionally redescribed in physical, biological, or related ways, by virtue of some version of type- or token-identity (type- or token-physicalism); but, again, it would be a mistake to suppose that all real properties designated by functional predicates were merely nonfunctional properties heuristically so designated – properties, that is, that (in something like Fodor's view) yield either true nomic universals of a physicalist sort or "true empirical generalizations" ("bridge statements" that are not nomologically cast but, by way of token-identities, appropriately link functional statements with physical laws). Wherever macro- and microphysical processes are not suitably matched by *type*-identities, or, congruently, wherever distinct macrophysical processes are not yet systematically linked by covering laws, we may expect functional "bridge statements" to facilitate our science. This is the unintended, prophetic role of that marvelously extravagant seventeenth-century doctrine known as occasionalism – and, nearer to home, the role instantiated by such devices as Maxwell's demon.[72] Among purely physical phenomena, functional, occasionalist, *and* informational idioms are taken to be purely heuristic, ultimately expendable representations of what, in principle, can be adequately redescribed and explained in the ideally homonomic vocabulary of physical science.

It is clear enough that the properties of systems *known* to be physical systems may be heuristically designated by functional predicates; but it is *not* clear that *every* real system is or is known to be a (purely) physical system (in that same sense) *or* that *every* entity is or is part of a real system. The standard prejudice favoring the latter view (that all real systems are physical systems and all real entities are parts of real systems) draws on the thesis that, since they all evolve from systems characterizable in principle in purely physical terms, all real phenomena belong to real systems and all real systems are purely physical systems (say, in Sellars's sense or in the sense J. J. C. Smart favors[73]); or, it draws on the thesis that those would-be functional properties (postulated in stages (b), (c), and (e) of our earlier declension) that behave in ways that violate strong extensional constraints are simply not real properties at all.[74]

Functionalism is a thesis that decisively affects the methodology of science – but only in the form in which the functional = the incarnate. This is the lesson of the declension of option (iii) within the space of (a): assuming, that is, the rejection of transparency, the symbiosis of self and world, the symbiosis of self and society, the scientific respectability of top-down explanation, and the coherence of insisting on the reality of properties (psychological, informational, cultural, linguistic, historical, praxical) that exhibit intentionality, particularly intensionalized forms of intentionality. On the argument, even if in the domain of purely physical phenomena functional predicates are heuristic designations of essentially nonfunctional properties, the admission of human selves, human societies, human cultures, human artifacts commits us to *real functional properties* of an "Intentional" sort (meaning, by that typographical device, intensionalized intentional properties).

The essential option, then, is this: on a functional "realism" that admits token-identities of physical and functional properties but denies *given* type-identities, the functional idiom is ultimately heuristic but ineliminable as long as suitable type-identities are not at hand; and, on a functional realism that rejects both token- and type-identities *for* those functional properties that are Intentional, the idiom cannot be merely heuristic, the properties in question are not reducible (either token- or type-wise), and we are bound to acknowledge incarnate properties. Fodor is undoubtedly the current champion of the first option; Davidson's version inadvertently exposes the inconsistency of affirming token-identity and denying the conceptual eligibility of type-identity;[75] Churchland's program shows the plausibility of moving from Fodor's sort of functionalism to more careful, possibly narrow gauge type-identities;[76] and Dennett's program shows what a fully reductive reading of functionalism would look like if it could be carried to completion (which, as it happens, Dennett never actually pursues). Tellingly, the second option is never discussed by functionalists any more than by straight-out reductionists and eliminationists.

All of this may be reduced to a single dictum: at a certain level of real emergence, phenomena obtain that exhibit incarnate properties; and at a certain level of emergence (possibly coextensive, possibly less extensive – depending on alternative restrictions) incarnate phenomena exhibit just those functional, or Intentional, properties that characterize selves, human societies, cultures, and cultural artifacts. *If* language is a real psychological aptitude (*contra* Sellars), then intentionality cannot be eliminated in the real world of human culture; and *if* language cannot be essentially regimented in extensional terms, then real *I*ntentionality cannot be eliminated or reduced. So the fate of an incarnatism strong enough to effect a fundamental adjustment in our conception of science depends on

1   the untenability, with respect to real linguistic (and lingual[77]) aptitudes, of both type- and token-identities of physically and functionally designated properties; and
2   the impossibility of regimenting the intensional complexities of linguistic (and lingual) phenomena congruently with the extensional treatment said to be appropriate to physical phenomena.

What is opposed in theses 1 and 2 is pointedly and most influentially advocated in Donald Davidson's union of (what he calls) anomalous monism and the intended extension to natural languages of Tarski's semantic conception of truth.[78] But the programs separately attacked in theses 1 and 2 are quite different from one another. Roughly speaking, advocates of the programs attacked in thesis 2 regard their own disciplines as autonomous sciences and are indifferent to identity claims. This is the common thread of so-called functional sciences – notably, the various forms of cognitive science pursued independently of neurophysiological and physical constraints (as in Zenon Pylyshyn's account[79]), analytic structuralisms (following the example of Ferdinand de Saussure and Claude Lévi-Strauss[80]), and Chomskyan linguistics (which opposes both cognitive science and structuralism[81]). Their essential objective is to demonstrate that the abstrac*ted* informational or functional properties of the autonomous domains they examine (natural language, kinship systems, cognition in general) pose no serious threat to any reasonably attenuated unity-of-science program. Advocates of the programs attacked in thesis 1 regard the functional idiom as ultimately heuristic and eliminable, treat token-identity as a stop-gap measure, and hold to a comparatively strong form of the unity of science (as, notably, do Sellars, Dennett, and the Churchlands). But, on the argument, the opponents of thesis 1 are too hasty and the opponents of thesis 2 never address the lacuna of their common project.

We must be cautious here. The admission of (a) in our declension of option (iii) – the denial of transparency, the advocacy of the realist–idealist symbiosis – is fully compatible with an ("internal") objectivism of science. On evidentiary grounds, we posit what we regard as the actual structure of the physical world, but we need not do so merely conventionally or arbitrarily. So, for instance, it need not be the case that "Entropy is an anthropomorphic concept [solely on the grounds that] it is a property, not of the physical system, but of the particular experiments you or I choose to perform on it." It *does* (on the argument) fall within the scope of (a), but in doing that it need not fail to meet the constraints of a reasonable realism regarding the physical world. It may be that entropic experiments are a function of arbitrary theoretical partitions of microphysical phase-states of a certain macrophysical space; but on the evidence, the experiments conducted may exhibit a certain suitable empirical invariance ranging over all such arbitrary partitions. This would of course accord with at least one general strategy for responsibly positing real entropic structures – the general strategy of inductivism (if inductivism may be relieved of its own tendency to favor transparency).[82]

In a word, we need not impute to the physical world we posit (a world we do posit as independent of our inquiry) the same Intentional features we impute to ourselves positing such a world – though we cannot deny that the determinate structures we impute to that world are artifacts of our Intentional effort, and though we cannot suppose that the cultural world in which we acquire the aptitude of science is similarly a world we can coherently posit as independent of our Intentional life. We must preserve the "internal" objectivism of science within these constraints,[83] but we must also preserve these constraints – or, we must preserve them to the extent that functionalist theories either ignore the reality of Intentional life or prematurely assure us that functional and Intentional concepts are no more than a mere *façon de parler* for discourse about sectors of the physical world still too complex for our explanatory powers. This is the proviso of the incarnatist idiom – which yields, along physicalist lines (type- or token-), wherever particular arguments compel it to, but insists as well on the coherence of an emergent order in which functional features may be conceptually abstracted from the incarnate attributes of that order.[84] The carelessness and concessiveness (by turns) of functionalism are due to its weak grasp of the full Intentional world of human culture. Functionalism is not mistaken in assuming that that world is hospitable to the physical sciences, but it is willing to go to extravagant lengths to ensure its welcome. Intentionality in all its variety is, therefore, the *functional* feature best placed to decide the contest between functionalism and incarnatism.

This means that not all forms of intentionality (or Intentionality) are inimical to the extensionalist programs of the physical sciences; effectively, then, they are not all inimical to the conceptual eligibility of reductive and eliminative programs regarding the human sciences. But they are also not bound to yield to them where the evidence is unfavorable. The issue is an entirely open one; and, to gain a proper grip on it, we need to have before us a fair table of the principal forms of intentionality. They may be construed as encumbering all those who, with Quine, simply repudiate the pertinence of the very notion in the context of science. But to collect them specifically as forms of intentionality will (as we shall see) prove decidedly unorthodox; for we need to sort them without presuming to explicate psychological phenomena only (Brentano) or what is phenomenologically accessible only among the data of the psychological (Husserl).[85] The forms to be tallied constitute, rather, the ground that naturalistic, phenomenological, and deconstructive inquiries have in common (or so the evolving argument insists). The tally itself is a table of themes with respect to which all the standard disputes centering on reductionism, eliminativism, extensionalism, the unity of science and similar programs may be collected. It is hardly a table of findings or irresistible solutions. But the sheer quantity of its distinctions accuses the partisans of the programs just mentioned of a certain philosophical lassitude at least.

Here, then, is the tally, deployed in the form of a plausible progression of themes, each of which (on arguments already supplied or not difficult to fathom) is not eliminable, is not reducible to another, and is not able to be completely neutralized or reduced by way of conceptual strategies available to the programs being contested. The themes are also not ordered, here, in any decisively compelling way, but their progression is dialectically suggestive. Furthermore, to each of the dimensions tallied there corresponds a distinct vocabulary of functional predicates – regarding the use of which questions of realist or heuristic import directly apply. It is extraordinary that such tallies as the following are so rare, given the obvious stakes at risk.[86]

1 *Intransparency* (or opacity): the indissoluble symbiosis of self and world; the inseparability of realist and idealist elements in the very cognizability of the world; our thesis (a) affirmed of science and inquiry in general (not restricted, of course, to propositional structures, in Quine's manner[87]).

2 *Presence*: the salience of determinate and distributed things within

the space of dimension 1; *Erscheinungen*, in a sense close to that featured in Hegel[88] but lacking in any presumption of reliable self-disclosure (*Anwesenheit*) or ontic fixity (sympathetic, say, with Heidegger's and Derrida's familiar criticisms[89]).

3 *Historicity* (or historicality): the open-ended, diachronic shift, and the continuity and discontinuity of shift, of the schemata of dimension 2 within 1 taken in a spirit close to Heidegger's themes of *Temporalität* (or *Zeitlichkeit*) and *Historizität* shorn of their privileged linkage to the ontology of *Dasein*;[90] therefore, also, *contextuality* or the temporal continuum or succession of contexts.

4 *Lebensweltigkeit* (or connectedness or co-presence): the encountered, coherent linkage of the distributed elements and distinctions of dimension 2 within the constraints of 3; roughly, the sense of potential or opening system subtended within what Husserl and Gadamer treat as the family of individual horizons linked within the common horizon of a historical interval.[91]

5 *Praxicality* (or agency): the reflexive specification of the presence of self and selves within the space of dimension 4 as the *conditio sine qua non* enabling, orienting, and predisposing all the schemata collecting the phenomena of 2 but not actually constituting them (contrary, say, to Husserl's solipsistic tendencies, sympathetic to Heidegger's dialectical use of *Zuhandenheit* and *Vorhandenheit*, and opposed to the disjunction of practical and theoretical life, as in Marx's socioeconomic speculations and studies[92]).

6 *Non-closure* (or the absence of system or totalization): what, somewhat paradoxically, is emphasized (in Derrida's attack on structuralisms[93]) as the radical alterity or presence-of-the-absence of procedures for effectively systematizing all possible classificatory schemes (totalizing);[94] the global bearing of the productive interventions of human selves within the space of dimension 4 on whatever may be specified regarding the determinate elements and distinctions of any scheme of 1.

7 *Constitution*: the upshot, for the distributed phenomena of a domain, of rejecting the pertinence of all forms of the correspondence theory of truth; that is, the impossibility of determining what, under the conditions of dimension 6, may be sorted as the determinate (eliminable) effect of 5-like interventions (technology) on the apparent structure of anything discriminated within 4; notably, the structure of experience and of a putative world disclosed through experience (the theme common to Kant, Nietzsche, Husserl, and Heidegger, if all presumption of cognitive privilege be discounted[95]).

8 *Artifactuality*: the presence of sub-sets of the members of the world of dimension 4 that are, as a causal or productive consequence of determinate 5-like interventions, not merely as a formal consequence of 7, constructed, formed, made, created, produced, or done by selves in ways that inextricably involve (by incarnation) 2-like elements; notably, artworks, linguistic utterances, actions, and related cultural phenomena.[96]

9 *Directedness$_1$* (or aboutness): the structure of phenomena, in accord with dimensions 7 and 8, in virtue of which such phenomena intrinsically possess or exhibit purposes, ends, reference, *Zweckmässigkeit*,[97] conative or volitional or cognitive powers; properties, therefore, that incorporate those classically noted by Brentano and Husserl but, now, not merely or even necessarily psychological, possibly linguistic or lingual or not, possibly relational or nonrelational, possibly propositional or nonprop-ositional,[98] possibly biological or electronic, possibly merely artifactual or cultural.

10 *Directedness$_2$* (or noematicity): the capacity of phenomena (charac-teristically psychological or artifactual) exhibiting dimension 9 to exhibit the further property of possessing propositional content or structure (roughly, in accord with Husserl's account, on the usual alternative interpretations[99]), phenomena paradigmatically linguistic or by extension, lingual, or by further extension (by way of heuristic modeling) to include the behavior and mental life of sub-linguistic animals or by direct extension (by nonheuristic modeling) to include the artifactual and the cultural in general.

11 *Rulelikeness* (or normicity): conformity to rules and norms, directly attributed to the regularities of societal ensembles or aggregates or to their aggregated behavior and underlying struc-tures; or, derivatively, conformity attributed to individual behavior, thought, work, production, and creativity, insofar as they instantiate regularities first and properly assigned societal phenomena, or to their products or artifacts produced; regularities in accord with dimension 4 (possibly along the lines of Husserl's notion of a *Lebenswelt* or, more projectively, in accord with Wittgenstein's notion of a *Lebensform*, or perhaps by an eclectic combination of the two influenced in alternative respects by Hegelian, Marxist or Foucauldian themes, as in Habermas, Bourdieu, and Deleuze[100]); phenomena that involve the symbiosis of self and society.

12 *Intensionality* (or nonextensionality): the possession of determinate significance, meaning, import, semiotic function (as in represen-

tation, exemplification, reference[101]), semantic content and the like; the capacity of anything within the space of dimension 4, relative to 5 (alternatively, of anything exhibiting dimension 9 or 10 to acquire or be assigned significance, whether linguistic or lingual, whether in a realist or heuristic sense, whether propositional or nonpropositional, whether construed as abstractly functional or incarnate, whether extensionally irreducible or reducible (as, famously, in quarrels involving Davidson's extension of Tarski's Convention T[102]).

The range of phenomena collected or collectible under dimensions 1–12 is staggering. The most important emergent dimension resulting from combining particular categories among 1–11 with 12 is what we have termed *the Intentional: intensionalized intentional phenomena viewed (at least provisionally) as the incarnated, essential, indissoluble mark of the cultural.* For our present purpose, it is enough to note that the Intentional constitutes the principal challenge to the adequacy of the unity-of-science interpretation of the human sciences, the challenge to all reductive, eliminative, physicalist, extensionalist, (or merely functionalist) programs regarding the description and explanation of the phenomena of human existence – regardless of what may obtain for any and all sub-human domains; although, on the argument, *all discrimination with regard to sub-human domains is itself a function of reflexive thought and activity at the human level.* All such discrimination implicates the reality of the human world. It follows, therefore, that the opponents of intentionality exhibit, by their own reductive projects, the (provisional) ubiquity of intentionality: in particular, the ubiquity of the Intentional.

To acknowledge that the phenomenon of science is a datum for science is, effectively, to admit that there are real functional properties, that such properties are and must be incarnate, that they must include intentional properties and, in particular, Intentional properties, and that the assignment of real nonfunctional (hence, nonintentional and non-Intentional) properties to the encountered phenomena of the physical world symbiotically entails the reality of entities exhibiting Intentional properties. If the argument supporting these entailments be admitted, the irony follows at once that the various reductive programs we have already surveyed are all self-defeating – without in the least disallowing the viability of suitably limited extensionalized claims of the sorts they favor globally.

More to the point: to acknowledge science as a datum for science is to admit that the Sellarsian and Quinean strategies we have been tracking through so many guises and disguises are all thunderously silent on the essential and ineluctable issues.

VII

Perhaps a final word should be risked here to collect the sprawl of our discussion of the functional and the intentional a little more neatly. We have scanned two tallies of two quite different sorts: one, a set of minimal features of the functional as it obtains in certain sciences and inquiries, where it appears it would take more than a strenuous effort to eliminate; the other, a declension of the principal forms of intentionality that any comprehensive overview of the space of human existence would have to acknowledge initially, however it finally disposed of such items. Short of effective counterstrategies, the intentional, as a sub-set of the functional, has application as *real, causally efficacious, not reducible or eliminable in purely physicalist terms,* and *indissolubly complex, incarnate, and emergent with regard to some physical or biological sector.* The supreme exemplar of intentional phenomena, by no means the only specimen, appears in the human use of language, in speech and thought. There is reason to think that if the intentional is real, it cannot be restricted to explicit uses of the linguistic: for one thing, the constraint of language may obtain in a "lingual" way only (as in music and dance, watchmaking, practicing dentistry and the like); for another, prelinguistic infants exhibit intentionality even in acquiring a first language (discounting nativism of course); for another, machines programmed by linguistically apt humans may be said to perform intentionally, and artifacts and artworks produced by humans may be said to possess intentional features as a result of human interventions; finally, sublinguistic creatures appear to behave in ways so similar to the human that, though the analysis of their mental and behavioral life may be modeled on the human, their own life may be said to exhibit actual intentional features.

Once matters are viewed this way, it seems reasonable to sort the principal ingredient strands of the structure of the intentional, as we have done by assembling the second tally. Here, we are guided by considerations of comprehensiveness (in the light of the accumulating literature) and of difficulties of reduction within the range of the intentional itself. There are, after all, two sorts of reduction to be considered: one, that by which any and all forms of the intentional are thought to be conceptually neutralized (for the purposese of strongly extensionalized or physicalized models of the sciences) by way of elimination, reduction, supercedence, or other strategies; the other, that by which one form, strand, structure of intentionality is demonstrably subsumable under, identical with, or tantamount to, another such form or strand or structure. The second tally is meant to provide a very reasonable first pass at a set of distinctions that, within the range of the intentional, appear to set significant limits to reductions of the second sort, once the first is disposed of.

Whether the functional and the intentional should be treated as coextensive depends on a number of considerations, including the narrowness with which the intentional is thought to be best applied and the sense in which we understand ourselves to be modeling natural phenomena. For example, when we speak of the genetic code and apply the notion distributively (as, in a recent medical breakthrough, in pinpointing a possible cause of Alzheimer's disease or as, in general, in detailing the normal development and differentiation of the cells of a human organism from the single cell that is a fertilized ovum), we speak inescapably in interlevel terms. There, we construe the *functional* (even teleological) but not purposive or intentional process as specifiable in terms of the "directive" power of the code somehow incarnate in the ovum's compositional material, ranging over "normal" developments marked only at a suitable macroscopic level. The reductive treatment of the functional in this context clearly depends on our ability to redescribe the macroscopic biological features we wish to account for (the development of a full-fledged human being, say) in ways suitably restricted to whatever are the terms in which law-like processes themselves restricted to the genetic could effectively service pertinent explanations (without recourse to a functional idiom at the genetic level). It is quite possible that we may never know how to eliminate the functional completely and that we may never know how to demonstrate that it is in principle and in an empirically responsible way eliminable. Certainly, if the intentional is ineliminable, then, trivially, the functional is ineliminable. But the proper restriction of the scope of the intentional is a matter of some finesse and diverging intuitions (as one can see merely by comparing the views of Dretske and Searle and Husserl for example). Only the central phenomena need be confirmed in order to concede the force of our second tally.

The intentional is clearly focused on the conscious, cognitive, cognitively informed, cognition-like, and productive aptitudes of human agents. But, as with our accommodation of the lingual, the prelinguistic, the artifactual, the sublinguistic, we may urge that there are strong reasons of continuity and analogy for ascribing intentionality in ways and under circumstances not canonically favored in such accounts as those of Brentano and Husserl. For example, a reasonable grasp of the social space of human existence – in effect, of the symbiosis of the psychological and the social (or societal), which we have considered at some length in terms of the linkage between our distinctions (b) and (c) of a previous tally – pretty well obliges us, *if* we admit the intentional, to acknowledge (*contra* Husserl, particularly) that the scope of the intentional must be extended to societal structures (involving Wittgensteinian practices, Gadamerian interpretive horizons, Marxist forms of production, for instance) that cannot be reduced to, or thought to be generated from, mere psychological or egological powers. By way of a similar argument, if the intentional is extended to institutional phenomena, it can hardly fail to be extended as well to the artifacts, art, technology, linguistic and lingual products of social existence. Construing matters thus, therefore, provides a most plausible rationale for articulating the various sub-structures of intentionality (that exhibit their own contributory forms or versions) without fear of any charge of having made a merely arbitrary use of entrenched distinctions. In a word, the inadequacy of any account of intentionality (Husserl's for instance) is simply the mirror of the inadequacy of some analysis of a part of the real world.

Given the peculiar dearth of discussions of philosophically motivating the full admission of the intentional and of the need to admit the full complexity of intentional phenomena, we should spell out a bit more explicitly the strategy of the foregoing argument. First of all, the deep prejudice favoring excluding or completely neutralizing the intentional had to be exposed – distributively. Here, one may think of Quine's rather cavalier (completely unvindicated) maneuver to box the troublesome features of the intentional in simply by leaving intentional expressions unanalyzed within larger expressions, themselves somehow *taken* to be unaffected by the included element – at least as far as extensionalist concerns are involved; or, one may think of Dennett's interpretation of the intentional as a magical device for altering reality merely by redescribing it; or, one may think of Sellars's treatment of intentionality's intensional complexities as the mere artifactual consequence of the idiom we use in discourse about the real world – an idiom that, somehow, is not itself robustly part of the real world; or, one may think of Dretske's treatment of the intentional as a series of nested restrictions of various sorts upon the informational, which is itself thoroughly extensional – hence, of the limited impact of intentionality within any comprehensive theory of reality and science; or, one may think Davidson's extended use of Tarski's semantic conception of truth, in application to natural languages, which enables us to construe intentional and intensional complications as informally tolerated within a schema already in place, that need not in principle exhibit any such features.

Secondly, the "intentional" and "intensional" appeared to be best characterized in terms of salient specimens – and appeared to call for a measure of informality in applying these epithets. There is hardly any point in insisting on an essential conceptual homogeneity where analysis leads us to depart in heterodox ways from the canonical picture favored by Brentano and Husserl. So, for instance, intentionality certainly centers on the thought and activity of individual human agents and is certainly focused in terms of aboutness or directedness addressed to propositional and nonpropositional objects, linguistically and lingually. But if, in reviewing the *environing* conditions within which and within which alone the intentional (thus construed) obtains, we find that there are other structures that must be admitted (for example, those involved in societally complex phenomena: the sharing of a language or a tradition or a culture), which cannot be satisfactorily analyzed in terms of the received canon and which cannot be eliminated or neutralized any more effectively than can the usual exemplars – by way of those strategies that would eliminate or neutralize the latter – we do then have good reason to extend the range of the "intentional" (if that is the right way of putting the point) to cover these additional phenomena, even at the risk of a certain loss of convenience and simplicity. The same argument holds for intensionality. The exemplary specimens of the intensional clearly feature, as salient, substitutivity – in sentential contexts – of semantically specified linguistic expressions. But within the world of human culture, within the world of linguistically apt creatures, it is hardly unreasonable to extend the use of "intensional" to cover the artworks, artifacts, institutions, histories, and actions of such creatures. And, of course, once these enlargements are in place, particularly given that the cultural is ubiquitous for man despite its peculiar instabilities and variety, given that the relationship between human culture and physical nature remains so problematic and yet so essential to man's understanding of himself, we may anticipate that there is bound to be a need to explore what may be distinctive or characteristic of the cultural as such. The cultural may then be seen to involve the enlarged sense of the intentional and to exhibit structures invariably articulated in terms of an equally enlarged sense of the intensional (for instance, as representational, presentational, referential, expressive, symbolic, significant, meaningful, purposive, propositional, semantic). We may collect such properties as Intentional ($=_{df}$ cultural) properties, although at this stage of the investigation the move is little more than a notational convenience. We must always bear in mind that our classifications are attentive to saliences only; they make no claims of an essentialist sort.

The extension of the intensional to the proto-cultural achievements of primates and cetaceans – below the level of a fully-fledged language – remains for instance an entirely feasible and worthwhile option.

Thirdly, we had to make provision for, but also to sort, heuristic and realist uses of functional and intentional predicates. Intentional predicates, we found, form a sub-set of functional predicates. The functional, however, is often treated as a purely heuristic distinction. When it is, it usually designates either a role or function deliberately assigned something capable in principle of being described and analyzed without reference to that assignment (as in Sellars's usage) or a causal aptitude not otherwise relationally examined in terms of independent causes and effects (as in Armstrong's and Churchland's usage). Hence, the functional, whether or not treated as coextensive or identical with the intentional, may be recovered as an actual or real structure by construing it as incarnate, complex, logically monadic, emergent with respect to the physical or biological, and causally efficacious. It no longer serves then as a mere place-holder for whatever *else* might enter into that causal nexus. The functional and the intentional cannot be causally efficacious *and* separable from the physical or biological, unless ontic dualism is a tenable doctrine. This is why it is reasonable to construe the intentional as indissolubly *complex, incarnate, monadic*. These are the minimal features of the intentional, consistent with admitting its reality and causal efficacy and consistent with avoiding dualism: once in place, they mark the inadequacy of all compositional, relational, reductive, and eliminative analyses of the pertinent phenomena. They do not, however, preclude alternative manifestations of monadic complexity (as of the propositional and nonpropositional content of emotional and volitional states); and they do not preclude relational complications (as in searching for one's dog as opposed to the Fountain of Youth, or of referring to one's brother as opposed to Sherlock Holmes). The essential point is that the relational can always be reconciled with the monadic, but the monadic cannot be admitted where the relational is taken as fundamental; we must, therefore, adjust the notion of intentionality to the saliencies of familiar experience.

Finally, within those saliencies, we needed to collect a reasonably ample array of distinct structures that combine with one another in familiar ways within the range of cultural life, but that are not obviously the same as others already admitted, and that incrementally incorporate much or most of whatever might be featured in an account of the human world.

Here, there is no pretense of collecting every pertinent distinction. On the contrary, the array provided confirms the endlessness and idiosyncrasy of the task. What we cannot fail to grasp, however, is the stubborn proliferation of the alternative structures we are here calling intentional or Intentional. Once we admit that artworks, histories, actions, rules, rituals, institutions, as well as linguistic acts are Intentional – and whyever should we not? – the possibility of falling back to the usual canons becomes very dim indeed. Beyond defeating the confident claim of the homogeneity of the intentional, tallies of the sort here collected serve only to give a sense of the large coherence that a theory of culture could supply and of certain strategic lines of investigation that such a theory would require or favor. We can say nothing about these as yet, just as we can say nothing about the nature of the actual (ontic) complexity of Intentional phenomena. Those are matters that would require an entirely fresh beginning.

Notes

1   See Wilfrid Sellars, "Philosophy and the Scientific Image of Man," *Science, Perception and Reality* (London: Routledge and Kegan Paul, 1963).

2   Paul M. Churchland, *Matter and Consciousness* (Cambridge, Mass.: MIT Press, 1984), pp. 21, 32.

3   See for instance Donald Davidson, "Mental Events," *Essays on Actions and Events* (Oxford: Clarendon Press, 1980). It should be clear that Davidson's (and Jerry Fodor's) less-than-eliminative theories are very much committed to the same general thesis as Churchland's: Davidson hopes to achieve by token identities what Churchland means to achieve by straighout elimination. We shall touch further on the issue below.

4   Paul Churchland, *Matter and Consciousness*, pp. 43–9.

5   See ch. 12, below; also Jerry A. Fodor, *The Modularity of Mind* (Cambridge, Mass.: MIT Press, 1983), pp. 112–17.

6   See ch. 5, above.

7   The point has been tellingly remarked in Jerry A. Fodor, *Psychological Explanation* (New York: Random House, 1968), pp. 100–6.

8   See Hilary Putnam, "Minds and Machines," *Philosophical Papers*, vol. 2 (Cambridge: Cambridge University Press, 1975).

9   Paul Churchland, *Matter and Consciousness*, p. 12.

10   Ibid., pp. 20–1.

11   I have explored the concept of the incarnate at some length: see Joseph Margolis, *Art and Philosophy* (Atlantic Highlands, NJ: Humanities Press, 1980), *Culture and Cultural Entities* (Dordrecht: D. Reidel, 1984), ch. 1, and *Philosophy of Psychology* (Englewood Cliffs, NJ: Prentice-Hall, 1984). The point at stake goes some distance toward explaining the Marxist use of the term "material" – meaning "real even where not physically reducible."

12   This suggests of course the underlying rationale (though not necessarily the validity of familiar speculations regarding the use of imagery in attempting to

achieve remission in certain types of cancer. See for instance D. C. Simonton, S. Simonton, and I. Creighton, *Getting Well Again* (Los Angeles: Tarcher, 1978).

13 Paul de Man, "Criticism and Crisis," *Blindness and Insight*, 2nd, rev. edn (Minneapolis: University of Minnesota Press, 1983), pp. 18, 19.

14 See for example Gerald Graff, *Literature against Itself: Literary Ideas in Modern Society* (Chicago: University of Chicago Press, 1979), ch. 6.

15 Sellars, "Philosophy and the Scientific Image of Man," *Science, Perception and Reality*, p. 40.

16 Ibid., pp. 38–9.

17 Sellars, "The Language of Theories," ibid., p. 126.

18 Ibid., p. 109, n. 3.

19 De Man, "Criticism and Crisis," *Blindness and Insight*, p. 15.

20 Sellars, "Philosophy and the Scientific Image of Man," *Science, Perception and Reality*, p. 40.

21 Sellars, "The Language of Theories," ibid., p. 121.

22 Daniel C. Dennett, *Content and Consciousness* (London: Routledge and Kegan Paul, 1969), p. 190 (Dennett's italics).

23 Ibid., p. 48.

24 Ibid., pp. 46, 52.

25 Ibid., pp. 82, 83.

26 Ibid., p. 80.

27 Ibid.

28 Stephen P. Stich, *From Folk Psychology to Cognitive Science* (Cambridge, Mass.: MIT Press, 1983), p. 247, n. 2 (ch. 1).

29 Paul Churchland, *Matter and Consciousness*, p. 16.

30 Ibid. (italics added).

31 Fodor, *Psychological Explanation*, pp. 51, 55. As a "mentalist," Fodor obviously does not share Churchland's notion of functionalism.

32 The thesis is also very close to (though not quite the same as) that espoused by D. M. Armstrong, in *A Materialist Theory of the Mind* (London: Routledge and Kegan Paul, 1968).

33 Paul Churchland, *Matter and Consciousness*, p. 37.

34 Ibid.

35 Ibid.

36 Ibid.

37 Ibid., pp. 41–2.

38 Ibid., p. 43.

39 See Davidson, "Mental Events," *Essays on Actions and Events*.

40 Ibid.

41 Fodor, *Psychological Explanation*, pp. xix–xx.

42 See Herbert Feigl, *The "Mental" and the "Physical": The Essay and a Postscript* (Minneapolis: University of Minnesota Press, 1967).

43 See Ernst Mayr, *The Growth of Biological Thought* (Cambridge, Mass.: Harvard University Press, 1982).

44   Thomas S. Kuhn, *The Structure of Scientific Revolutions*, 2nd, enlarged edn (Chicago: University Press, 1970), pp. 126, 119–22. Kuhn actually says, "Until that scholastic paradigm was invented, there were no pendulums, but only swinging stones, for the scientist to see. Pendulums were brought into existence by something very like a paradigm-induced gestalt switch" (p. 120). In fact, in the Postscript, Kuhn explicitly remarks (a matter he hedges on later), "Notice now that two groups, the members of which have systematically different sensations on receipt of the same stimuli, do *in some sense* live in different worlds. We posit the existence of stimuli to explain our perceptions of the world, and we posit their immutability to avoid both individual and social solipsism. About neither posit have I the slightest reservation" (p. 193).

45   Ibid., pp. 121, 122, 126.

46   Ibid., p. 206.

47   See P. K. Feyerabend, *Philosophical Papers*, vol. 1 (Cambridge: Cambridge University Press, 1981), pt. 1, particularly ch. 7.

48   See Karl R. Popper, *Conjectures and Refutations: The Growth of Scientific Knowledge*, 2nd edn (New York: Harper and Row, 1965), ch. 10, section 3, and *Realism and the Aim of Science*, ed. W. W. Bartley, III (Totowa, NJ: Rowman and Littlefield, 1963), pp. xxv–xxvii, 56–62. I cannot see that Popper extricates himself from the essential difficulty of his view of verisimilitude. He seems to feel (now) that it was unfortunate to have attempted a formal definition of the notion, but the doubt really goes to the heart of his progressivism (cf. p. 58). Cf. also, Kuhn, *The Structure of Scientific Revolutions*, pp. 206–7.

49   Ludwik Fleck, *Genesis and Development of a Scientific Fact*, ed. Thaddeus J. Trenn and Robert K. Merton, tr. Fred Bradley and Thaddeus J. Trenn (Chicago: University of Chicago Press, 1979), p. 68 (in the context of ch. 3); pp. 118–19, 83 (in the context of ch. 2); cf. Prologue.

50   Ibid., p. 46.

51   Ibid., pp. 83, 98, 101.

52   Ibid., pp. 101–2.

53   See Gyorgy Markus, *Language and Production: A Critique of the Paradigms* (Dordrecht: D. Reidel, 1986).

54   See ch. 4, above.

55   See Nancy Cartwright, *How the Laws of Physics Lie* (Oxford: Clarendon Press, 1983). The attempt to disjoin methodology and ontology (or ontic commitment) surfaces in a forceful way in Adolf Grünbaum's criticism of the hermeneutic conception of science. Independently of the fate of that conception, Grünbaum's disjunction must fail. See Adolf Grünbaum, Introduction to *The Foundations of Psychoanalysis* (Berkeley, University of California Press, 1984), section 5. See also ch. 11, below.

56   Daniel C. Dennett, "Artificial Intelligence as Philosophy and as Psychology," *Brainstorms* (Montgomery, Vt: Bradford Books, 1978), pp. 122, 124.

57 Ibid., p. 124.
58 Paul Ricoeur pointedly explores the "hermeneutics of suspicion" or the "hermeneutics of 'false consciousness'" in *The Conflict of Interpretations*, ed. Don Ihde (Evanston, Ill.: Northwestern University Press, 1974).
59 See Michel Foucault, *Power/Knowledge: Selected Interviews and other Writings 1972–1977*, ed. Colin Gordon, tr. Colin Gordon et al. (New York: Pantheon, 1980), particularly chs 5–6.
60 See ch. 5, above.
61 See Thomas Nagel, "What Is It Like to Be a Bat?" *Philosophical Review*, LXXXIII (1974).
62 See Roderick M. Chisholm, *Perceiving: A Philosophical Study* (Ithaca, NY: Cornell University Press, 1957), particularly chs 6, 10.
63 See Margolis, *Culture and Cultural Entities*, ch. 3. A fair sense of the naturalness of a strongly realist intentional idiom in characterizing the primates may be found in David Premack, *Gavagai!* (Cambridge, Mass.: MIT Press, 1986).
64 I have explored the theme, for the arts, in *Art and Philosophy*, and have extended it for cultural life in general, in *Culture and Cultural Entities*.
65 See ch. 6, above.
66 Robert Cummins, *The Nature of Psychological Explanation* (Cambridge, Mass.: MIT Press, 1983), p. 29.
67 Ibid., pp. 29–30 (italics added). See also section 1.2.
68 Jerry A. Fodor, *The Language of Thought* (New York: Thomas Y. Crowell, 1975), p. 20 (italics added).
69 Ibid., pp. 19, 20. Cf. Davidson, "Mental Events," *Essays on Actions and Events*.
70 Ibid., ch. 2.
71 We shall return to this theme in ch. 12, below.
72 See Feigl, *The "Mental" and the "Physical"*. There is a suggestive discussion along these lines in K. S. Trincher, "Information and Biological Thermodynamics," in Libor Kubát and Jiří Zeman (eds.), *Entropy and Information in Science and Philosophy* (Amsterdam: Elsevier, 1975).
73 J. J. C. Smart may be one of the most explicit advocates of this doctrine; see his *Philosophy and Scientific Realism* (London: Routledge and Kegan Paul, 1963).
74 This is the implicit view of Quine and the explicit view of Dennett. Wherever token- (but not type-) identities are favored, as by Fodor and Davidson, it is more difficult to be sure that the putative properties involved are denied reality; it is more likely that the functional idiom is regarded as "folk-ish," heuristically fruitful, but distinctly dispensable. Functional "realisms" of this sort are fashionably (but not accurately) said to be not reductionistic.
75 See Davidson, "Mental Events," *Essays on Actions and Events*; also Joseph Margolis, "Prospects for an Extensionalist Psychology of Action," *Journal for the Theory of Social Behavior*, XI (1981).

76 The truth is that this appears most tellingly in the work of Churchland's wife. See Patricia M. Churchland, *Neurophilosophy* (Cambridge, Mass.: MIT Press, 1986).

77 I have taken a deliberate liberty here with the import of "lingual" as a translation of Ricoeur's *langagier* and Gadamer's *Sprachlichkeit* (transformed into an adjective). In my usage, the lingual presupposes the specifically linguistic, but it need not explicitly involve the use of speech or linguistically formed thinking – as for instance in the dance. See Paul Ricoeur, "Hermeneutics and the Critique of Ideology," *Hermeneutics and the Human Sciences*, tr. and ed. John B. Thompson (Cambridge: Cambridge University Press, 1981), p. 78. Cf. also Thompson's "Notes on Editing and Translating," ibid., p. 28.

78 See Davidson, "Mental Events," *Essays on Actions and Events*; and ch. 1, above.

79 See ch. 1, above.

80 See Joseph Margolis, "'The savage mind totalizes,'" *Man and World*, XVII (1984).

81 See Margolis, "Cognitivism and the Problem of Explaining Human Intelligence," *Culture and Cultural Entities*.

82 The remark appears in E. T. Jaynes, "Biggs vs. Boltzmann Entropies, *American Journal of Physics*, XXX (1965), and is cited in Adolf Grünbaum, "Is the Coarse-Grained Entropy of Classical Statistical Mechanics an Anthropomorphism?" in Kubát and Zeman, *Entropy and Information in Science and Philosophy*. Grünbaum understandably answers his own question with a no, but he does not discuss the general import of (a). Cf. ch. 11, below.

83 This, as we shall see in ch. 10, below, marks the essential weakness of Roy Bhaskar's attempt to recover the objectivity of the human sciences.

84 See Joseph Margolis, *Pragmatism without Foundations: Reconciling Realism and Relativism* (Oxford: Basil Blackwell, 1986), particularly ch. 11.

85 See ch. 7, above.

86 The terms supplied are all terms of art, though most will be familiar.

87 See W. V. Quine, "Two Dogmas of Empiricism," *From a Logical Point of View* (Cambridge, Mass.: Harvard University Press, 1953).

88 See G. W. F. Hegel, *Phenomenology of Spirit*, tr. A. V. Miller (Oxford: Oxford University Press, 1977), Introduction and Division A.

89 See Jacques Derrida, *Of Grammatology*, tr. Gayatri Spivak Chakravorty (Baltimore: Johns Hopkins University Press, 1976).

90 See Martin Heidegger, Introduction to *Being and Time*, tr. John Macquarrie and Edward Robinson (New York: Harper and Row, 1962).

91 See Hans-Georg Gadamer, *Truth and Method*, tr. Garrett Barden and John Cumming from 2nd German edn (New York: Seabury Press, 1975), pt II, section II.

92 Marx's "Theses on Feuerbach" rightly serve as the focus of Marx's emphasis here.

93 Derrida, *Of Grammatology*, pt II.
94 The counterpart of Derrida's program is Michel Foucault's, also within the Nietzschean tradition, of construing historical change as dependent in subterranean ways on forces that gradually form those thereby altered reflexive powers by which their own provisional structure can be identified. This fixes the sense, also, in which the Foucaldian thesis is, by attenuation, Marxist as well as Nietzschean. See Foucault, *Power/Knowledge*, ch. 5.
95 See Margolis, *Pragmatism without Foundations*.
96 I have given an extended account of such phenomena in *Art and Philosophy*, particularly chs 2–3; and in *Culture and Cultural Entities*, ch. 1.
97 See Charles Taylor, *The Explanation of Behavior* (London: Routledge and Kegan Paul, 1964), pt I; also Marjorie Grene, *The Understanding of Nature* (Dordrecht: D. Reidel, 1974).
98 See ch. 7, above.
99 See ch. 7, above.
100 See Gilles Deleuze and Felix Guattari, *Anti-Oedipus*, tr. Robert Hurley et al. (Minneapolis: University of Minnesota Press, 1983); and Pierre Bourdieu, *Outline of a Theory of Practice*, tr. Richard Nice (Cambridge: Cambridge University Press, 1977).
101 See for instance Arthur C. Danto, *The Transfiguration of the Commonplace* (Cambridge, Mass.: Harvard University Press, 1971); and Nelson Goodman, *Languages of Art* (Indianapolis: Bobbs-Merrill, 1968).
102 See ch. 1, above.

10

Emergence and the Unity of Science

To have collected a great variety of functional and intentional distinctions of which at least a significant number may be taken in a robustly realist sense is to confront the ineluctable threat of emergence. For the question arises of whether such *properties* can be reconciled with the explanatory powers of those sciences that have no particular use for the intentional. It is, however, a question that suffers from its own complexities. We shall require an entirely fresh beginning to bring these rather different issues within common boundaries, in a way that permits a measure of conceptual commensuration.

There are many senses in which things are said to emerge or be emergent: a fox may emerge or be on the point of emerging from its earth, come into view from its hidden shelter. But in terms central to the inquiries of science, to speak of emergent phenomena signifies

1 the presence or occurrence of what is novel relative to some putative system, and/or
2 the occurrence or existence of what cannot be construed as an element within, or an attribute of any element within, such a system.

It is obviously possible that what satisfies condition (1) may be a mere contingency of inquiry or of a limited sector inquired into: for example, one may fail to discover green orchids either because one has not yet examined the flora of the Cincinnati Conservatory or because there are none to be found in Nutley, New Jersey. It is only when the phenomena of condition 1 are construed in terms of the considerations of condition 2 that the central conceptual issue of *emergence* arises. The core idea, then, is of things characterized as emergent relative to a given system that serves as ground zero (and that, so identified, contains or exhibits no emergent phenomena). Again, the core idea is of something whose nature or properties cannot be assigned to the phenomena within the ground-zero system, whose nature or properties are not analyzable in terms suitable and sufficient for characterizing the nature or properties of what does belong within that system. It pays to approach this core notion somewhat gingerly, so that we need not be prematurely explicit about what a system is or about what may be taken to fall inside or outside a system. The burden of precision rests with those who claim or deny emergence.

Nevertheless, our idea of emergence is still insufficiently well formed. It would, for example, hardly do to hold that the flowering of green orchids was (in the pertinent sense) emergent with respect, say, to Spassky's and Fisher's playing their famous chess match, or *vice versa*: some as yet unspecified constraint of causal relevance is normally thought to link a given system with what is said to emerge relative to that system. Things are said to emerge with respect to a system because of some principled explanatory inadequacy or *pertinent* limitation linking the two (consistent *with* explanatory success) – or because the phenomena in question have a nature or properties answering to such inadequacy or such limitation. Here, again, a certain informality and vagueness are initially advisable, so as not to preclude alternative varieties of explanatory difficulty. One thinks of the emergence of language with respect to physical nature, the emergence of *Homo sapiens* among the primates, the emergence of living organisms within an inanimate world, the emergence of the totalitarian state at a certain time of history. Emergence is a relational ascription linking certain phenomena to certain systems, in virtue of principled inadequacies or limitations in explaining the nature or properties of the emergent phenomena in terms of the causal or generative powers imputed to this ground-zero system. Questions of explanatory relevance, therefore, are paramount to such ascriptions: the familiar assumption that the order of nature forms one interlocking system confirms the sense in which, at some level of understanding reality, the question of emergence is ubiquitously pertinent. We sometimes ask for instance whether, if indeed we do inhabit *one* world, there is or must in principle be one comprehensive theory that could explain everything – or, what the structure of such a theory might or must be like. It is worth noting that hardly anyone would claim that the mere acknowledgement of one world entails that a single omnicompetent science must be possible, a science that can explain all natural phenomena in terms of a set of covering laws ranging over fundamental (even if not ultimate) physical phenomena.

Questions of emergence, then, arise in specially structured contexts. A sector of nature must be construed as in some sense a system; the phenomena in question must be relevantly linked to a given such system: in particular, they must be thought to be of a sort that could be explained in just those terms deemed suitable and sufficient for explaining whatever does fall within the system; and, if they are emergent, they will fall outside the system, though, conceivably, within a more inclusive system – which would confirm the judgment that they were pertinently linked to the phenomena of the more restricted system. This way of viewing things is intended to be neutral between alternative versions of realism and between alternative measures of tolerance regarding divergent theories fitted, within the constraints of realism, to what is taken to be one and the same systematic order of nature. But it is reasonably clear that, in the absence of ramified systems, the question of emergence does not arise at all or, where it appears to arise, at once loses its point. Within the context of science, the notion of system is undoubtedly a complex and multivocal one, so there is a corresponding variety of senses in which the question of emergence can be posed. Whatever else a system may be, it can be said to *close*, or impose a measure of *closure* on, a given domain of inquiry. As it happens, the varieties of emergence and closure are easier to manage than the notion of a system *tout court*. Nevertheless, the question of emergence is of particular interest wherever biological, psychological, social, linguistic, and cultural phenomena are to be explained; for it is in these areas that the nature and admissibility of emergence directly affect the human sciences and their relationship to the physical and life sciences. The issue is hardly ever examined in a detailed way that favors a strong sense of emergence. This is what we shall seek to provide here.

I

By and large, for what are taken to be physical systems, two principal versions of emergence have been identified: emergence$_F$ and emergence$_B$. These are conveniently so dubbed in virtue of the views of authors prominently associated with the formulation of each; and each has had an important historical role in the formulation of the unity-of-science program.[1] "Emergence$_F$." signifies the sense of emergence proposed by Herbert Feigl. It is essentially taken from the well-known discussion of emergence offered by Paul Meehl and Wilfrid Sellars – which, as it happens, also adumbrates the alternative notion, "emergence$_B$." Meehl and Sellars were primarily concerned to save the sense in which "Whether

or not there are any emergents [in their sense] is an empirical question"
– in particular, whether or not there are so-called "raw feels," "the
qualia of feeling and sensation."[2]

On their view, the notion of emergence is captured fairly well by the
following terminological distinction:

Physical₁: an event or entity is *physical₁* if it belongs in the space-
time network.

Physical₂: an event or entity is *physical₂* if it is definable in terms
of theoretical primitives adequate to describe completely the actual
states though not necessarily the potentialities of the universe
before the appearance of life.[3]

The interesting conclusion Meehl and Sellars wished to support was that
"raw feels" (if there were any) were not *physical₂* phenomena, though
they were *physical₁*; also, that whether there were "non-epiphenomenal
emergents" (that is, phenomena that were *not physical₂* but were *physical₁*)
was itself a meaningful and empirically eligible question.[4] They did not
at this point reject as empirically false the claim that there were such
emergent phenomena. Feigl adopts their distinction and ventures the
further thesis (strengthened in his Postscript) "that *physical₂* laws will
prove sufficient."[5] Feigl's concern, drawn directly from theirs, was that
the admission of emergent phenomena in the sense given would entail
the insufficiency of *physical₂* laws to explain the phenomena of the
physical₁ domain (where "*physical₁*" is "practically synonymous with
'scientific', i.e., with being an essential part of the coherent and adequate
descriptive and explanatory account of the spatio-temporal-causal
world").[6] The reason is that even though the (emergent) phenomena in
question would be law-like, the laws themselves, not being of the
physical₂ sort, would effectively challenge the adequacy of *physical₂* laws
with respect to the entire *physical₁* domain (and even within the putatively
physical₂ domain). They would, therefore, threaten the closure and
homonomic nature of the physical world.[7]

Feigl argues that both "current physical theory" and emergentism are
committed to physical determinism ("that degree of precise and specific
in-principle-predictability that even modern quantum physics would
allow as regards the macro- and some of the micro-processes in
organisms"); and yet, "once mental states have emerged, their very
occurrence is supposed to alter the functional relations between the
neurophysiological (*physical₁*) variables in a manner in principle susceptible to confirmation."[8] "Mental states or raw feels," he says, "be they

regarded as states of an interacting substantial mind (or soul) or as values of emergent scientific variables, would in any case entail a breach in $physical_2$ determinism."[9] This would mean that even the "system of neurophysiological events" would be "*open*" in a radically new sense, not merely in the sense of being subject to, say, "extraneural" or "extradermal" events (anything, that is, in the $physical_1$ world) but in the sense that we could not count on confirming the law-like regularities of putatively $physical_2$ phenomena in terms only of $physical_2$ variables. Feigl is not altogether clear here, because he seems to speak of "nomological danglers" in several ways: first, against epiphenomenal doctrines – which deny "the causal efficacy of raw feels" and introduce "peculiar lawlike relations between cerebral events and mental events," "relations which connect intersubjectively confirmable events with events which *ex hypothesi* are in principle not intersubjectively and independently confirmable";[10] and, second (following Meehl and Sellars), against non-epiphenomenal but emergentist views of mental events (events empirically confirmable in a law-like way) – since such views would deny that the $physical_1$ world was *closed* in a law-like way in $physical_2$ terms. It is the identity theory, Feigl believes, which ensures that the second possibility – emergence in a strong sense – does not obtain. "Successor" ($physical_2$) concepts supplied for "all phenomenal concepts" will (Feigl believes) eliminate "nomological danglers," in the sense that "nothing important [will be] omitted" – "spatio-temporal-causal features ... essential for the world's description, explanation, prediction, and retrodiction (as much as whatever degree of fundamental determinism or statistical regularity permits)" will be provided in $physical_2$ terms.[11] In this sense, emergence and reduction are contradictories: emergence entails the denial of the adequacy of $physical_2$ concepts with respect to $physical_1$ phenomena; and reduction (in particular, in Feigl's case, physicalism[12]) entails the denial of emergence. This gives the full meaning of emergence$_F$.

Feigl's account, however, implicitly directs us to the sense in which emergence could be said to obtain *within* the competence of a science restricted to $physical_2$ terms. One has only to consider that the $physical_2$ world may be of such complexity that what is emergent with respect to one "level" of complexity may fall within a higher-order "level" of complexity – where the array of all such "levels" falls within a $physical_2$ system. The denial of emergence$_F$ does not as such countenance such a condition, though Feigl is aware of the possibility (he certainly does not preclude it) and though Meehl and Sellars are obviously interested in making provision for it. This, in fact, is precisely the point of their acknowledging "theories of emergent qualities with emergent laws" –

that is, of emergence that is not merely epiphenomenal, of a state of affairs "in which certain characteristics supervene upon other characteristics but in which the lower level characteristics [are] not adequate to explain the occurrences on their level."[13] The importance of this provision is plain: if emergence can occur within a *physical$_2$* system, in the sense that it can occur with respect to hierarchically ordered sub-systems within such a system, then, assuming with both Meehl and Sellars and Feigl that whether causal explanation is closed with regard to *physical$_2$* variables is an empirical question, we cannot preclude *a priori* the emergence of non-*physical$_2$* phenomena.

Certainly, one of the most sanguine, strategic, and cautious defenses of the adequacy of a relatively conservative *physical$_2$* science may be found in the speculations of Donald Davidson. Davidson's anomalous monism rejects psychophysical laws, which, construed in a realist manner, would (in effect) entail emergence. He does not oppose the use of a psychological *vocabulary* in identifying or describing phenomena that are essentially physical, since this amounts to no more than a convenient *façon de parler*. (We need not concern ourselves, here, with the consistency of Davidson's argument, which as it happens is demonstrably defective but which does not as such affect the coherence of physicalism.[14]) Davidson's recommendation runs as follows:

> Let us call a description of the form "the event that is *M*" or an open sentence of the form "event *x* is *M*" a mental description or a *mental open sentence* if and only if the expression that replaces "*M*" contains at least one mental verb essentially. (Essentially, so as to rule out cases where the description or open sentence is logically equivalent to one not containing mental vocabulary.) Now we may say that an event is mental if and only if it has a mental description, or (the description operator not being primitive) if there is a mental open sentence true of it alone. Physical events are those picked out by descriptions or open sentences that contain only the physical vocabulary essentially. It is less important to characterize a physical vocabulary because relative to the mental, it is, so to speak, recessive in determining whether a description is mental or physical.[15]

In these terms, Davidson supports "a version of the identity theory that denies that there can be strict laws connecting the mental and the physical," in effect, a version that affirms the thesis "that the mental is nomologically irreducible."[16] An essential part of his argument is that the physical sciences are *homonomic and* have the required scope to accommodate what is usually identified in mental terms. A generalization is homonomic, Davidson holds, "whose positive instances give us reason to believe the generalization itself could be improved by adding further provisos and conditions stated in the same general vocabulary as the original generalization." Furthermore,

> Within the physical sciences we do find homonomic generalizations, generalizations such that if the evidence supports them, we then have reason to believe they may be sharpened indefinitely by drawing upon further physical concepts: there is a theoretical asymptote of perfect coherence with all the evidence, perfect predictability (under the terms of the system), total explanation (again under the terms of the system). Or perhaps the ultimate theory is probabilistic, and the asymptote is less than perfection; but in that case there will be no better to be had.[17]

Now, it is certainly plausible that mental events do not (or do not ever appreciably) affect the behavior of the sun, so there may well be "homonomic generalizations" *within* the physical sciences, as Davidson says. But, *if* the identity theory fails, *if* the mental vocabulary cannot be construed as nothing more than a *façon de parler* (so that the mental cannot be eliminated altogether), and *if* (as Davidson acknowledges) the mental is causally efficacious, *then the physical sciences, ranging systematically over the entire domain of physical phenomena, cannot be homonomic as such.* This is precisely what Feigl feared: there may well be, on empirical grounds, non-*physical₂* phenomena that are not merely epiphenomenal. Davidson offers no argument for his case – only an argument to the effect that what he calls anomalous monism (incorporating a token-identity claim, not a type-identity claim) is consistent.[18] In effect, Davidson helps to clarify an important sense in which the physical sciences are or are not *closed*: he trades on the notion of emergence, though he does not actually note the extent to which and manner in which he subscribes to the restrictions of a specifically *physical₂* vocabulary.

One of the most explicit statements of the admissibility of emergence *within* the scope of a *physical₂* science is provided by Mario Bunge. The relevant notion – what may be dubbed emergence$_B$ – falls quite strictly within the scope of the denial of emergence$_F$, but it does so by providing a ramified account of a hierarchy of "levels" within that domain. Feigl had taken emergence to be the denial of *physical₂* reduction; Bunge concedes a sense of emergence that is not incompatible with *physical₂* reduction if we understand reduction in terms of an explanatory *vocabulary* and not in terms of the denial of what Meehl and Sellars mean by "emergent *laws*." In effect, Bunge reconciles emergence and reduction *within* the range of Feigl's thesis – though contrary to a very strong unity-of-science program. Alternatively, Bunge's formulation is neutral to the adequacy of the unity-of-science program, though he personally believes the evidence is against it.

On Bunge's view, "an emergent thing (or just *emergent*) is one possessing properties that none of its components possess."[19] So construed, emergence$_B$ addresses part–whole relations in an ontologically neutral way, even though Bunge's motivation is clearly to support the thesis that all natural phenomena fall within the scope of physical explanation. On Bunge's view, natural phenomena are such that they cannot all be explained in terms of the law-like relations that hold *only among their components*, by way of a transitive regress; in Meehl and Sellars's terms, there must (to capture Bunge's views) be "emergent laws," laws connecting the properties of "emergent things" and the properties of what obtains at the level of their components. The issue at stake is precisely the one that exercised Rudolf Carnap: the translatability of psychological terms by physical terms (Carnap was more sanguine that most physicalists) does not entail the "deducibility" of (say) psychological laws from the laws of "inorganic physics."[20]

Bunge's account depends on the notion of "levels" within single systems and is fixed by the following postulates:

Postulate 1   Some of the properties of every system are emergent;
Postulate 2   Every emergent property of a system can be explained in terms of properties of its components and of the couplings amongst these;
Postulate 3   Every thing belongs to some level or other;
Postulate 4   Every complex thing belongs to a given level that has self-assembled from things of the preceding level.[21]

Here, the account appears (thus far at least) to be entirely congruent with Oppenheim and Putnam's well-known précis of the unity-of-science program. It seems fair to say that, for Oppenheim and Putnam, the working hypothesis of a hierarchy of "levels" is construed heuristically, though the underlying motivation is strongly realist.[22] Bunge is not committed, by adherence to emergence$_B$, to the so-called "strong" (or "strongest") sense of the unity of science – the "unity of laws" – in accord with which, that is, "the laws of science become reduced to the laws of some one discipline" (ultimately, a "unitary science," "an all-comprehensive explanatory system").[23] Bunge is not committed to Oppenheim and Putnam's sense of the *micro-reduction* of one branch of science, B_2, to another, B_1 – because, although he appears to accept the condition (Oppenheim and Putnam's) that "the objects in the universe of discourse of B_2 are wholes which possess a decomposition into proper parts all of which belong to the universe of discourse of B_1," he does not believe that we can, "in principle, dispense with the laws of B_2 and explain the relevant observations [as of the emergent properties of a high-level system B_2, 'self-assembled' from the things of the next lower level, B_1] by using B_1 [only]."[24] His reason is simply that, at some emergent levels and for some phenomena, the requisite explanatory theories will involve "suitably enriching [lower-level theories] with new assumptions and data" and that the reduction of laws is a logically distinct question. In fact, Oppenheim and Putnam note both that, "at a given time," Dalton's "chemical theory of molecules might not be reducible to the best available theory of atoms ... if the latter theory ignores the existence of the electrical properties of atoms," and that "Such hypotheses and models as those of Crick and Watson, and of Delbrück, are at present far from sufficient for a complete micro-reduction of the major biological generalization, e.g., evolution and general genetic theory (including the problem of the control of development). But they constitute an encouraging start towards the ultimate goal. ..."[25]

Actually, Oppenheim and Putnam provide the essential clue to any demurrer of Bunge's sort: not "all reductions are micro-reductions"; "it seems very doubtful," they concede, "that a branch B_2 could be reduced to a branch B_1, if the things in the universe of discourse of B_2 *are not themselves in* the universe of discourse of B_1 and also do not possess a decomposition into parts in the universe of B_1."[26] The special sciences may formulate laws that are not expressed (and are not known to be expressible) in terms of the parts of phenomena in accord with Bunge's Postulate 4 – even within the scope of a *physical₂* world. Furthermore, *physical₂* reduction does not entail micro-reduction; in particular *physical₂* reduction does not entail the transitive reduction of the laws of hierarchically ordered levels of natural phenomena, even if those levels accord with Bunge's Postulate 4. This helps to expose, for example, the logical blunder of B. F. Skinner's attempt to reduce the putative laws of behavior in terms of environmental variables by construing intervening cognitive variables (which Skinner really supposes to be merely heuristic or even fictional) in terms of environmental variables: there is no reason to believe that the reduction of functional distinctions (even if valid) yields transitive micro-reductions.[27] Jerry Fodor explicitly acknowledges the general point in a strong sense, although it is clear that he, too, is sanguine about the adequacy of explaining emergent psychological phenomena in terms of "physical mechanisms" (specifically, along the lines of Davidson's anomalous monism).[28] Here Fodor tends to agree with Bunge, favoring physical reduction, opposing the strong micro-reductive program of the unity of science, and reconciling emergence and reduction (apparently) within a $physical_2$ schema.[29]

II

The advantage of our review of emergence within physicalist terms is just that it prepares us for emergence (for example, with respect to biology) that may not be expressible in terms of emergence$_B$, that may require that $physical_1$ terms not be coextensive with $physical_2$ terms (in Feigl's sense) – hence, that may exhibit emergence$_F$; *and* for the possibility of emergence (for example, with respect to psychology and the human or cultural sciences) that meets the condition of emergence$_F$, is not, *a fortiori*, expressible in terms of emergence$_B$, *and yet does not entail any form of Cartesian dualism.* Such a possibility – conceding causal interaction – entails a radical rejection of the unity-of-science program, hence, a rejection of the homonomic nature of the laws of at least large subsectors of the physical world or (alternatively) of the *closed* nature of the physical world. Broadly speaking, then, two possibilities present themselves:

1 that biological phenomena, reasonably identified as $physical_1$, cannot be accommodated in a descriptive or explanatory respect in $physical_2$ terms; and
2 that linguistic and cultural phenomena cannot be sufficiently characterized in $physical_1$ terms, phenomena which, since they concern matters of fact and causal efficacy, should nevertheless fall within the competence of some science, or which, even if generously characterized in $physical_1$ terms, cannot be explained in terms of laws in accord with emergence$_B$, in the sense both of the strong unity-of-science program and of the view (favored by Bunge) that opposes the micro-reduction (or at least the wholesale micro-reduction) of putative emergent laws.

The issue is a complex one involving what we have earlier termed incarnate phenomena.[30]

One consideration of importance concerns the peculiar informality (even vacuity) and elasticity of the notion of the "physical" or "$physical_1$."

Often it appears that *whatever* is subject to scientific description and explanation falls within the scope of the physical – even if, as it now seems reasonable to argue, the formal properties of a science are not (and perhaps cannot be) reliably fixed. The irony is that the very concept of what is properly a *scientific* matter has, for historical reasons, been peculiarly intimately linked with the prospects of physicalism (of Carnap's sort, for instance – though not merely of Carnap's sort – and of the unity-of-science program). Evidence of this linkage is rather interestingly provided by Noam Chomsky, since Chomsky is concerned to counteract the threat the explanation of linguistic phenomena appears to pose to the unity-of-science program. So Chomsky says,

> let us assume that it makes sense to say, as we normally do, that each person knows his or her language, that you and I know English for example, that this knowledge is in part shared among us and represented somehow in our minds, ultimately in our brains, in structures that we can hope to characterize abstractly, and in principle quite concretely, in terms of physical mechanisms.[31]

Chomsky speculates that the pertinent mechanisms may be those of Cartesian automata or of "contemporary natural science"; or "perhaps principles now unknown enter into the functioning of the human or animal minds, in which case the notion of 'physical body' must be extended, as has often happened in the past, to incorporate entities and principles of hitherto unrecognized character."[32] Nevertheless, he confines his own studies exclusively to abstract, purely formal regularities.

Chomsky's entirely plausible suggestion (whether, ultimately, true or false) marks the inevitable informality with which the term "physical" is used *and* the impossibility of drawing a sharp demarcation line between "*physical$_2$*" distinctions and other putatively *physical* distinctions (biological, for instance) not reducible in *physical$_2$* terms: the concepts of the *physical* and *physical$_2$* simply evolve historically and without sharp boundaries. So quarrels about emergence$_F$ must be relativized to particular phases of the history of science. Another way of putting the point is this: "emergence$_F$" signifies, in historically relativized terms, what would entail the *denial* that nature forms a comprehensive *closed* system, a system under homonomic laws – in the "weakest sense" of the unity of science (the unity of "all the terms of science"), whether in accord with Carnap's notion or Kemeny and Oppenheim's or in some alternative sense, *or* in a sense clearly more informal than that of even the "weakest sense" of the unity of science.[33] The advantage of the "weakest sense" doctrine is precisely that it provides a conceptual grip on what may be meant by extending (as with Chomsky) the meaning of "physical" or by extending by micro-reductive means the scope of $physical_2$ explanation itself. By contrast with "emergence$_F$," "emergence$_B$" signifies only what falls *within* a putatively *closed* system, whether or not such a system accords at least with the requirements of the unity-of-science program, as long as it accords at least with a compositional hierarchy of increasingly complex physical levels. So the two notions of emergence are not easily unified, though they bear on common issues.

We do not really know what the usually announced constraints on "physical" and "$physical_2$" amount to. The positivist philosophy of science, particularly in Carnap's hands, treated first-person sensory or phenomenal reports, expressed in protocol sentences, *as* translatable in physicalist terms; but Carnap never supplied the required reduction (and never gave the idea up, though he gave up foundationalist claims about them).[34] Feigl is notoriously vague and sanguine about the $physical_2$ reduction (or replacement by "successor" concepts) of "all phenomenal concepts" which ensure the "anchoring [of science] in the data" and which fix the sense of "physical" (or "$physical_1$").[35] Here, Feigl follows Sellars, but Sellars merely treats the phenomenal (without any argument whatsoever) – or, better, phenomenal reports – as functionally eliminable as far as the scientific characterization of nature is concerned (the "scientific image").[36] Along similarly motivated lines, Davidson simply announces the token-identity of the mental and the physical, without any attempt at a theoretical analysis of the mental *in virtue of which* the identity might have been justified (but which would surely then have reintroduced the problem of type-identity – and with it the prospect of psychophysical laws).[37] Again, Jerry Fodor accepts Davidson's direction here, without any independent analysis of the mental.[38]

John Searle's recent attack on reductive programs exhibits, we may observe, the reverse difficulty. Searle appears to favor an emergent interpretation of mental properties and processes: they are said to be "macro-level" properties *of* the brain. Searle affirms both "naive physicalism" ("all that exists in the world are physical particles with their properties and relations") and "naive mentalism" ("there really are mental states; some of them are conscious; many have intentionality; they all have subjectivity; and many of them function causally in determining physical events in the world"); Searle further affirms that "naive mentalism and naive physicalism are perfectly consistent with each other."[39] But on that view either the vocabulary of mental phenomena is no more than a mere *façon de parler* (say, along the lines of Davidson's account – which Searle opposes) or mental properties are taken to be physical despite the fact that they are *not* demonstrably type- or token-identical with anything that may be independently identified as physical. In that impoverished sense, they are "emergent": *"all mental phenomena [just] are caused by processes going on in the brain ... and they just are features of the brain (and perhaps the rest of the central nervous system)."*[40]

The trouble is that Searle merely enunciates what he believes. More perspicuously put, Searle never explains what, regarding the brain's physical nature and composition, could account for the "subjective" and "intentional" features of mental properties – which are physical. He utterly lacks a clear conception of physical emergence (say, emergence$_B$) in virtue of which he might offer a systematized defense of his claim. Also, he lacks what may be called a sense of "ontic adequation," a principled sense of how to analyze the "macro-level" (the mental properties) *of* brain activity, so that their very ascription may be conceptually reconciled with the usual ascription of physical properties to the brain – so that we may grasp the very fact that they "just are" physical properties.[41] The same conceptual lacuna appears, as we have seen, in Chomsky's extension of the "physical." It's no good merely affirming that the mental is the physical. We must supply grounds for (1) a type-identity or (2) a token-identity or (3) a "successor" substitution (in Sellars's sense) or (4) a causal account along the lines of emergence$_B$ (or some suitable substitute) or (5) a functional analysis of the mental in accord with the required adequation between our ascriptions and the nature of what we make our ascriptions of.[42]

The fact remains that the very scope of the *physical$_2$* and the demarcation between biological (and psychological) and non-biological (*physical$_1$*) concepts cannot be drawn even provisionally, if we cannot say what the conceptual status of first-person psychological reports are. So we actually find in the standard literature opposed to the strong doctrine of emergence$_F$ no clear sense in which either phenomenal experience or linguistic (or linguistically qualified) behavior and states are or are not reasonably construed as physical, *physical$_1$*, *physical$_2$*, or *physical$_1$* but not *physical$_2$*; and, without that distinction, the prospects of denying emergence$_F$ *or* of affirming a comprehensive emergence$_B$ (whether in terms of micro-reductions or not) remain entirely moot, particularly at the levels at which biological, psychological, linguistic, and cultural phenomena are conceded.

This is a most remarkable lacuna, given the vigorous history of reductionism. *If* we cannot establish a secure linkage between what, on the usual theory, gives an empirical standing to science itself and what are at least the strategically placed *compositional* parts of systematic wholes known to be subject to covering laws, then not only the unity-of-science program but also the larger notion of a hierarchy of compositionally specified levels will prove seriously defective – just at that point at which the adequacy of emergence$_B$ will need to be gauged. We do not know precisely how to understand emergence$_F$ because we do not know what the defensible relationship is between the phenomenal experience and linguistic reports of scientists and the putatively *physical$_2$* phenomena they are considering: the usual theory supposes, without much in the way of argument, that the former either fall within the scope of the latter or can be eliminated in a suitably canonical account. But, *if* mental and cultural phenomena are real and causally efficacious, then the very point of denying emergence$_F$ remains altogether untested if we have not examined these recalcitrant data. Furthermore, if we cannot settle this question, then we cannot pretend to have a proper basis for testing the adequacy of emergence$_B$ just where the thesis would be most controversial and significant.

Apart from eliminating the mental, and apart from construing the functional as (ultimately) heuristic or fictional,[43] the principal strategies for ensuring the adequacy of *physical$_2$* accounts require type-identity, token-identity, or "successor" substitutions (along the lines of Feigl's or Sellars's accounts). The enlargement of the extension of "physical" or "physical$_1$" terms to include the mental *is not* suited to any *physical$_2$* program (in particular, the unity program) unless that extension is (at least) expressly reconciled with emergence$_B$. This marks the essential weakness of Chomsky's recommendation as well as John Searle's (offered on entirely different grounds).[44]

III

Several further distinctions suggest themselves. First of all, *emergence cannot be convincingly restricted in epistemological terms.* This goes directly counter to Hempel's view: "emergence of a characteristic," he says, "is not an ontological trait inherent in some phenomena; rather it is indicative of the scope of our knowledge at a given time; thus it has no absolute, but a relative character; and what is emergent with respect to the theories available today may lose its emergent status tomorrow."[45] It is certainly true, as Hempel says, that ascriptions of emergence are bound to change with the state of our science; but that cannot affect the concept of emergence with respect to a realist science: emergence must be an ontological trait – or (less argumentatively) a trait empirically accessible in the nature of things (if justifiably ascribed). In fact, it could be not unfairly said that Hempel's way of construing emergence signifies that ascriptions of emergence are simply admissions of ignorance; in a suitably developed science (in which phenomena are subsumed under laws in accord with pertinent micro-theories), emergence as such would be regularly eliminated.[46] The intended implication is clearly that emergent = scientifically unexplained. (Call this emergence$_H$.) But the thesis seems to go directly contrary to Hempel's own empirical and realist orientation[47] as well as to the obvious intention of serious disputants about the status of biology as a science of emergent phenomena. Thus, David Hull, writing under a strong Hempelian bias toward the unity of science, still cannot quite bring himself to claim to be able to decide the "ontological" issue – even after he has exposed the characteristic weaknesses of a number of the best-known emergentist views.[48]

Biology challenges (but does not necessarily defeat) both the denial of emergence$_F$ and the affirmation of emergence$_B$, but any serious reading of the relevant data utterly precludes emergence$_H$. Perhaps the central theoretical concern posed by biological phenomena, congruent with that denial and that affirmation (which, despite Bunge's intended use, is compatible with both a strong micro-reductive program and one that opposes micro-reduction) is that of hierarchical organization. (This bears on a second point to be made below.) The evidence is apparently not yet in to decide the matter empirically; but it seems fair to say that a version of emergence$_B$ is quite promising, even if it should be the case (which is uncertain) that strong micro-reductive claims will have to be abandoned. It is easy to formulate a conception of hierarchical organiz-ation that *could* not fail to facilitate a strong micro-reductive reading, though it is not at all clear that biological and higher-level phenomena would support it empirically: one has only to assume "a linear order of parts, sub-parts, sub-sub-parts, etc." in accord (in effect) with Bunge's Postulate 4.[49] Robert Causey recommends avoiding such an assumption, particularly in the context of biological hierarchies. He also notes that, in dealing with organisms, the behavior of given systems may depend on *parts* (such as water and hormones) that are not proper parts of the system in the sense intended. So he treats a "hierarchical structure" as "one which involves at least two structural descriptions, namely, a description of the whole as composed of certain kinds of parts, plus a description of at least one of these parts as a structure composed of certain sub-parts."[50]

It needs to be added, of course, that reference is often made to "dynamic systems" or "open systems" – particularly where such "systems" are living organisms or societies, active evolutionary aggregates of various sorts, markets, or the like, and where what is being considered are measures of equilibria or optimizations (or even constellations that exhibit stable growth of various sorts far from equilibrium) with respect to given inputs and outputs linking a would-be system with some enclosing ecological niche or environment, *and* where the processing capacities assigned a system may, over time, be, in part at least, an artifact of the interaction between a system and its environing world. There need be no contradiction, here, between this entirely realistic use of the term "system" and what we have intended in speaking of systems with respect to the unity program: open systems are not (yet) systems in the sense here favored, though they obviously have their distinctive dynamics and may well belong to closed systems. Popper, for example, reporting the work of Ilya Prigogine, mentions "open systems" and has the following to say: "*open systems in a state far from equilibrium* show no tendency towards increasing disorder, even though they produce entropy. But they can export this entropy into their environment, and can increase rather than decrease their internal order. They can develop structural properties, and thereby do the very opposite of turning into an equilibrium state in which nothing exciting can happen to them any longer." Popper adds that "Prigogine's work may be looked upon as a piece of exciting physicalist reduction" or at least as opening "the way to understanding the reason why the creativeness of life does not *contradict* the laws of physics."[51]

Now, generally, but especially in biological systems, hierarchies exhibit the property that "the parts of a structure are under the influence of special boundary conditions created by the structure."[52] Causey shows that, in itself, within the tolerance of the adjusted definition and conceding additional complexities (for instance, that, as is usual in living systems, certain parts cannot be studied except within certain larger parts of the system; or, as in genetics, reference may have to be made in an indissolubly interlevel way to genes and "gross phenotypic characteristics"), this is not incompatible with the micro-reductive program.[53] Be that as it may (the matter is controversial), explanatory accounts formulated in such terms appear to be strongly compatible with the denial of emergence$_F$ and even promisingly congruent with emergence$_B$.

It is of course the "special boundary conditions" that threaten micro-reduction and, conceivably, emergence$_B$: this is probably also close to what Feigl (and Meehl and Sellars) had in mind in considering evidence for emergence$_F$. The matter is unclear at the moment since, even though a fair picture of the function of such boundary conditions is at hand, *we have no clear idea of the origin or conditions under which such conditions obtain.* This accounts for Polanyi's resistance to the reduction of the hierarchical behavior of living cells in terms of the laws governing the motions of molecules – even though the molecules within living systems obey physical laws.[54] H. H. Pattee notes, in one of the most careful and balanced accounts of biological hierarchy, that Francis Crick's micro-reductive confidence is simply premature:[55] the critical question of the "hierarchical interfaces" between the law-like functioning of molecular structures and the functioning of cells and higher-order systems under what appear to be the "unlikely constraints" of the special boundary conditions of living systems – *and* the origin of the latter – is still apparently unresolved.[56] In fact, on Pattee's view, this *is* the central problem of the origin, evolution, development, and characteristic functioning of living systems, the problem of "time-dependent boundary conditions . . . not imposed by an outside agent, but . . . inseparable from the dynamics of the system" itself, by means of which a given "collection of elements [for example, an aggregate of molecules, can] impose *variable* constraints on the motion of individual elements" – having "an effect which is like modifying the laws of motion themselves."[57] Hempel agrees that, thus construed, biological and psychological phenomena *are* largely emergent (in effect, emergent$_F$): although he also takes that characterization to be

> trivial, for the description of various biological [and psychological] phenomena requires terms which are *not contained* in the vocabulary of present-day physics and chemistry; hence we cannot expect that all specifically biological phenomena are explainable, i.e. deductively inferable, by means of *present-day* physico-chemical theories of the basis of initial conditions which themselves *are described in exclusively physico-chemical terms.*[58]

This, of course, is the strong unity-of-science position.

A second useful point (alluded to above) is simply that *complex systems – particularly at the biological, psychological, or cultural levels – may not be hierarchical at all (or, hierarchical in accord with a principle similar to that of emergence$_B$).* This, of course, would, if true, defeat the denial of emergence$_F$ (in its original sense), *a fortiori* the strong micro-reductive reading of emergence$_B$. It might, however, *not* entail the denial of a looser version of physicalism, not confined to *physical$_2$* notions but also not involving any form of vitalism or dualism – for example, along the lines of Bertalanffy's general systems approach, which calls for isomorphism of laws in different physical fields, in which strong micro-reduction does not obtain. (Call this thesis emergence$_{VB}$, intended to be compatible with a distinctive form of the unity of science.) Apparently, it fails specifically to accommodate such important facts as that "The 'system' features of the [Jacob–Monod] operon theory ... do not seem to have any analogues or isomorphic counterparts outside the area of *biological* control systems."[59]

Herbert Simon has formulated a very influential conception of hierarchical systems, which he interprets in a way favoring a strong micro-reductive reading.[60] More narrowly viewed, Simon claims that "we do not need to postulate processes more sophisticated than those involved in organic evolution to explain how enormous problem mazes [even at the human level] are cut down to quite reasonable size. ... [H]uman problem solving, from the most blundering to the most insightful, involves nothing more than varying mixtures of trial and error and selectivity."[61] Furthermore, on Simon's view, "complex systems" are, generally, "hierarchic systems" or can be so construed. "If a complex structure," he says, "is completely unredundant – if no aspect of its structure can be inferred from any other – then it is its own simplest description. We can exhibit it, but we cannot describe it by a simpler structure." Otherwise, "many [it seems, in the argument, perhaps all other] complex systems have a nearly decomposable, hierarchic structure [which enables] us to describe, and even to 'see' such systems and their parts." Alternatively, "If there are important systems in the world that are complex without being hierarchic, *they may to a considerable extent escape our observation and our understanding.*"[62] Human problem-solving is, then, on Simon's view, intelligible, just because on "a growing body of evidence" it is a form of "means–end analysis" (trial and error and selectivity, as remarked above) involving "continual translation between the state and process descriptions of the same complex reality" (so construed).[63]

This is a breathtaking generalization, but it suffers from some obvious difficulties. In the first place, the modeling of *segments* of intelligent human processing by hierarchically organized machines may be formally apt, without having any realist force at all; secondly, such modeling does not increase the likelihood that open-ended human *capacities* can be so modeled; thirdly, on Simon's own view though clearly prejudicially, *we ourselves* would escape our own observation and understanding *if we were not hierarchically structured*; and, fourthly, Simon offers no direct empirical evidence to show that, at the psychological, social, linguistic, or cultural level of behavior, human beings do actually function *as* hierarchical systems or, more accurately, systems reducible to hierarchical systems favoring micro-reductive accounts or accounts in accord with emergence$_R$.[64]

The key notion in Simon's view is that of "nearly decomposable systems" – which clearly favors micro-reduction. For instance, it is pretty well governed by the rule, "In organic substances [apparently to include living systems], intermolecular forces will generally be weaker than molecular forces, and molecular forces weaker than nuclear forces."[65] But, judging from Pattee's remarks, this is either flatly false, very doubtful, or completely uninformative about biological or higher-level systems. (Simon could of course intend his remarks in Hempel's sense.) Nevertheless, it is on this basis that Simon suggests that a "system [like a rare gas, may be described] as decomposable into the sub-systems comprised of the individual particles." Then, for other gases, for denser gases, for actual gases, "we can treat the decomposable case as a limit and as a first approximation; [and] as a second approximation, we may move to a theory [bearing on living organisms and higher-level systems] of *nearly decomposable systems*, in which the interactions among the subsystems are weak but not negligible."[66] So Simon sees biological and higher-level systems as continuous, in hierarchically relevant respects, with the phenomena of inorganic levels – hence, as providing (in effect) a reasonably economical adjustment, *via* the notion of the "nearly decomposable," of emergence$_B$ (and, more or less in accord with the view attributed to Causey: namely, a micro-reductive reading of emergence$_B$). The hierarchical structures of complex systems, then, are characteristically nearly decomposable – that is, decomposable in the limit; correspondingly, with a measure of tolerance, versions of emergence$_B$ that are not of the micro-reducible sort are in the limit micro-reducible.

This brings us to a third fundamental point: *the characteristic functioning of complex or hierarchical systems, particularly of the biological and higher-level sort, are not merely functionally describable – that is, abstract, indifferent to material realization or the like.* To suppose that they are merely abstract is the special thesis known as functionalism.[67] Where the thesis acknowledges that the relevant phenomena are real, abstract, not identical with physical phenomena, it risks Cartesian dualism (with which, in any case, it is formally compatible). Where functionally described phenomena are, taken "token-wise," identical with physical phenomena (as in Davidson's and Fodor's views), functionalism is a *mere façon de parler* – normally introduced (as with Davidson) in order to disallow psychophysical laws or similar breaches of physically homonomic systems. Independently of versions of the identity theory, functionalism remains a *mere façon de parler* – as, for instance, with Sellars, Stephen Körner, May Brodbeck, and Daniel Dennett[68] – by essentially insisting on the closed nature of the physical world. In addition, functionalism is sometimes pressed into service in support of functionally characterized laws said to be not reducible to physical laws, on the grounds that there are no "natural-kind terms" (pertinent to emergence$_B$ or the strong unity-of-science program) to which the functionally identified regularities can be reduced. This is Fodor's view, for instance, in contrast to Davidson's.[69] But neither identity nor token-identity is (as such) a nomological notion, and Fodor fails to supply a reasonably developed account of natural-kind terms with regard to which his thesis is plausible.[70] The more serious objections to functionalism lead us back to the issue of biological and higher-level hierarchies: these must be construed (to avoid Cartesian dualism on the one hand and a merely heuristic or fictional use on the other) as complex systems with respect to which functionally described properties *are abstracted from incarnate properties*, properties indissolubly characterized as either emergent$_B$ (precluding micro-reduction) or emergent in some as yet unspecified sense more extreme than emergent$_B$.[71] The coding powers of living cells, for instance, are either incarnate in this sense or else the micro-reductive reading of emergence$_B$ is not likely to be weakened by biological phenomena. (For our present purpose, it is perhaps sufficient to say that *incarnate* properties prove, *qua* emergent, conceptually simple or indissoluble with respect to accounts that press to reduce them compositionally to, or permit them to be analyzed without remainder into, ingredient properties including the physical, the *physical*$_1$, the *physical*$_2$, or to such properties *plus* [heuristically introduced] abstract or functional properties. Nevertheless, once *that* [emergent] simplicity is acknowledged, they may prove as complex [polyadically] as you please – at *their* level of emergence. It is an entirely open question, empirically, whether and to what extent biological and higher-order emergents conform to the putative laws of the *physical*$_2$ world; but conceptually, on the emergentist thesis, we may only "abstract" *physical*$_2$-like "aspects" of [indissoluble] emergent properties – since they are not "composed" of, or additively specified from, *physical*$_2$ properties.)

A fourth and final point is this: *the real structures of certain sectors of the order of nature – specifically, linguistically and culturally qualified phenomena – cannot exist independently of the structures of the cognizing powers of those who share the culture in which they appear.* This is an easily overlooked but extremely subtle and strategic point: the effect of denying it is to retreat to reductionism or dualism. It is entirely reasonable to *assume* that the physical world, structured in whatever way it is, exists independently of our inquiry and knowledge, even if what *we* hold to be knowledge of that world is ineluctably subject to conceptual orientations that we cannot count on ever becoming completely aware of reflexively.[72] But the world of human culture exists only insofar as culturally apt aggregates of humans knowledgeably participate in the practices of the culture they share. The assumptions it is reasonable to make about the independence of physical nature and human culture are quite different, therefore, even though (rejecting the functionalism described, and avoiding Cartesian dualism or analogous doctrines) cultural, linguistic, historical, psychological properties must be incarnate in the sense sketched (and, correspondingly, culturally emergent entities – persons, artworks, words and sentences, for instance – must be indissolubly "embodied" in some way in suitable physical or biological entities[73]). This may be fairly construed as a sympathetic reading of Wittgenstein's powerful notion of "forms of life" (along lines, of course, that Wittgenstein himself would never have been drawn to). But, without risking the details of any such special ontology, it should be clear that the mere admission of the reality and causal efficacy of cultural phenomena and linguistically qualified behavior and mental states poses an extraordinary burden on the defensibility of the denial of emergence$_F$ and of the affirmation of emergence$_B$ (whether of the micro-reductive or non-micro-reductive sort). For that reason, the admission cannot fail to affect the profound question of the "bifurcation" of the physical and human sciences.[74] The reason is elementary. Once the kind of *complex* phenomena that a nondualistic and nonheuristic treatment of linguistic and cultural phenomena would introduce is conceded, we cannot fail to consider whether we are or we are not conceptually obliged to admit that a crucial sector of the real world *is both complex and not hierarchically structured in a way restricted to emergence$_B$*. (This of course is not to deny that such phenomena *are* hierarchically structured, only that they may not – and perhaps cannot – be in the sense favored by Bunge, Causey, Simon, Crick, Hempel and the army of reductively oriented theorists attracted to views like theirs.)

IV

The crucial issue is a double one. For there to be a form of emergence that satisfies (in the sense originally intended) emergence$_F$, that *a fortiori* does not satisfy emergence$_B$ – in either the strong unity-of-science sense or in a sense opposed to micro-reduction, or even in the sense (favored alternatively by Pattee and Simon) that could support only an idealized tendency toward the alternative readings of emergence$_B$, or even in a sense (favored by Fodor) that conceded no more than the heuristic usefulness of functional regularities or laws not known to be reducible in either sense of emergence$_B$ – the *real* phenomena in question would

(1) have to be complex but not hierarchically composed in accord with the conditions of emergence$_B$; and (2) have to possess properties of a kind that could not be explained as a result of the composition of any lower-level elements of systems that do accord with emergence$_B$ (*a fortiori*, that could not be explained in any micro-reductive way). Now, then, the best candidates for such phenomena are *human persons, artworks, words and sentences*, and the like – that is, culturally emergent entities. (Call the intended form of emergence, emergence$_M$.[75]) It is not enough to say that such phenomena (thus construed) are *sui generis*. The same could be said in a strong sense in accord with Bunge's reading of emergence$_B$. The distinction required is that culturally emergent entities and their distinctive properties are such that the proper parts of systems in accord with emergence$_B$ are *not* the proper parts of such systems (for instance, the proper parts of a piece of marble are not the proper parts of Michelangelo's *Pietá*, manifest in the marble); the salient parts or proper parts (if the term has any sense at all, here) of such systems are not characterizable in terms of the "self-assembly" of the elements of any lower-level systems restricted to emergence$_B$ (for instance, the parts of a sentence are not expressible in terms of any merely acoustical array or the like), and the generation of such emergent systems cannot be accounted for in terms of any law-like regularities ranging over systems confined within the scope of emergence$_B$.

These constraints, admittedly, threaten to produce a discontinuity in the generation of natural systems. But the discontinuity is really only apparent, and its resolution provides a decisive set of distinguishing features regarding phenomena in accord with emergence$_M$. The key lies with the fourth distinction offered above. Interestingly, Roy Bhaskar has, in two different, strongly realist and strongly anti-inductivist accounts of the theory of science, managed to formulate diametrically opposed (or, at least, what look like diametrically opposed) views of this fourth consideration – both of which turn out to be inadequate. Charity may recommend a way of reconciling the intent of both statements (which need not be resisted); but they are instructive precisely because they reinforce the importance of the point itself. In the earlier account, Bhaskar insists in various ways,

> Without the concept of real strata apart from our knowledge of strata we could not make sense of what the scientist ... is trying to do: viz. to discover the reasons why the individuals which he has identified (at a particular level of reality) and whose behavior he has described tend to behave the way they do. ... Now if

changing knowledge of strata is to be possible the strata must not change with our knowledge of them. Thus the concept of real strata apart from our knowledge of them is necessary if both the ideas of scientific structure and scientific change, which are central to recent critical philosophy of science, are to be intelligibly sustained.[76]

Bhaskar adds that, "in the transition from knowledge of any one stratum to knowledge of the next, knowledge of three levels of the objective world order is progressively obtained: of relations between events, of causal laws and of natural kinds."[77] This cannot be right if emergence$_M$ obtains; and, if it obtains, it obtains at the level of (human) cultural phenomena.

The point is that, *if* Cartesian dualism, reductionism, and a merely heuristic or fictional reading of cultural and psychological phenomena are rejected, *then the meaning and conditions of scientific realism cannot be the same for the physical and the human sciences.* For the physical sciences, some presupposition in accord with Bhaskar's claim is reasonable; but, for the human sciences, it is not only unreasonable, it is impossible: languages, traditions, practices, institutions, histories and the like must, in some sense, be assigned societies; but there are no societies (or cultures that they sustain) except insofar as aggregates of individual human agents behave in ways reasonably construed as sharing (in Wittgenstein's sense) a form of life. It is true that the practices of actual societies are, in a fair sense, independent of the behavior of any individual member of that society (although even this may be too gross-grained a concession); but an observed society cannot be said to have (culturally distinctive) properties independent of the cognitive properties of the *reflexive observers* (*a fortiori*, the participants) *of that society. The realism of the human sciences is essentially consensual.*

This is what Bhaskar appears, but only appears, to acknowledge in his second account, where he considers restrictions on the adequacy of the first. Assuming "society [to be] *sui generis* real," he says, the following "ontological limitations" will have to be acknowledged:

(1) Social structures, unlike natural structures, do not exist independently of the activities they govern; (2) Social structures, unlike natural structures, do not exist independently of the agents' conceptions of what they are doing in their activity; (3) Social structures, unlike natural structures, may be only relatively enduring (so that the tendencies they ground may not be universal in the sense of space-time invariant). These [he adds] all indicate real

differences in the possible objects of knowledge in the case of the natural and social sciences.[78]

Uncertainty about the force of what Bhaskar means by these distinctive rests on whether he means by (1)–(3) to emphasize differences in the *real* phenomena of the natural and social sciences (which, of course, he does mean) or whether he means, in emphasizing that, to emphasize a difference *also* in the sense and conditions of *realism* itself applied to the natural and social sciences (which it seems he does not mean). He sorts the differences in the *objects* of both kinds of science, but he does not sort the differences in the *observing scientist's* relationship to the domains observed. Nevertheless, he does concede, somewhat obliquely, possibly fundamental differences in the causal processes examined by the two kinds of science. Bhaskar does not consider the peculiarly consensual basis (ontological as well as epistemological) of whatever the human sciences may judge to be the real properties of the cultures they examine – at first reflexively and then, by extension, of historically removed or alien cultures; also, Bhaskar does not consider the strong (Wittgensteinian) sense in which the very sharing of language, the very exercise of spontaneous linguistic aptitudes – the precondition of human culture and its paradigmatic manifestation – entails not only "the internal complexity and the interdependence of social structures"[79] but also the *reflexively consensual* sense in which the reality of social structures is accounted for and confirmed. In disputing Peter Winch's remarks about would-be objective studies of the Azande, for instance, Bhaskar makes the telling and telltale observation: "it is important to note the limits of all Viconian arguments for hermeneutics: *what we do not make, we have no privileged understanding of.* And we make neither society nor ourselves. It follows from this that the social structure, motivation and the tacit knowledge (skills and competences) we employ in social interaction and the transformation of nature may all be more or less opaque to our undertsanding."[80] The point (which Bhaskar appears to miss) is not that knowledge of nature presupposes knowledge and understanding of ourselves (socially) but rather that there is no scientific understanding of *ourselves* at all (*a fortiori*, of any other society) unless (more or less in the Wittgensteinian sense) *our* understanding of the prevailing practices of our own society is generally reliable (which, of course, is fully compatible with, even the precondition of, anyone's being mistaken about any particular detail). The human sciences are and cannot fail to be inherently hermeneutic.

The quarrel should be construed dialectically, of course. *If* this final point is denied, then it is entirely possible that some form of emergence$_B$

will prove adequate for the social sciences as well as the natural sciences. But, if the emergence of the phenomena of human culture is inseparable from the power (of aggregated human agents) to generate such phenomena *and*, in being able to do that, to understand the phenomena they share and generate, then there is good reason to believe that culturally emergent phenomena (emergent$_M$) are *sui generis* in a sense utterly unlike that in which emergent$_B$ phenomena can be. This is not the place to examine the intensional complexities of human thought and action and practice, the force of admitting that causal processes may not always behave extensionally, the consequence that causal processes may not always behave nomologically.[81] But these complications would mean, minimally, that cultural entities – persons (as distinct from but not opposed to members of *Homo sapiens*) and what they do and produce – are *not* characterizable (*contra* Bhaskar) in "natural kind" terms congruent (in the emergence$_B$ sense) with the natural kinds that, in the physical and life sciences, may be said to be "self-assembled" from the elements of some next-lower level. The reality of culturally emergent phenomena is accounted for in terms of their possessing, in addition to whatever properties are assignable to natural entities within the hierarchy of levels pertinent to the physical and life sciences, *properties "sui generis" to the (incarnate) complexity of the culturally emergent level of reality. If* language is the condition *sine qua non* of human culture and its paradigm manifestation, *if* it is causally efficacious (in the sense, say, in which speaking is efficacious), *and if linguistic phenomena cannot be accounted for compositionally in terms of any sub-linguistic phenomena*, then the bifurcation of the physical and human sciences cannot be resisted – and some form of emergence$_M$ cannot be denied.

This seems to be a reasonably clean and straightforward argument – which does not depend on any particular theory of the peculiar properties of cultural phenomena. On the contrary, it invites a careful study of those properties. It obliges us to recognize only that, if linguistic behavior is real, and if language cannot be analyzed in terms of the hierarchical composition of emergent$_B$ phenomena, then quite radical consequences are bound to follow with regard to the nature of those sciences that study such phenomena and the nature of the phenomena studied. It remains to add a word about the apparent discontinuity within the emergent systems of nature.

Part of the appeal of emergence$_B$ lies with the elegantly simple way in which emergence and reduction are linked, in which a strong unity-of-science program is seen to be a limiting case of a completely comprehensive theory of causal processes in nature. On the theory, an adequate explanatory vocabulary may be generated (in empirically suitable ways) in accord with a single rule that ensures the conceptual continuity of all emergent systems. Emergence$_M$ breaks that continuity (1) by positing entities that are not self-assembled in accord with Bunge's Postulate 4; and (2) by positing properties possessed by such entities that, though manifest in causal processes, cannot be featured in laws limited by the constraints of the denial of emergence$_F$ or the affirmation of emergence$_B$.

There is and can be one and only one resolution of this difficulty consistent with the causal continuities of nature; it must be that the real structures of the cultural world are inseparable from the cognitively reflexive, consensual life of those creatures (ourselves) that discern such structures. In a profound sense, we find ourselves in a uniquely emergent *world* (the world of cultural phenomena) because *we find ourselves* in finding the structures of the natural world. Our understanding (our mode of understanding) the natural world already *presupposes and entails* the reality of human culture; and that mode of understanding *presupposes and entails* our own understanding of our own cultural world. The hierarchical structures of the natural world (in accord, say, with emergence$_B$) do not as such presuppose or entail the world of human culture; but our *science*, our ability to know and understand such structures, presupposes and entails that world. It cannot be thought to be generated merely by some compositional assembly from the elements of an inferior level. Perhaps the emergence of the human species can be accounted for by reference to a causal schema very much like that of emergence$_B$. Nevertheless, within some critical phase of the development of the biologically grounded capacities of human beings – in a way that, quite frankly, no one understands – structures begin to emerge, and continue to develop from what thus emerges, that depend primarily on the incipient reflexive capacities of the species. What emerges – the linguistically, intentionally, historically, purposively complex phenomena of human culture – may well be linked to some extremely small, otherwise unpromising, biological change; their mode of organization is undoubtedly constrained by physical and biological regularities in which (like language) they must be incarnate in some way. But that is hardly a reason for supposing that their own compositional structure, their own hierarchical order, is continuous with some mere extension or adjustment of compositional regularities captured by emergence$_B$. The discontinuities of cultural emergence are due to the cognitively reflective power of whatever biologically emergent structures first made that world possible. The discontinuities signify only the *sui generis* properties of the world of cognitively reflexive lives: the power of such lives is itself (reflexively) seen to be incarnate in some (probably minimal) biological capacity linked by straightforward causal continuities to the rest of nature. The entities and the properties of the entities of the cultural world must be embodied or incarnate (in some way) in the entities and properties of the physical and biological worlds – which is what ensures their causal efficacy and reality and their escape from Cartesian dualism.

There can be no doubt that many puzzles remain to be faced regarding the nature of human cultures, human languages, human histories, human existence itself. Nevertheless, if the account given is a reasonable one, then we have provided a fair sense in which *the concept of natural emergence need not be restricted to the processes of emergence normally admitted within the scope of the physical and life sciences and yet not, for that reason, irreconcilable with the regularities of those sciences.* The recent history of pursuing this question confirms how rarely such an undertaking is attempted.[82] Needless, to say, explanations regarding emergent$_M$ phenomena will have to be managed top–down. The phenomena in question are, on the hypothesis advanced, *sui generis*: they obtain only at the reflexive level of linguistically and culturally apt humans. Alternatively put, it is the failure to resist emergence$_F$ that commits us, discontinuously, to top–down explanations geared to emergence$_M$. But that just is the thesis of the bifurcation of the sciences.

V

One final consideration suggests itself. The essential theme of emergence$_M$ holds that cultural phenomena are *incarnate* – "simple" as regards the range of phenomena of emergence$_B$ and the denial of emergence$_F$, indissolubly complex as regards conceptually abstractable but not actually separable physical and functional or Intentional "features." As incarnate, at the level at which they are discerned, cultural phenomena may of course be successfully analyzed in ways independent of the continuities of emergence$_B$ and the denial of emergence$_F$. The only way in which to reconcile this discontinuity in realist terms is to admit that the phenomena in question are discernible only reflexively and are real only in the space of such reflexive attention – at the level at which human investigators *and* the artifacts of their culture actually emerge. Unless we eliminate the human altogether (assuming the reductions of emergence$_B$ and of the denial of emergence$_F$ to fail), the human sciences cannot but function top–down: the relevant phenomena first obtain and are first discerned at the (ontic) level at which inquiry itself obtains; their description and explanation proceed by scanning, top–down, *which* incarnate processes best account for those selected; physical and biological continuities are never breached; the discontinuity of the cultural reflects only our confinement to the top–down restrictions of the (cultural) nature of science itself; and the realism of the cultural is captured by the indissoluble *complexity* of the incarnate – which lies in the structure of its properties and does not depend on admitting causal forces that have a provenance other than the physical or material.

Clearly, the strategy requires

1   that we reject treating incarnate properties as, somehow, the "addition" or "composition" of physical and functional properties; and

2   that we disallow any and all counterstrategies that characterize the descriptive and explanatory role of intentional, psychological, cognitive, cultural and similar properties in ways incompatible with requirement 1.

The policy seems innocuous enough, but it actually has far-reaching implications. Of course, it means to oblige any countermove to emergence$_M$ to show cause, *first*, why requirement 1 should not be conceded *before* entertaining any reductive or eliminative program opposed by requirement 2. An example will clarify what is at stake.

Robert Cummins has explicitly endorsed certain essential claims of all known eliminative (psychological) programs favoring AI within the terms of requirement 2. Cummins provisionally treats *intentionality* as equivalent to or as entailing the full cognitive capacities of humans as envisaged by "folk" scientists: on that view intentionality is not to be ascribed to anything that does not exhibit a "subjective" capacity to "understand" its own intentional states – states said to exhibit, at least to a first approximation, propositional content.[83] Cummins then introduces the potentially subversive (eliminative) counterpart notions, *cognition and UNDERSTANDING, as what, in a computer simulation of intentionality, would count as the noncognitional analogue of whatever in the "folk" account would constitute genuine cognition, understanding, subjectivity or the like.[84] Cummins then professes (or confesses) that *he* cannot offer, beyond the intuitions suggested, any account of what must be added to *cognition and UNDERSTANDING to constitute cognition and understanding. He also claims that there is no convincing reason for denying that, where a computer *cognitively simulates the cognitive capacity of humans, we should *eliminate* "cognition" and "understanding" (in the folk vocabulary) and install "*cognitive" and "UNDERSTANDING" as sufficient for the needs of whatever (regarding human cognition) *is* recoverable within the relevant sciences. The successor notions would then not be reductive in a crass way and the justification would be empirical – would rest on the impending success of computer simulation. Such a "computationalism" *may* "fail as an explanatory strategy even if its truth is experimentally beyond serious doubt." "I think," Cummins cheerfully affirms, "the experimental evidence is already significant, if not overwhelming. It seems quite possible that we will shortly be in the position of knowing that the strong thesis [which we have yet to introduce] is probably true without having more than the vaguest idea of how that is possible."[85]

Now, the important point is that Cummins would be entirely justified *if the simulation were empirically adequate and successful.* So his maneuver helps to force a confrontation in the essential contest regarding emergence. The trouble is, in spite of his insistence on construing the contest empirically, he begs the question in a hopelessly tendentious way.

Consider two decisive doctrines that Cummins introduces linking (folk) cognition and (AI) *cognition:

Strong Thesis: intentional characterization can be explicated via semantic interpretation, i.e., an intentionally characterized capacity can be instantiated as an information-processing system – e.g., a system of discursive capacities.[86]

Imitation Thesis: Exercise of any intentionally characterized capacity C_I can be imitated by exercise of a computational capacity C_C.[87]

There are technical quibbles one could press, but it would be better to make all the concessions Cummins wants (short of the conclusion itself) in the interest of clarifying what is at stake. Regarding both theses, we are to understand that reference to "interpretation," "explication," "instantiation," "imitation" (or "simulation"), "computation," and "discursive capacities" does *not* smuggle into the *explanans* (the AI analysis) *any* cognitive (folk) assumptions. What Cummins wishes to defend is reasonably captured by the following: "When I say that exercise of C can be imitated by exercise of C', I mean that C and C' are input–output isomorphic (within idealization)."[88] Now, the usual objection to this is that it does not capture what is intuitively meant by cognition or intentionality; whereas our objection is that the "Strong Thesis" and the "Imitation Thesis" are simply false or at least not confirmed – certainly not empirically supported in the sanguine way Cummins claims they are.

In particular, Cummins's account presumes favorable answers to a number of questions implicated in construing the contest as he does, that are so strong that, once granted, the issue cannot but be settled in favor of *cognition and UNDERSTANDING. The point is that Cummins begs the essential question – but, in begging it, identifies it and, in identifying it, fixes the sense in which the computationalist (the eliminationist who approaches cognition by way of AI simulation) has never really faced the *empirical* issue that he presumes can be favorably settled. His strategy – a perfectly reasonable one – is fairly caught by the familiar advice: if it looks like a duck and walks like a duck and talks like a duck, then it is a duck. This is *not* intended reductively but eliminatively. That is, it is not that anything that looks like a duck is a duck, but that there is no other way to judge whether anything is a duck but by how it looks and. ... So Cummins gamely concedes, speculating about "what the capacity for intentional states is *for*" – meaning by that, what in the "folk" or "subjective" sense intentionality is supposed to signify – that "The easy [folk] answer is that the capacity for intentional states is necessary for the capacity to communicate, to deliberate, to premeditate, and a host of others, for each of these is analyzable into intentionally characterized capacities"; and that [from] another

> [the AI, "imitation"] perspective ... the capabilities required by [or necessary to the folk conception] seem almost epiphenomenal ... because we don't know why, from a [computer] programmer's point of view, a system should need to be able to represent the semantics of its own states in order to communicate, deliberate, etc. Intentionality seems essential when we analyze the capacity to communicate, but when we analyze the capacity for intentional states, we can't see how it *could* be essential or even important.[89]

First of all, Cummins confuses and conflates the imitation of a determinate *disposition* and the imitation of a *capacity*; and he fails to distinguish between the imitation of a finite *segment* of cognitionally pertinent behavior (conceding the imitation of behavior to count as the imitation of the central states governing such behavior – for the sake of simplicity) and the imitation of either a disposition or a capacity. "A capacity is specified," he says, "by giving a special law linking precipitating conditions to manifestations – i.e., by specifying 'input–output' conditions. What makes a capacity cognitive is that the outputs are cognitions."[90]

But this is both false and misleading. What Cummins offers may capture the sense of how the capacity of a computer is (or could be) specified – where there is *no* difference between its capacities and its dispositions (and no difference between how a segment of its behavior can be explained and how its dispositions or capacities can be explained). But that is certainly not how any familiar human capacity is specified. Think only of Shakespeare's capacity to write great theatre: certainly, Shakespeare was not always disposed to produce great plays and in fact did not always do so. Furthermore, it is a travesty of the problem Cummins poses to claim that we specify a cognitive capacity by specifying "a special law" linking "input" and "output." (In fact, we usually "specify" such a capacity by specifying only its fruits – for instance, *Hamlet* rather than *Timon of Athens*.) We know no such laws in the human case; and because we know no such laws, it is reasonably clear that we do not, from the "folk" perspective, treat intentional or cognitive capacities *as a system of any kind*. We certainly do not do so where we lack the required "special law" (even if we are doctrinally disposed to imagine that such capacities must fall within a system). So, at one stroke, both the "Strong Thesis" and the "Imitation Thesis" are invalidated – empirically.

Secondly, Cummins explains in the following way what he means by his special law:

> cognitive capacities are *inferentially characterizable capacities* (ICC's hereafter): the transition law specifying a cognitive capacity is a rule of inference. ... To explain an inferentially characterized capacity is to explain the capacity to conform to the characteristic inferential pattern. More precisely, to explain an inferentially characterized capacity is to explain how it happens that the output is characteristically interpretable as a sentence that is inferable in a specified way from the sentence interpreting the input that precipitated it. ... [T]he instantiation of an ICC in [some system] S must demystify the epistemological successes which are the exercises of the ICC in S, and ultimately this means that drawing the right conclusion must be (interpret) a physical transition characteristic of physical systems like S.[91]

(The resemblance to Stephen Stich's account is noteworthy. It may not be unfair to think of Cummins's strategy as something very much like the result of applying Stich's strategy to Dennett's homuncular program.)

Cummins is entirely within his right in treating all this as explicating *cognition rather than cognition (in the folk sense): the sentences by which the inputs and outputs of the "system" are interpreted are not themselves represented in the system itself, as a condition of the system's success: it merely "executes" its moves. It merely "E-represents" these moves because it executes them: "physical instantiation is sufficient for [such] representation"; it does not represent them in the "internal manual" sense of the folk theorists (such as Fodor and Chomsky).[92] But this means very plainly that a systematic isomorphism between some sub-set of the discrete physical states of a human being (as well as of a computer) and the putative sub-states of *any* cognitive capacity first identified in the folk theorist's sense can always be empirically produced. But this is simply not true. It is in fact just the assumption that needs to be converted into an empirically confirmed conclusion if the computationalist is to make his case. *To assume it is to assume his case already.* To assume it is to vindicate substituting *cognition and UNDERSTANDING for cognition and understanding.

No, the argument against the computationalist's elimination of the human goes this way. First, we take ourselves to be the paradigms of cognition and understanding, but we allow the computationalist's thesis as a possibility. Next, we specify intentionality and cognitive capacity in the full cultural (folk) sense, and we oblige the computationalist to show us how (if he dares) to analyze *such capacities* (not those of the computer) in terms of a closed system of inputs and outputs that, by some generous idealization, could be made to function isomorphically with the actual *cognizing* power of the human specimen. (This, incidentally, also touches on the weakness of Chomsky's nativism, since Chomsky restricts his account to the parsing of idealized sentences. He does not consider the actual accessing of language by cognitively apt humans in the flesh, *by* which the presumed parsing is activated.[93]) The required isomorphism must be empirically managed, and the would-be isomorph (on the human side) must first be suitably analyzed by way of an extensional program of the sort suited to a computer and capable of being instantiated physically in the way in which a computer program can be. Cummins himself says as much: "intentionally characterized capacities are computationally instantiated," can always be mapped in the form of a "flow chart."[94] Now, there is no known demonstration of this sort to be had. On the argument, if Cummins provided one, we should have to give up resisting. But, by an equal bit of fair play, *we* may insist that the computationalist hold his peace until he has met the full challenge – which, apparently, he willingly accepts. To have failed in every such effort is just the point of advocating emergence$_M$ and of accepting all its difficult consequences. It is the very heart of the bifurcationist argument.

Beyond the local skirmish regarding computationalism, Cummins's strategy shows very clearly how it is that the extension of the unity model (or at least the extension of its principal tenets) essentially centers on carpentering the distinctive *properties* of the human sciences so that the extensionalism, nomologicality, and physicalism favored by the unity canon are straightforwardly applied to the otherwise emergent$_F$ sciences of the cultural world. This is the focus of ultimate convergence among the immense variety of would-be such extensions – including all pertinent versions of computationalism (Cummins's, Dennett's, Stich's, Pylyshyn's, Marr's, Dretske's), all versions of nativism (Chomsky's, Fodor's), all versions of eliminationalism or of heuristic functionalism (Sellars's, Quine's, Davidson's, Feyerabend's, the Churchlands'), all versions of possible-worlds semantics (Stalnecker's, Lewis's), of analytic structuralism (Lévi-Strauss's, Hjelmslev's).[95] It is the *analysis* of such properties (as Cummins rightly observes) rather than the direct *subsumption* of the phenomena of the cultural world under causal laws (as Popper and Grünbaum divergently suppose – call their common view "methodologism") that determines the fate of the attack on the emergence of the human beyond the clutches of what, canonically, is thought to be adequate for the *physical*$_2$ world. Hence, it is not surprising that the analysis of the intentional should be as controversial (as we have already seen) as the analysis of the *physical*$_1$ is vague, uncertain, elastic, and completely unsystematic.

Notes

1   See Joseph Margolis, *Persons and Minds* (Dordrecht: D. Reidel, 1978), ch. 12.

2   P. E. Meehl and Wilfrid Sellars, "The Concept of Emergence," in Herbert Feigl and Michael Scriven (eds), *Minnesota Studies in the Philosophy of Science*, vol. 1 (Minneapolis: University of Minnesota Press, 1956), pp. 252, 247. The argument is directed against Stephen C. Pepper, "Emergence," *Journal of Philosophy*, XXIII (1926).

3   Meehl and Sellars, "The Concept of Emergence," in Feigl and Scriven, *Minnesota Studies in the Philosophy of Science*, vol. 1, p. 252.

4   Ibid.

5   Herbert Feigl, *The "Mental" and the "Physical": The Essay and a Postcript* (Minneapolis: University of Minnesota Press, 1967), p. 10.

6   Ibid.

7   See Donald Davidson, "Mental Events," *Essays on Actions and Events* (Oxford: Clarendon Press, 1980).

8   Ibid., pp. 10, 12.

9   Ibid., p. 11.

10   Ibid., p. 61.

11 Ibid., pp. 144, 139–40.
12 See Joseph Margolis, *Philosophy of Psychology* (Englewood Cliffs, NJ: Prentice-Hall, 1984) and *Persons and Minds*.
13 Meehl and Sellars, "The Concept of Emergence," in Feigl and Scriven, *Minnesota Studies in the Philosophy of Science*, vol. 1, pp. 243, 241.
14 See Joseph Margolis, *Philosophy of Psychology*, ch. 2, and "Prospects for an Extensionalist Theory of Action," *Journal for the Theory of Social Behavior*, II (1981).
15 Davidson, "Mental Events," *Essays on Actions and Events*, p. 211.
16 Ibid., pp. 212, 216.
17 Ibid., p. 219.
18 Though, as remarked, it actually is not. The physicist Eugene P. Wigner has offered two extremely instructive arguments bearing at once on homonomic prospects and the unity-of-science problem. "The first is that if one entity [he is thinking of physical entities] is influenced by another entity, in all known cases the latter one is also influenced by the former. . . . More generally, we do not know any case in which the influence is entirely one-sided. Since matter clearly influences the content of our consciousness, it is natural to assume that the opposite influence also exists, thus demanding a modification of the presently accepted laws of nature which disregard this influence. The second argument . . . is that all extensions of physics to new sets of phenomena were accompanied by drastic changes in the theory. In fact, most were accompanied by drastic changes of the entities for which the laws of physics were supposed to establish regularities" – "Physics and the Explanation of Life," in R. J. Seeger and R. S. Cohen (eds), *Philosophical Foundations of Science* (Dordrecht: D. Reidel, 1974), pp. 128–9. To support Wigner here, of course, is not to endorse his views on biological emergence itself.
19 Mario Bunge, "Emergence and the Mind," *Neuroscience*, II (1977), 502.
20 Rudolf Carnap, "Psychology in Physical Language," tr. George Schick, in A. J. Ayer (ed.), *Logical Positivism* (Glencoe, Ill.: Free Press, 1959), pp. 166–7. See also Joseph Margolis, "Schlick and Carnap on the Problem of Psychology," in Eugene T. Gadol (ed.), *Rationality and Science* (Vienna: Springer, 1982).
21 Bunge, "Emergence and the Mind," *Neuroscience*, II, 503–4.
22 Paul Oppenheim and Hilary Putnam, "Unity of Science as a Working Hypothesis," in Herbert Feigl, Michael Scriven, and Grover Maxwell (eds), *Minnesota Studies in the Philosophy of Science*, vol. 2 (Minneapolis: University of Minnesota Press, 1958), pp. 28–9.
23 Ibid., p. 4.
24 Ibid., pp. 6–7.
25 Bunge, "Emergence and the Mind," *Neuroscience*, II, 509; Oppenheim and Putnam, "Unity of Science as a Working Hypothesis," in Feigl et al., *Minnesota Studies in the Philosophy of Science*, vol. 2, pp. 13, 22.
26 Ibid. (Oppenheim and Putnam), p. 8 (italics added).

27 See B. F. Skinner, *Science and Human Behavior* (New York: Macmillan, 1953).

28 Jerry A. Fodor, *The Language of Thought* (New York: Thomas Y. Crowell, 1975), p. 19.

29 Fodor, "Introduction: Two Kinds of Reductionism," *The Language of Thought*; and Mario Bunge, *Causation and Modern Science*, 3rd, rev. edn (New York: Dover, 1979), pp. 290–1. See further ch. 9, above.

30 See ch. 9, above.

31 Noam Chomsky, *Rules and Representations* (New York: Columbia University Press, 1980), p. 4.

32 Ibid., p. 6; cf. pp. 31, 39, 187, 257 (n. 21).

33 Oppenheim and Putnam, "Unity of Science as a Working Hypothesis," in Feigl et al., *Minnesota Studies in the Philosophy of Science*, vol. 2, p. 3. See Rudolf Carnap, "Logical Foundations of the Unity of Science," in Otto Neurath et al. (eds), *International Encyclopedia of Unified Science*, vol. I (Chicago: University of Chicago Press, 1955); J. G. Kemeny and P. Oppenheim, "On Reduction," *Philosophical Studies*, VII (1956).

34 Carnap, "Psychology in Physical Language," in Ayer, *Logical Positivism*. See also Otto Neurath, "Protocol Sentences," tr. George Schick, in Ayer, *Logical Positivism*.

35 Feigl, *The "Mental" and the "Physical"*, pp. 144, 87.

36 Wilfrid Sellars, "Philosophy and the Scientific Image of Man" and "Empiricism and the Philosophy of Mind," *Science, Perception and Reality* (London: Routledge and Kegan Paul, 1963). See also Margolis, *Persons and Minds*, chs 1, 6.

37 Davidson, "Mental Events," *Essays on Actions and Events*. Cf. Hilary Putnam, "Reflections on Goodman's *Ways of Worldmaking*," *Philosophical Papers*, vol. 3 (Cambridge: Cambridge University Press, 1983), pp. 158–60.

38 Fodor, *The Language of Thought*, particularly p. 18.

39 John Searle, *Minds, Brains and Science* (Cambridge, Mass.: Harvard University Press, 1984), pp. 26–7; cf. also ch. 4.

40 Ibid., pp. 18–19.

41 See Joseph Margolis, "Constraints on the Metaphysics of Culture," *Review of Metaphysics*, XXXIX (1986).

42 There is an interesting adumbration of the linkage between causal accounts and analysis (sympathetic with (3) and (4)) in Robert Cummins, *The Nature of Psychological Explanation* (Cambridge, Mass.: MIT Press, 1983), chs 1–2. It is overly committed, however, to a very strong sense of system.

43 See ch. 9, above.

44 See ch. 12, below.

45 Carl G. Hempel, "Studies in the Logic of Explanation," *Aspects of Scientific Explanation* (New York: Free Press, 1965), p. 263.

46 Cf. for example ibid., p. 259.

47 See for example Carl G. Hempel, *Philosophy of Natural Science* (Englewood Cliffs, NJ: Prentice-Hall, 1966), ch. 8.

48 David Hull, Introduction to *Philosophy of Biological Science* (Englewood Cliffs, NJ: Prentice-Hall, 1974). See also ch. 5 for the discussion and convincing rejection of prominent emergentist views: including those of Michael Polanyi, "Life's Irreducible Structure," *Science*, CLX (1968); Walter Elsasser, *Atom and Organism* (Princeton, NJ: Princeton University Press, 1966); and G. G. Simpson, *Principles of Animal Taxonomy* (New York: Columbia University Press, 1961). Cf. J. Bronowski, "New Concepts in the Evolution of Complexity," in Seeger and Cohen, *Philosophical Foundations of Science*.

49 Robert L. Causey, *Unity of Science* (Dordrecht: D. Reidel, 1977), p. 138. See also Kenneth F. Schaffner, "The Watson–Crick Model and Reductionism," in Marjorie Grene and Everett Mendelsohn (eds), *Topics in the Philosophy of Biology* (Dordrecht: D. Reidel, 1976).

50 Causey, *Unity of Science*, pp. 138–9.

51 Karl R. Popper, *The Open Universe*, ed. W. W. Bartley, III (Totowa, NJ: Rowman and Littlefield, 1982), pp. 173–4. See also, Ilya Prigogine, *From Being to Becoming: Time and Complexity in the Physical Sciences* (San Francisco: W. H. Freeman, 1980).

52 Ibid., p. 140.

53 Ibid., pp. 139–42. See also Robert L. Causey, "Polanyi on Structure and Reduction," *Synthese*, XX (1969); and Kenneth F. Schaffner, "The Peripherality of Reductionism in the Development of Molecular Biology," *Journal of the History of Biology*, VII (1974).

54 Polanyi, "Life's Irreducible Structure," *Science*, CLX.

55 H. H. Pattee, "The Problem of Biological Hierarchy," in C. H. Waddington (ed.), *Towards a Theoretical Biology*, 3rd ser. (Edinburgh: Edinburgh University Press, 1970), pp. 121–4. See also Francis Crick, *Of Molecules and Men* (Seattle: University of Washington Press, 1966); and F. J. Ayala and T. Dobzhansky (eds); *Studies in the Philosophy of Biology* (Berkeley, Calif.: University of California Press, 1974).

56 Pattee, "The Problem of Biological Hierarchy," in Waddington, *Towards a Theoretical Biology*, 3rd ser., pp. 119, 117.

57 Ibid., pp. 120, 127.

58 Hempel, "Studies in the Logic of Explanation," *Aspects of Scientific Explanation*, p. 263 (italics added).

59 Kenneth F. Schaffner, "The Unity of Science and Theory Construction in Molecular Biology," in Seeger and Cohen, *Philosophical Foundations of Science*, p. 515. Cf. L. von Betalanffy, *General Systems Theory* (New York: George Braziller, 1968), pp. 48–9, cited by Schaffner; also, F. Jacob and J. Monod, "Genetic Regulatory Mechanisms in the Synthesis of Proteins," *Journal of Molecular Biology*, III (1961).

60 Herbert A. Simon, "The Architecture of Complexity," *Proceedings of the American Philosophical Society*, CVI (1962); repr. in Herbert A. Simon, *The Sciences of the Artificial* (Cambridge, Mass.: MIT Press, 1969). Note especially p. 87 (in *The Sciences of the Artificial*).

61 Ibid., p. 97.

62 Ibid., pp. 110, 108 (italics added).

63 Ibid., p. 112.

64 I have explored these issues at greater length in *Philosophy of Psychology* and *Persons and Minds*.

65 Simon, "The Architecture of Complexity," *The Sciences of the Artificial*, p. 99.

66 Ibid., pp. 99–100.

67 See ch. 9, above. Further on functionalism, see Margolis, *Philosophy of Psychology*, ch. 3. In the context of psychology and artificial intelligence, the thesis is perhaps most clearly formulated in Hilary Putnam, "Minds and Machines," *Philosophical Papers*, vol. 2 (Cambridge: Cambridge University Press, 1975). Putnam appears to have given up this position. See also Ned Block, "Troubles with Functionalism," in C. W. Savage (ed.), *Minnesota Studies in the Philosophy of Science*, vol. 9 (Minneapolis: University of Minnesota Press, 1973).

68 See Stephen Körner, *Experience and Theory* (London: Routledge and Kegan Paul, 1966); May Brodbeck, "Mental and Physical: Identity versus Sameness," in P. K. Feyerabend and G. Maxwell (eds), *Mind, Matter, and Method* (Minneapolis: University of Minnesota Press, 1966); Daniel C. Dennett, *Content and Consciousness* (London: Routledge and Kegan Paul, 1969) and *Brainstorms* (Montgomery, Vt: Bradford Books, 1978). Further on Dennett, see Margolis, *Philosophy of Psychology*, ch. 5.

69 Fodor, *The Language of Thought*.

70 See Causey, *Unity of Science*, pp. 142–51.

71 See ch. 9, above. The meaning of "incarnate" is explored further in Joseph Margolis, "Nature, Culture, and Persons," *Culture and Cultural Entities* (Dordrecht: D. Reidel, 1983) and *Philosophy of Psychology*, ch. 4.

72 See Joseph Margolis, *Pragmatism without Foundations: Reconciling Realism and Relativism* (Oxford: Basil Blackwell, 1986).

73 The concepts of "embodiment" and "incarnate" properties are jointly developed in Joseph Margolis: *Art and Philosophy* (Atlantic Highlands, NJ: Humanities Press, 1980), pt 1; "Nature, Culture, and Persons," *Culture and Cultural Entities*; and *Persons and Minds*, chs 1, 6, 12.

74 See Chomsky, *Rules and Representations*, ch. 1; and Donald Hockney, "The Bifurcation of Scientific Theories and Indeterminacy of Translation," *Philosophy of Science*, XLII (1975). The term is already used by Oppenheim and Putnam, "Unity of Science as a Working Hypothesis," in Feigl et al., *Minnesota Studies in the Philosophy of Science*, vol. 2.

75 The topic is systematically explored in Margolis, *Persons and Minds* and *Culture and Cultural Entities*.

76 Roy Bhaskar, *A Realist Theory of Science* (London: Leeds Books, 1975), p. 170.

77 Ibid., p. 171.

78 Roy Bhaskar, *The Possibility of Naturalism* (Atlantic Highlands, NJ: Humanities Press, 1979), pp. 48–9.

79 Ibid., p. 49.
80 Ibid., p. 180; cf. p. 204.
81 See further Margolis, *Culture and Cultural Entities*.
82 We shall return to the question of systems in ch. 12, with special attention to functionalist (that is, primarily noncausal) views of the human sciences. Such views are intended to enlarge the scope of the unity model so as to accommodate these sciences more suitably than programs such as emergence$_B$ appear able to afford.
83 Cummins, *The Nature of Psychological Explanation*. The argument is pretty well confined to ch. 3.
84 Cf. ibid., pp. 53–62, 77–82.
85 Ibid., p. 102.
86 Ibid., p. 89.
87 Ibid., p. 104.
88 Ibid.
89 Ibid., pp. 116–17.
90 Ibid., p. 53.
91 Ibid., pp. 53–5. Cf. John Haugeland, "The Nature and Plausibility of Cognitivism," *Behavioral and Brain Sciences*, II (1978).
92 Cummins, *The Nature of Psychological Explanation*, p. 46.
93 See ch. 12, below.
94 Cummins, *The Nature of Psychological Explanation*, pp. 90, 78.
95 On structuralism, see Joseph Margolis, " 'The savage mind totalizes,' " *Man and World*, XVII (1984).

11

Context, History, and the Human Condition

I

There is a curious similarity between the doctrine of original sin and the general thrust of contemporary theories of language: in both, men are said to be born into a condition that is a native defect of the race, which, however, grants them the wit to understand their plight reflexively, but which they cannot overcome. The linguistic thesis holds that language "fits" the world in some sense and, in doing that, distorts the world it fits in a way we cannot possibly specify or correct. The theme is a larger one than that of the historicity of human existence, but it accommodates it in a way not altogether unlike the way in which the doctrine of sin accommodates the finitude of political and personal projects. In the Christian vision, it is of course possible to be the beneficiary of divine grace, which, because of a convenient revelation, even sinful man has an inkling of. But, within the scope of the theory of language, there is no corresponding, external source of redemption – unless, by a failure of analogy, by way of denying the ineffable gap between language and the world, or of arbitrarily affirming a preestablished harmony in spite of the gap.

The question of the distorting match of language – or, better, of the reasonable (second-order) guess that a creature so obviously unaware of the subterranean interests that shape its own orientation and sense of identity, or so obviously incapable of confirming the match in a way unaffected by the very forces of language and the world – has, perhaps surprisingly, proved as absorbing a question as that of sin itself. In a sense, it is a more touching question, because, with respect to the ultimate realism of language, we have no hope of any direct cognitive redemption. Nevertheless, at least two of the most influential spirits of our age have been distinctly preoccupied with it: the reception of their views has, in the most multifarious ways, obliged both their admirers and detractors to attend carefully to the conceptual puzzle of *context* and *history*. To have been born into a languaged world without benefit of cognitive transparency entails relativizing our science in a radical way.

Nietzsche is the more mordant of the two. "[M]an wants the truth," he says,

> he desires the agreeable life-preserving consequences of truth, but he is indifferent to pure knowledge, which has no consequences; he is even hostile to possibly damaging and destructive truths. And, moreover, what about these conventions of language? Are they really the products of knowledge, of the sense of truth? Do the designations and the things coincide? Is language the adequate expression of all realities? Only through forgetfulness can man ever achieve the illusion of possessing a "truth" in the sense just designated. ... What, then, is truth? A mobile army of metaphors, metonyms, and anthropomorphism – in short, a sum of human relations, which have been enhanced, transposed, and embellished poetically and rhetorically, and which after long use seem firm, canonical, and obligatory to a people: truths are illusions about which one has forgotten that this is what they are; metaphors which are worn out and without sensuous power; coins which have lost their pictures and now matter only as metal, no longer as coins.[1]

There is only one conceivable consequence of acknowledging the justice of Nietzsche's charge: we will and must be perpetually *suspicious* of the effective but unperceived ways in which our use of our own language shapes our sensibilities and convictions and behavior. This may play itself out in a double way: either we can drone on forever about the ineffable equality of linguistic sin, ignoring altogether what, under that condition, occupies particular men at particular times; or, in whatever regard we suppose we can be critical within the pale of linguistic damnation, we can "suspiciously" toil on linguistically as before, knowing full well that all our appraisals and reforms obtain "east of Eden."

The important point to notice is that the theme of linguistic (or conceptual) suspicion *has no systematic or methodological or epistemic or differential bearing at all on how, rationally, we judge particular questions* – except of course to collect them all "suspiciously." The same is true of the doctrine of sin, once we take seriously how utterly alien and unfathomable is the mind of God. Nietzsche's assessments, therefore, of what accounts *differentially* for the distortions *of human history* are quite uninformed (though obviously not uninspired) by his ulterior grasp of the ultimate puzzle of language. There simply is no connection between the two that a human agent could possibly claim. This explains why, for instance, in a true Nietzschean moment, Jacques Derrida introduces the "neographism" *différance*, which "is literally neither a word nor a concept" but a surd that (in the Saussurean idiom Derrida both features and attacks) "*is* [tricked as we are by the copula into appearing to affirm what cannot be affirmed or denied] the non-full, non-simple, structured and differentiating origin of [all linguistic and conceptual] differences."[2] *Whatever* the persuasiveness of Nietzsche's and Derrida's exposés of (the unequal targets of) Christianity and Saussure and Husserl, the force of their *comparative* remarks has no privileged link to their respective surds (which remain unutterable); although it is true that, even as *all* discourse is committed (equally, before the god of correspondence) to the spontaneous fit of words and things, *some* theorists actually presume to affirm a foundationalist or essentialist privilege – a first-order realism that violates our appreciation of the originating puzzle. These, of course, may (on the hypothesis) be exposed in the strongest sense, but the entailed judgment conveys no further advantage within the range of (equally suspicious) comparative findings. What *is* true is that, since the Nietzschean view of the correspondence puzzle is so insistent on the falsity and systematic delusion of language, there is no reserve left for ensuring even a modicum of communicative rigor for any of our smallest sublunar purposes. This leads directly to an exaggerated vigilance regarding the distorting power of the least utterance and, correspondingly, to a remarkable and unexplained precision in our being able to fix each distortion.

Wittgenstein is the other voice, of course. But his is a strikingly different one: first, because, in the *Tractatus*, Wittgenstein clearly believed he had succeeded in solving the problem that (in effect) he shared with Nietzsche; and in his later work, notably in the *Philosophical Investigations*, he repudiated not merely the solutions of the first book but the very intelligibility of such questions and such answers. Admitting the force of *that* maneuver, however, has obliged us to turn in a fresh way to the import of history and context and, through that, to begin to formulate a quite revolutionary conception of the entire undertaking of rational inquiry. In the Preface to the *Tractatus*, Wittgenstein says very plainly,

> what can be said at all can be said clearly, and what we cannot talk about we must pass over in silence. Thus the aim of the book is to draw a limit to thought, or rather – not to thought, but to

the expression of thoughts: for in order to be able to draw a limit to thought, we should have to find both sides of the limit thinkable (i.e., we should have to be able to think what cannot be thought). It will therefore only be in language that the limit can be drawn, and what lies on the other side of the limit will simply be nonsense.[3]

He adds quite straightforwardly, "the *truth* of the thoughts that are here communicated seems to me unassailable and definitive. I therefore believe myself to have found, on all essential points, the final solution of the problems."[4]

To appreciate how extraordinary was Wittgenstein's claim, one must piece together at least a few (possibly alternative sets of) specimen propositions from the *Tractatus* itself – for example, the following: "The world is the totality of facts, not of things" (1.1); "The facts in logical space are the world" (1.13); "Logic deals with every possibility and all possibilities are its facts" (2.0121); "If I know an object I also know all its possible occurrences in states of affairs. (Every one of these possibilities must be part of the nature of the object)" (2.0123); "There must be objects, if the world is to have an unalterable form" (2.026); "Objects are what is unalterable and subsistent; their configuration is what is changing and unstable" (2.0271); "The existence and non-existence of states of affairs is reality" (2.06); "States of affairs are independent of one another" (2.061); "We picture facts to ourselves" (2.1); "There must be something identical in a picture and what it depicts, to enable the one to be a picture of the other at all" (2.161); "A picture depicts reality by representing a possibility of existence and non-existence of states of affairs" (2.201). Now, Wittgenstein evidently believed that he had constructed a syntactically explicit schema, based entirely on the formulability of atomic propositions and truth-functional or extensional ways of constructing all other possible propositions from them – so that all possible states of affairs become accessible in a formal sense. Apart from what he took to be Russell's muddled Introduction to the *Tractatus*,[5] Wittgenstein notes, at the end of his Preface, "how little is achieved when these problems are solved."[6] He evidently meant this in the spirit of his ethical and mystical views (characteristically ignored in the early analytic reception of his book). But he shows not the slightest interest here in the serious possibility that the aptness and scope of the very syntax of a language intended to facilitate the formulation of statements about the world *might be a function of our interaction with the world* – for instance, in terms of the contingent history of our natural language, our nonlinguistic experience, or the dimly grasped pecularities of human nature.

In the *Investigations*, Wittgenstein simply ignores or rejects the entire enterprise of deciding the extensional reducibility of natural languages. It would certainly have been theoretically possible, as indeed Carnap (and Neurath) made clear, for linguistic events and the facts they describe to obtain "within the world"; so an (extensionally) idealized language along the lines Wittgenstein *appeared* to favor (but in an important sense did not) need not have been viewed as totalizing all possibilities relative to the empirical description of the actual world. In particular, Carnap distinguished rather sharply between the bearing of Wittgenstein's views (in the *Tractatus*) on a formal and uninterpreted language schema and a language that had a determinate cognitive content.[7] Hence, in a double sense, Wittgenstein's project in the *Tractatus* was radically different from both Nietzsche's and Carnap's. Nietzsche raised the fatal correspondence question about actual historical languages and the humans who used them; and Carnap divided Wittgenstein's concern so that, where it could be said to totalize all possibilities, the (syntactically idealized) language schema had no content at all; and, where a scientifically disciplined language had a determinate content, the question of painstaking extensional translation precluded altogether the pointless extravagance of talking of totalized possibilities.

Wittgenstein's project is something of a mystery, then, since he seems to have been irritated by the Carnapian line of interpretation. His effort perhaps was a thought-experiment of a grand sort, a Schopenhauerean speculation *sub specie aeternitatis*, since how, after all, *could* contingent empirical difficulties be expected to bear on specifying the properties of a language (synchronically) totalized for all possible propositions? The very sense of the adequacy of the *Tractatus* is, for Wittgenstein, the mystical. But now the reason is clear: the picture of a finitely totalized world constrained everywhere by logical necessity is itself mystical. This is certainly close to what Wittgenstein intends, as the following final propositions make clear: "the only necessity that exists is logical necessity" (6.375); "*How* things are in the world is a matter of complete indifference for what is higher. God does not reveal himself *in* the world" (6.432); "It is not *how* things are in the world that is mystical, but *that* it exists (6.44); "To view the world sub specie aeterni is to view it as a whole – a limited whole. Feeling the world as a limited whole – it is this that is mystical" (6.45); "There are, indeed, things that cannot be put into words. They *make themselves manifest*. They are what is mystical" (6.522). Hence, Wittgenstein thinks the unthinkable and (also) repudiates the pretense that that can be done: "My propositions," he says, "serve as elucidations in the following way: anyone who understands me eventually recognizes them as nonsensical, when he has used them

– as steps – to climb up beyond them" (6.54). He has, therefore, dared to think of the identity that links "both sides of the limit" – which he denies is possible.

Here, Wittgenstein seems to join Nietzsche – but he only seems to. Of course, they share the same question. Nietzsche bores subversively from within the only language we have, to reach his unqualified exposé. And Wittgenstein gathers his reassuring thesis about the necessary structure of all truth-bearing utterances from a supposed vantage outside the legitimate domain of the other aptitude – apparently by way of a deeper use of language. Nietzsche is entirely aware that there is no other language but the natural language we are spawned in: suspicion is our consequent function; but the exercise of suspicion requires something more than suspicion (which is denied), and Nietzsche himself is too ardent to be satisfied with his own *aporia*. Wittgenstein thinks and states what surpasses the limits of what he insists we can think and state. He takes his discovery to be achieved in an eternal moment not bound by the limits of language or world: hence, he states in a worldly language truths that cannot be there contained; but the strict pride of reason that his own mysticism subtends cannot pretend that the resulting propositions are actually intelligible.

Ultimately, therefore, Wittgenstein was dissatisfied with the *Tractatus*. Everyone knows that. But what needs to be grasped – what is most instructive about his review of the human condition (what links the "early" and "later" Wittgenstein[8]) – is that, in the *Investigations*, Wittgenstein considers only *human* language, the forms of human discourse *within* the contingent world that the language of the *Tractatus* (God's language, one may suppose, the language of the "eternal life" the author of the *Tractatus* apparently believed he had entered[9]) utterly disregards. (There is *no* reason, however, to think that in a sense favorable to his mystical concern Wittgenstein ever repudiated the "adequacy" of the Tractarian view.)

At least three fundamental constraints on human discourse are regularly featured in the *Investigations* and related texts – constraints that gain their full force *only* by way of linkage with the impossible project of the *Tractatus*: first of all, the world of natural discourse is a thoroughly contingent world in the sense the language of the *Tractatus* invites us to *ignore* (6.432); secondly, in spite of this and in spite of the fact that, as Wittgenstein came to realize, his present "philosophical remarks" (*philosophische Bemerkungen*) could not, "against their natural inclination," be forced into an order that *resembled* that of the *Tractatus*, were in effect only an "album" ("sketches of landscapes"),[10] there does remain a reliable but contingent order of life in which doubts can be resolved, in which rules can be applied to the extension of practices, in which a measure of certainty can be achieved – all without appeal to any unchanging or necessary structures; and, thirdly, the reliability with which we manage to carry on with our inquiries and practices holds under conditions of ineliminable *indeterminacy* with respect to those very practices.

The essential key to his radical change of emphasis rests with the notion of "forms of life." In a sense, the entire ("later") account can be drawn from this single remark: "What has to be accepted, the given, is – so one could say – forms of life."[11] In the first place, the notion of a "form of life" entails the rejection or irrelevance of the correspondence problem of the *Tractatus* for an understanding of the actual achievement of human languages. Secondly, the notion of a "form of life" preempts the need to raise the correspondence worry in the context of a functioning natural language. Although it could hardly be Wittgenstein's conceptual preference, we might say that our never "exiting"[12] from the forms of life which we have learned from infancy signifies that their contingency is indissolubly linked with the *en bloc* realism (biologized or pragmatized, if one may put things this way) of every viable form of life. And, thirdly, introducing the notion of a form of life is *meant to be contrasted* with the idiom of the eternal fixity, finite boundedness, logical necessity, and totalized order of the world of the *Tractatus*. Thus, Wittgenstein remarks, "You must bear in mind that the language-game [for instance, with regard to knowing what to expect in different circumstances] is so to say something unpredictable. I mean: it is not based on grounds. It is not reasonable (or unreasonable). It is there – like our life."[13] Think of this in the context, for instance, in which, actually recalling the correspondence problem, Wittgenstein says, "Like everything metaphysical the harmony between thought and reality is to be found in the grammar of the language. ... Reality is not a property still missing in what [for instance] is expected and which accedes to it when one's expectation comes about."[14] *Now*, however, the linkage between language and reality is specified *within a form of life* – not in accord with the preferences of an idealized language Carnap might have favored and not in any way intended to bring it into line with the structures the *Tractatus* had imposed.[15] Furthermore, *the natural order of things is sufficient for human functioning*, contrary to Carnap's fears and (also) contrary to the constraints of the Tractarian minima. The point, here, is that the standard Tractarian picture is false to the right reading of the *Tractatus*.

But, if this is so, then, according to Wittgenstein, *all* uncertainties, all doubts, all indeterminacies, all pluralities come to rest in the *given* – the forms of life. This identifies the sense of the practical and procedural adequacy, the smooth functioning, of natural languages and natural cultures. "'Obeying a rule' is a practice," says Wittgenstein, "*not* an *interpretation*."[16] An *interpretation* (he thinks) would lead to a regress or the impossibility of any form of effective certainty – which, in an important sense, *is* the line of theorizing favored by the Nietzschean strain, running through Heidegger, Gadamer, Derrida, Michel Foucault, Paul de Man, and others.[17] Interpretation *presupposes* the mastery of a practice (on Wittgenstein's view) – rather than establishing it in the first place, or making it reasonable to *treat* behavior as a regular practice: "When I obey a rule, I do not choose. I obey the rule *blindly*."[18] What Wittgenstein favors here is the sufficiency of the spontaneous and tacit – but *not* singularly correct – continuation of a *natural* practice or *natural* application of a would-be rule. In effect, Wittgenstein retires the puzzle of the *Tractatus* by installing the primacy of "forms of life," because (1) the source or origin (the originary correspondence of word or thought and world) is then seen to be conceptually impenetrable or unrecoverable, and because (2) such a recovery is "no longer" needed to ensure the stability and smooth functioning of *any* form of natural inquiry. This yields the fair sense in which – granted the crudity of the choice of terms – Wittgenstein and the pragmatists agree. The Nietzscheans suffer by comparison, because (i) they persevere in exposing the correspondence obsession (finding it everywhere), and because (ii) they treat *any* and *every* science and inquiry as no more than an interpretation of the world governed by that same obsession. The result is that whatever of merit they say about particular claims and findings cannot but be completely arbitrary from their own perspective. Wittgenstein outflanks the *aporia*, and the Nietzscheans rejoice in it.

Friedrich Waismann has recorded an extremely helpful remark of Wittgenstein's, from the Vienna Circle period, regarding continuing with an arithmetic series (for instance, writing the square of a natural number down below the number): "A formula of algebra," says Wittgenstein, "corresponds to an induction, but it does not express the induction for the reason that the latter is inexpressible. . . . The rule cannot be expressed by a single, concrete configuration and hence not by the one written down [algebraically]. Generality shows itself in application. I have to *read* this generality *into* the configuration [as I might have done with the original numbers]."[19] Here, the distinction between what can be *said* or expressed and what can only be *shown* or applied undergoes the most profound sea-change: what needs to be grasped is the transition from the mystical contrast of the *Tractatus* to the (contexted but abstractly historical) praxical contrast of the *Investigations*. All the doubts, challenges, uncertainties of the human world come to an end – not in propositions that are ultimately unassailable *cognitively*, but in the *sub-cognitive* forms of life in which those piecemeal doubts and challenges first arise and acquire their proper function. So what is "given" – forms of life – is *not* given foundationally but only, as it were, viably. *Nothing* remains fixed. Nothing determinate may be safely presupposed. It may even be fair to see here a remote similarity to William James's empiricism and to Aron Gurwitsch's phenomenology – in the sense that neither cognizing subjects nor cognized world (or determinate objects) are (is) fixed *prior* to the flux of the forms of life. What proves salient there is neither necessary nor apodictically discerned. And, yet, what is salient *is* capable, rising from the tacit practices of a society's viable life, of *grounding* (but not *founding*) all the forms of normal inquiry.[20] This is the essential clue to the "internalist" and "externalist" versions of naturalism and phenomenology and the relationship between the matched versions of each.[21] The "internalist" versions concern distributed claims within the space of the salient; the "externalist" concern holist (nondistributed), noncognitive grounds for the realism of the other. The boldest theories of each address both aspects of the theory of the sciences, though internalist strategies require supplementary grounds for their own dialectical gains within the space of the other.[22] Deconstruction, on the other hand, shrinks the externalist theme to the mere exposure of the obsession with correspondence and totalizing and has no contribution at all regarding the internalist theme.[23] In fact, it cannot have an internalist theme, on pain of contradiction.

Two further remarks will round out this important but difficult line of reflection: first,

> "Do the same." But in saying this I must point to the rule. So its *application* must already have been learnt. For otherwise what meaning will its expression have for him? To guess the meaning of a rule, to grasp it intuitively, can surely mean nothing but: to guess its *application*. And that can't now mean: to guess the kind of application; the rule for it. Nor does guessing come in here . . . the "and so on" presupposes that one has already mastered a technique[24]

and, second, "The concept of a living being has the same indeterminacy as that of a language."[25] For Wittgenstein, then, sharing or mastering a *natural* practice does positively signify that we *can* settle questions of how to go on spontaneously, tacitly, without appeal to any foundations or prior rules or the like, without thereby invoking interpretations, *and* without there being any uniquely determinate solutions to the question of how to go on.[26] To deny the claim is to risk the coherence of actually doing so.[27]

II

It is the asymmetry between Nietzsche and Wittgenstein that is particularly useful to explore. The correspondence question that they share is conceptually unmanageable. Wittgenstein *discards* the project of the *Tractatus* (though not necessarily the mystical vision that made it possible): that, conceivably, might have been reconciled with the entirely different undertaking of the *Investigations*; there is, after all, the same serene confidence in both. Nietzsche takes it that the hopelessness of resolving the correspondence puzzle, which he clings to, *entails* the inescapable distortion of every form of human discourse. This is why he leads us to the "hermeneutics of suspicion" – although the critical theories of Marx and Freud show full well that that direction of the theory of interpretation neither requires nor necessarily favors Nietzsche's own extravagance about the correspondence issue. Paul Ricoeur has put the point in an admirably clear light: "Hermeneutics," he says, "seems to me to be animated by this double motivation: willingness to suspect, willingness to listen; vow of rigor, vow of obedience."[28] The hermeneutic question arises *within* the space of human discourse; its *double entendre* is the double nature of human communication itself. Hermeneutics is essentially internalist in spirit. But it *arises* in the first place because of the inherent indeterminacy of context and history: that is the condition on which the utterances and behavior of men *first become texts*, signs that require interpretation. "By hermeneutics we shall always understand," says Ricoeur, "the theory of the rules that preside over an exegesis – that is, over the interpretation of a particular text, or of a group of signs that may be viewed as a text."[29] But, he adds,

> there is no general hermeneutics, no universal canon for exegesis, but only disparate and opposed theories concerning the rules of interpretation. The hermeneutic field ... is internally at variance with itself. ... According to the one pole [of this internal opposition], hermeneutics is understood as the manifestation and restoration of a meaning addressed to me in the manner of a message, a proclamation, or as is sometimes said, a kerygma; according to the other pole, it is understood as a demystification, as a reduction of illusion. Psychoanalysis, at least on a first reading, aligns itself with the second understanding of hermeneutics.[30]

What makes interpretation problematic – *whether* for the recovery of intended meaning or for suspicion – is that that which invites interpretation, the *text*, displaces purely natural phenomena (which are merely possible and actual) with profoundly intentional (and linguistically intensionalized) phenomena that pose (in addition) questions of the *legitimacy* of linking texts to the world and to other humans in this way or that.[31] Wittgenstein, as we have already seen, makes interpretation (wherever it can be said to arise) conceptually dependent on particular forms of life being already in place – natural and tacit dispositions (at once praxical, evolutionary, pluralistic, nonfoundational) – that ensure a viable network of spontaneously apt behavior and responses, with respect to which (alone) reflexive theorizing and interpretative speculation have their point and hope of resolution. In this sense, Wittgenstein *cannot* be seriously mistaken: any alternative either returns us to foundationalism or some philosophy of cognitive privilege (or "presence") or else imposes on us the *aporia* of Nietzsche's (and, of course, Derrida's) preference. The point is that Wittgenstein had *detached* the *Investigations* from the conceptual project of the *Tractatus*, whereas Nietzsche elaborated his own critical and suspicious vision of man as the very *import* of his theory of language. "One would have to know," says Nietzsche, "what *being* is, in order to decide whether this or that is real (e.g., 'the facts of consciousness'); in the same way, what *certainty* is, what *knowledge* is, and the like. But since we do not know this, a critique of the faculty of knowledge is senseless: how should a tool be able to criticize itself when it can use only itself for the critique? It cannot even define itself."[32] The force of this global attack is *not* quite captured by Nietzsche's doctrine of "the innocence of becoming."[33] That signifies instead the effort, *within* the boundaries of human experience, to liberate man from every fixed or essentially "valid" metaphysics.

If Wittgenstein's account had installed the assurance of knowledge or cognitive certainty within the reliably effective practices of his forms of life – that is, distributively, for particular claims, according to rules – then he would indeed have reinstated a form of foundationalism (contrary to what, in effect, he shares with Nietzsche). But he doesn't do that. On the contrary, in *On Certainty*, he utterly opposes such a move. As he says (combatting G. E. Moore's views),

> If "I know etc." is conceived as a grammatical proposition, of course the "I" cannot be important. And it properly means "There is no such thing as a doubt in this case" or "The expression 'I do not know' makes no sense in this case." And of course it follows

from that that "I *know*" makes no sense either. "I know" is here a *logical* insight. Only realism can't be proved by means of it.[34]

"All testing," he says, "all confirmation and disconfirmation of a hypothesis takes place already within a system ... the system is not so much the point of departure [for our arguments], as the element in which arguments have their life."[35] And again, most perspicuously: "'An empirical proposition can be *tested*' (we say). ... What *counts* as its test? ... As if giving grounds did not come to an end sometime. But the end is not an ungrounded presupposition: it is an ungrounded way of acting."[36]

What Wittgenstein shows is that, if correspondence "does not have any clear application" cognitively, then, also, "since from 'I know it is so' there follows 'It is so,'" foundationalism is effectively entailed in confusing certainty and knowledge, or the absence of doubt and the "objectively" reliable grounds of truth:[37] "'knowledge' and 'certainty,'" he says, "belong to different *categories*. They are not two 'mental states'. ..."[38]

Wittgenstein's treatment of the expression "I know" as a "logical insight" deprives us, at a single stroke, of all presumption of apodicticity, without at all disallowing a functional certainty within the lives of those who share a form of life – *a fortiori*, without disallowing our *positing*, within that shared form, rules for the extension of given practices or even necessary relations systematized for selected such practices. In separating certainty and rules and necessity from foundational sources, Wittgenstein implicitly features the praxical grounding of context and history while disallowing their epistemic founding. Hence, cognitively considered, those sciences that explicitly accommodate the contingencies of context and history (the human sciences, notoriously) cannot but appear indecisive, indeterminate, or simply logically feeble when compared to the physical sciences – which, for their part, seem only minimally occupied with human contexts and human histories. But that is an utter illusion. The marvel of the physical sciences is their achievement despite the constraints of context and history – the constraints that are the internalist *minima* required by the externalist rejection of cognitive transparency. In fact, this fixes, more effectively than any *explication de texte*, the fatal allure of the apodictic for Husserl; for Husserl transforms what should have been a most remarkable externalist exposé of the epistemic presumptions of the classical forms of naturalism into a new form of internalist inquiry that suffers from its own much deeper foundationalist presumption.

What the dialectical contrast between Nietzsche and Wittgenstein shows, then, is this: (1) that the hermeneutics of suspicion must be reflexively linked to and dependent upon the hermeneutics of explication – otherwise, the Nietzschean *aporia* becomes unavoidable; and (2) that the hermeneutics of explication must itself depend on the tacit forms of life within (the practices of) which it finds its function – otherwise, objectionable forms of foundationalism become unavoidable. This is the master strategy that can be traced through every current effort to reconcile the natural and human sciences and studies.

There is room, of course, for a variety of theories. Wittgenstein assigns to the tacit practices of a viable society priority *over* interpretation: first, because he sees that there is no alternative to foundationalism that does not embed questions of truth, knowledge, and certainty in *non*cognitively specified practices – only a biologized account (as contemporary pragmatism has argued) is plausible here; and, second, because, eschewing foundationalism, he construes that priority only *en bloc*, never distributively. It is for this reason that he says, "In certain circumstances a man cannot make a *mistake* ('Can' is here used logically, and the proposition does not mean that a man cannot say anything false in those circumstances.) ... In order to make a mistake, a man must already judge in conformity with mankind."[39] Read thus, the hermeneutics of explication is to be construed either as an internal feature of the cognitive work of natural practices, in the face of the ubiquitous possibility of distributive error; or as the ubiquitous feature of determinate, specialized cognitive work arising out of the more tacit practices of a viable society. A certain tolerance for alternative views is understandable here. Furthermore, once we detach Nietzsche's and Wittgenstein's discussion of the cognitive concerns of human societies from the correspondence puzzle itself (which does more violence to Nietzsche than to Wittgenstein), our two theorists tend to move in somewhat different ways – though still with surprisingly convergent convictions. Of the two, Wittgenstein is more the master of the constraints of small-gauge contexts; and Nietzsche, of the constraints of the larger tendencies of human history. One sees this for instance in Wittgenstein's sensitive discussions of doubt and certainty regarding expectation and pain; and in Nietzsche's exposés of Christianity and the morality of the herd. But, ultimately, context and history are one and the same.

In fact, the charm of the original comparison rests entirely with grasping why it is that the bafflement of the correspondence puzzle itself leads to a strong contextualism and historicism – and with grasping what that ultimately means. The issue is straightforward enough but not without its subtlety.

III

Context is essentially an intensional distinction. There are no contexts in nature. Animals occupy their ecological niches, and causes exhibit salient regularities under given conditions; but context signifies a suitably characterized part of the world in which, thus characterized, whatever is rightly assigned, or assignable, meaning or semiotic function or the like has and can be perceived to have such (or that) meaning or function. Generally speaking, *it is texts that can be and must be assigned their proper contexts.* The paradigm of course is human speech; but by extension anything of cultural significance may be construed as a text: actions, artworks, artifacts, institutions, practices, rules, human feelings, thoughts – even persons themselves.[40] Ricoeur understandably (but quite unnecessarily) insists that "Without a specific investigation of writing, a theory of discourse is not yet a theory of the text."[41] He is concerned that a theory of texts focused only on speech (in some direct context of address) will fail to grasp the complexity of the interpretive problem. Others, such as Gadamer, are primarily concerned to emphasize the infinite, horizontally open, even prejudiced, project of interpreting a text and the consequent nonequivalence of an author's intention and the meaning of his text.[42] But, by and large, *a text is a human utterance*, something suitably generated by linguistically and culturally apt human agents, or what may be taken to be thus deposited or thus intensionally ordered or informed – *or* the very uttering being itself, reflexively affected by its own activity. Hence practices, which cannot be accounted for by the determinate acts of particular agents, are nevertheless due (on a reasonable conjecture) to the aggregated activity of the members of a given society.

In this sense, texts and only texts have histories. A history *is* the temporal career of a text: texts are intensionally significant utterances; and a history is the intensionally unified or coherent diachronic career of (or, assignable to) a text. (Intensionally ordered) texts are identified as such in their (intensically ordered) contexts; and in those contexts they have histories. Roland Barthes may well have provided the most felicitous brief characterization of the symbiosis of text and context, in spite of his notorious prose: "Every text," he says, is "itself the intertext of another text [and thus] belongs to the intertextual."[43] So texts form the contexts of texts. In rather different ways, this is the governing theme for instance of such diverse literary theorists as Michael Riffaterre, Harold Bloom, and Stanley Fish.[44] But it is also the governing theme of theorists attracted to Thomas Kuhn's insistence on the indissolubility of science and the history of science.[45] Ultimately, what this means is that texts exist, are found, are recognized, and are understood only in the world or space of human culture – (*only*) by persons, being naturally groomed within their respective societies to be apt for producing and understanding themselves and their texts.

This is the affirmative face of what Nietzsche and Wittgenstein share in their grasp of the insolubility of the correspondence puzzle: the point of convergence between the themes of "the will to power" and the "forms of life." Nietzsche's notion, of course, is the more problematic; perhaps it is close to the notion of the Heraclitean flux viewed in terms of the generation and replacement of provisionally functioning networks of categories of understanding, drawn from sources below (and, there, first organizing) the cognitive powers of man himself: "*This world*," says Nietzsche, "*is the will to power – and* nothing besides! And you yourselves are also this will to power – and nothing besides!"[46] Nietzsche grants the effective hegemony of particular conceptual schemes – shifting from age to age, organizing the historical focus and range of human intelligence. This is why Foucault is essentially a Nietzschean historian. But Nietzsche has no patience for exploring the small dynamics of the process; he is too intent on exposing its would-be eternal (therefore, false) truths. Thus he proclaims, "Dionysus versus the 'Crucified': there you have the antithesis."[47] It is Wittgenstein who painstakingly explores the ubiquitous, minimal, ineliminable, structuring context of human culture: the forms of life, within the generic range of which particular contexts of tacit practices are (somewhat indeterminately) identified. Cultural phenomena, texts, have no identity except in context; their meaning and significative import cannot be grasped or interpreted except in context.[48] Still, the Nietzschean (or Foucauldian) insistence on the "normalization" of conceptual frameworks *is* an insistence on the radical nature of context and history – in effect, an insistence on the textualized status of whatever we inquire into.[49]

Now then, *if* the correspondence thesis fails, then no realism with respect to either natural or cultural phenomena can claim a cognitively privileged access: this is what is meant by repudiating foundationalism, essentialism, logocentrism and the like and of insisting on the indissoluble unity of the realist and idealist aspects of human inquiry. (It should be said, however, that, although it entails the correspondence question, the question of foundationalism need not explicitly broach the other.) But, if the point be granted, then, first of all, although there are, in (what we presuppose) physical nature (to be), no contexts (because there is no intensionality and no language), the cognitively accessible natural world *is* texted and contexted by the categories of human inquiry: ineliminably, in the sense that *we* cannot exit from the lingual nature of inquiry itself – contingently, in terms of the historical fortunes of particular conceptual schemata. So nature has a history – within human inquiry; and, relative to human efforts at explanation, it is a kind of text. In fact, this is the very theme of Kuhn's distinction between normal and revolutionary science. But, secondly, if inquiry is to proceed at all, it must have some stability: concepts must be reliably applicable to both nature and culture – to culture as it emerges from and rests upon physical nature, to nature as it is explored by culturally emergent agents.

This is the source of the profundity of Wittgenstein's notion of the forms of life: Wittgenstein provides a template for innumerable accounts of the tacit or "natural" condition within which the epistemically explicit probings of human societies (must come to) depend on the signal fact that the members of viable such systems cannot meaningfully fail to know how to "go on" with their linguistic and cultural practices. Every form of realism presupposes the condition, but no form of realism can establish it or be established by it. Nietzsche, of course, implicitly acknowledges the point, since he actually *uses* language to expose its ulterior pretense; but his *aporia* oppresses us to the extent that (in his spirit) we fancy we may claim not to trust, even provisionally, any of the ascriptions of ordinary discourse.

One sees the difficulty most clearly in the early essays of Paul de Man, who, among all the followers of Nietzsche, may well come closest to permitting Nietzsche's view of correspondence to *invade* (and thereby to disorder) the detailed discussion of literature and science. So de Man says, reviewing "the imaginary loci of the physicist and the *fictional* entities that occur in literary language," "One entirely misunderstands ... the priority of fiction over reality, of imagination over perception, if one considers it as the compensatory expression of a shortcoming, of a deficient sense of reality."[50] This *might* have been no more than a particularly florid way of restating the symbiosis of the realist and idealist aspects of language and inquiry; but de Man insists that "a work of fiction asserts, by its very existence, its separation from empirical reality."[51] And elsewhere he remarks – in a warning that surely centers on the question of context –

> Prior to any generalization about literature, literary texts have to be read, and the possibility of reading can never be taken for granted. It is an act of understanding that can never be observed, nor in any way prescribed or verified. A literary text is not a phenomenal event that can be granted any form of positive existence, whether as a fact of nature or as an act of the mind. It

leads to no transcendental perception, intuition, or knowledge but merely solicits an understanding that has to remain immanent because it poses the problem of its intelligibility in *its own terms*.[52]

So there is no remotely specifiable *public context* for accommodating the "immanence" of the literary work itself – which "invents fictional subjects to create the illusion of the reality of others"[53] and which can only be penetrated, apparently, by entering that peculiarly private "void" in which *it* (and not another work) invents what it invents.

Perhaps de Man has not really erased the contextual conditions for the very intelligibility of literature (here, its "priority" over science may be benign enough). But what he appears to be saying surely fixes the mad limit to which the Nietzschean rhetoric may be pressed – and points unmistakably (if unintentionally) to the need to retreat *in Wittgenstein's direction*. The *success* of a favorable correspondence thesis would have eliminated the need to speak of context altogether, because we would then have possessed a ubiquitous, single, all-embracing context: whatever our foundationalist or essentialist accounts would have correctly ident-ified. On the other hand, on the *failure* of the correspondence thesis, discourse itself would also fail – *if* there were innumerable "atomic" or solipsistic contexts. De Man flirts with the idea, because he is so terribly afraid that *any* concession to a functioning language (1) that *is* realist (in the biologized sense already sketched), (2) that links empirical reality *and* fiction (perception *and* imagination), and (3) that has *any* tendency toward system will have granted too much to be able (then) to sustain the subtleties of a "critical reading" of literature. But he is not merely wrong in this: he positively courts incoherence.

De Man's dangers inevitably point to the conceptual beauty and power of Wittgenstein's notion of "forms of life." Let us tabulate its advantages:

1 it escapes the impossible strictures of correspondence;
2 it substitutes a symbiosis of realist and idealist elements;
3 it repudiates foundationalism and totalizing;
4 it provides a rationale for improvisation and historical change;
5 it ensures a sense in which certainty and cognitive claim are reasonably grounded without invoking either a vicious regress or skepticism or a retreat to foundationalism itself;
6 it eliminates the need to rely on necessary or sufficient or necessary and sufficient conditions of discourse;
7 it links in a natural way individual behavior with the practices of a society;
8 it indicates the profound continuity of biology and culture; and

9 it clarifies the textual and contextual nature of the entire human world.

And in doing that, it does still more:

10 it shows how it is that any and all extensional regularities among our categories are ultimately controlled by the intensional distinctions we introduce *contextually*.

Or, to put this last point in its most powerful but least controversial way: *to introduce distinctions contextually, without any foundationalist privilege, is to treat conceptual distinctions intensionally; or, to treat extensional regularities as finite, provisional conveniences for given runs of cases originally marked in intensionally irreducible ways.* Ultimately, this is why it is so puzzling to find that the most strenuously devoted extensionalists among analytic philosophers – Carnap, Quine, Sellars, Davidson – are also so implacably opposed to foundationalism: in effect, they give away with one hand what they insist on taking back with the other. This also explains why the concept of a context is simply the concept of a history: *context is the punctuated locus of a history with respect to which the dynamic stabilities of a language so function; and history is the diachronic movement of the human aptitude to change from one scheme of contexts to another.*

The intentionality of history (the *I*ntentionality of history, to invoke a distinction introduced earlier[55]) invites a small warning and an ample detour. In conceding the puzzles of context and history, we should not for that reason discredit the projects of science or even the projects of an extensionalist regimentation of science: they obtain, as well as they can, *within* the shifting, plural contexts of shifting history; they are themselves contexted, historicized. This is the rightly remembered (originally radical) lesson of Kuhn's study of paradigm shifts, even though Kuhn's thesis gradually drifts back through rearguard adjustments (indefensibly) to as much as it can salvage of an earlier progressivism in science. There *is* room for assessing progress, induction, falsification, comprehensiveness of theory and the like; but there is no longer any room for progressivism, inductivism, falsificationism, verisimilitude, totalizing, cognitive privilege. In abandoning these, we are forced to concede a deeper sense of "context" and "history" – namely, just that sense that, alone, is consistent with the abandonment of these other doctrines.

But, within the space of that concession, we are also obliged to acknowledge the *sui generis* features of human history (history proper)

– of processes that cannot be adequately captured in terms of physical or clock time and cannot be equated with what is usualy meant by "natural history." The essential clue is straightfoward enough. *Human history is Intentional, and natural history is not* – even though accounts of natural histories, together with the categories by the use of which they are identified, are clearly subject (as, *a fortiori*, are those regarding human history) to the Intentional life of human investigators.

The inadmissible (or at least completely undefended) conflating of the two is nowhere more explicit, in the current literature, than in Adolf Grünbaum's *The Foundations of Psychoanalysis*.[56] For, for one thing, Grünbaum does not satisfactorily address the question of the relationship between Intentionality and history (which clearly affects the ulterior issue of the unity or bifurcation of the sciences – *a fortiori*, the nature of psychoanalysis *as* a science); and, for another, Grünbaum's own intended investigation of the "*credentials* of psychoanalytic theory"[57] presupposes a resolution of the first question. (We shall have to take care to confine this detour to essentials.)

There is every reason to believe that when he advises us that "I shall argue ... that all of Freud's clinical arguments for his cornerstone theory of repression should be deemed to be fundamentally flawed,"[58] a large part of what Grünbaum offers in support presupposes a relatively clear canon of science fitted to the common methodological features of the natural and human sciences and, in particular, a canon suitably fitted to accommodate all pertinent Intentional, contextual, historical complications. This, it should be said at once, is *not* to deny that psychoanalytic claims are (in some fair sense) testable, or that Grünbaum's charge is unsupportable, or that Freud himself did not favor some version of the unity-of-science program. It is only to set reasonable limits on what we need to explicate here regarding the distinction between human history and natural history which could not fail to affect any assessment of what, methodologically, may be expected of psychoanalysis as a science (and as a science of one sort rather than another).

There are at least three distinct puzzles prominently associated with the analysis of context and human history that would need to be favorably resolved if a unity conception of psychoanalysis – or, at least, that much of it as is supported by Grünbaum – is to be vindicated at all. These are as follows.

(i)   The holism of the mental, the potentially radically open-ended dependence of any descriptive characterization of individuated mental or psychological phenomena on their being only relationally so distinguished within a holistic model of mental life.[59]

(ii)   The double interpretive problem of Intentional ascriptions: *either* the Intentional import of particular physical or biological phenomena must be *assigned* by the use of conceptual schemes not *first* (or perhaps ever) known to collect phenomena subsumable in a pertinently regular way under physical laws, or not known to yield any serviceably rule-like or law-like regularities linking physical (or biological) states and intentional (or Intentional) states *in either direction*;[60] *or* the attributive similarity or identity of intentional (or Intentional) properties itself depends on potentially radically variable interpretive schemata (also essentially Intentionalized and historicized) – ranging, say, between the Kuhnian and the Foucauldian visions – that have never been (perhaps cannot be) regularized (reductively or nonreductively) *in accord with any known physical laws.*[61]

(iii) Causal contexts that may not, on the admission of incarnate phenomena, be able to be uniformly characterized as extensional; or, as a result, causal regularities that may not be able to be regimented in a nomological way, or causal laws that may prove logically and methodologically quite different from physical laws.[62]

(These puzzles, of course, do not address the question of history directly.)

The fact is that, when he examines the "flaws" in Freud's "philosophy of science" (that is, in the early "Scientific Project" and in Freud's implicit views on clinical explanations) as well as in his metapsychology (his "ontology") – and in the use made of all of this among prominent psychoanalysts and hermeneutic theorists – Grünbaum attends to two distinct issues (only) bearing on (iii): first, the well-known "reasons-*versus*-causes" argument; and, second, the causal effectiveness of intentions. On the first, Grünbaum rehearses the by now well-established finding that reasons (really: *having* reasons) cannot be denied a causal role in psychological contexts; that the numerous theorists who have insisted otherwise – notably, Ricoeur, Roy Schafer, George Klein among the psychoanalytically minded – have utterly failed to make their case out.[63] Fair enough, but this finding alone goes no distance at all toward solving the methodologically crucial questions focused in (i) and (ii) that bear decisively on how we should understand (iii).

The issue is not (or not merely) the "metapsychological" or "ontological" question of reducing or replacing the mental by the physical, the biochemical, or the neurophysiological. Grünbaum makes a plausible case to the effect that Freud "explicitly deemed the metapsychology *epistemologically expendable* as compared to the clinical theory."[64] (Also, Grünbaum himself opposes a physicalist reduction of the psychological sciences.[65]) The issue is rather *how* to manage "epistemologically" the Intentional complexities *of the clinical material.* Grünbaum never quite says. The mere defeat of the "reasons-*versus*-causes" thesis cannot guarantee an answer; but the "epistemological" effect of superseding the metapsychology is completely opaque.

It seems unlikely, but it is true nevertheless, that Grünbaum simply abandons the question in favor of ensuring a biographically accurate account of Freud's own mature persuasion about the methodology of psychoanalysis. So, for instance, he offers the following summary assessment of George Klein's reading of Freud:

> Since Freud personally claimed scientific status for his enterprise [Klein declares, having only the superseded strategy of the "ontological reduction" of psychoanalytic phenomena in mind], even the mature Freud aspired to such a physicalistic reduction *in order to vindicate* just this avowal of scientificity. Oddly enough, Klein seems to have been unaware that Freud had long since repudiated his early reductively ontological hallmark of scientificity in favor of an epistemic or methodological one.[66]

Perhaps so, but Grünbaum nowhere shows us *how,* "epistemologically," we may bring the intentional complexities of *clinical* psychoanalysis into accord with the rigors of "scientificity." The issue hardly surfaces at all.

But, if it does not, then Grünbaum has as yet no firm basis on which to specify *the "epistemological" standing of psychoanalysis: on the strength of which the "flaws" in Freud's clinical hypotheses (at least in part – that is, apart from mere empirical error and inadvertence and the like) count as failures to meet the methodological requirements of all "bona fide" sciences.* The truth is that Grünbaum is content to show, as far as intentionality is concerned, that

(a) having reasons, as in deliberate behavior, may rightly be counted as causally efficacious (against the "reasons-*versus*-causes" thesis); and

(b) "repressed ideation," "unconscious motives" and the like may also be counted as causally efficacious, though they are not (in Grünbaum's sense) "intentional" or "intended" or else they are intended "in *only a Pickwickian or metaphorical sense."*[67]

But neither (a) nor (b) is addressed to the issues (i)–(iii). And yet

Grünbaum specifically chides Habermas and Gadamer for having drawn "a *pseudo*contrast between the nomothetic [paradigmatically: the physical] and the human sciences": "For that major physical theory [the theory of classical electrodynamics] features laws that embody a far more fundamental dependence on the history and/or context of the object of knowledge than was ever contemplated in even the most exhaustive of psychoanalytic explanatory narratives or in any recapitulation of human history."[68] With the best will in the world, it cannot but be hopeless to press the point if, as we have just learned, Grünbuam simply does not consider the puzzles of intentionality – (i)–(iii) – in the context of the human sciences; and, of course, the claim (made just above) is patently false.

What does Grünbaum offer to support his charge? He rejects the claim that "explanations in physics are generically based on context-free ahistorical laws";[69] and he cites the electrodynamic laws as instances of physical laws that are neither context-free nor ahistorical: "As against Habermas, I submit," he says,

> that these electrodynamic laws exhibit context-dependence with a vengeance by making the field produced by a charge for any one time dependent on the particular infinite past history of the charge. And, to the detriment of Gadamer, these laws are based on replicable experiments but resoundingly belie the thesis that "no place can be left for the historicality of experience in science."[70]

The argument is worth reporting in some detail:

> Consider an electrically charged particle having an arbitrary velocity and acceleration. We are concerned with the laws governing the electric and magnetic fields produced by this point charge throughout space at any one fixed time t. In this theory, the influence of the charge on any other test charge in space is postulated to be propagated with the finite velocity of light rather than instantaneously, as in Newton's action-at-a-distance theory of gravitation. But this *non*instantaneous feature of the propagation of the electrodynamic influence contributes to an important consequence, as follows: At any space point P, the electric and magnetic fields at a given time t depend on the position, velocity, and acceleration that the charge had at an earlier time t_o. That earlier time has the value $t - r/c$, where r is the distance traversed by the influence arriving at P at time t after having traveled from the charge to P with the velocity of light. . . . It follows [from similar considerations

involving increasing the distance r] that at ANY ONE INSTANT t, the electric and magnetic fields produced throughout infinite space by a charge moving with arbitrary acceleration depend on its own PARTICULAR ENTIRE INFINITE PAST KINEMATIC HISTORY![71]

Grünbaum would appear to be right in what he says – *at least* in the sense that time, read in terms of the relation "earlier than" and "later than" *when fixed psychologically*, behaves nonextensionally, cannot be rendered in truth-functional terms. In *that* sense, "natural histories" *are* context-dependent and historically complex. This is the point of construing natural histories as "texted." *But the phenomena they order are not themselves intensionally (or Intentionally) complex, even including anisotropic time; and their own intensionality is apparently exhausted by this assigned feature of physical time (directionality essentially linked to human time).* Human history – the phenomena of psychoanalysis, preeminently – encompassing states and events that, in addition to the directionality of time (which undoubtedly links all inquiry – and physical time – to the directional life of human investigators), includes the Intentionally complex phenomena (favored by the "hermeneuts") that pose the puzzles of (i)–(iii). *Human history*, then, contrary to what Grünbaum implies, commits us to construing *the directionality of time as an incarnate feature of phenomena that accord with (i)–(iii)*: if so, then the "historicity" and "contextedness" of the electrodynamic laws are themselves a limited abstraction *from* the historicity and contextedness of time-directed human existence. The irony remains that Grünbaum himself treats the anisotropy of physical time as entirely unrelated to the directionality of "psychological time."[72] On that thesis, admitting that anisotropic time treated as a feature of the phenomena objectively studied rather than as an abstraction from the time of human inquiry may be managed extensionally, Grünbaum would no longer be able to support his claim about history and context in the challenging way he does, and natural history and human history would have to be seen as entirely distinct.[73] *If* anisotropic time is conceptually distinct from psychological time and *if* it forms the basis for the claim that the electrodynamic laws are "intentional," "context-dependent," and replicable (in Grünbaum's sense), then what Grünbaum says about their "historical" nature remains irrelevant to the analysis of the methodology of psychoanalysis unless the *phenomena* of psychoanalysis *do not* introduce any more profound features of the intentional and context-dependent than is captured by anisotropic time. *But they do introduce such features.* First of all, psychoanalytic phenomena, even on Grünbaum's account of alternative kinds of time, entail "psychological time" – which *cannot* be captured by anisotropic time; secondly, such psychoanalytic phenomena as the repression syndrome *do* introduce intensional (or Intentional) complexities that must be satisfactorily regimented *before* we are even able to claim replicability or to claim that the replicability they exhibit conforms with the "nomothetic" model or any unity view that disallows the bifurcation of the natural and human sciences. Thus we conclude that Grünbaum fails to make his case, fails to address it directly, even subverts it by his admission of the fundamental difference between anisotropic and psychological time.

So much, then, for our detour.

VI

Clearly, those who would eliminate selves and persons (Daniel Dennett, for instance) would not regard the conceptual distinction between natural history and human history as an insurmountable barrier to their reductive and extensional programs. But it is notorious that they are often unaware of the complexity of the challenge. For example, one may claim that reference involving proper names is doubly affected by Intentional (or historical) complications: first, because of the speaker's "location" in referring; second, because actual referents are so designated relative to the speaker's (or the speaker's community's) "location."[74]

Under the conditions acknowledged, there would be no coherent inquiries or practices of any sort if nothing like Wittgenstein's forms of life obtained; that is, if there were no nonfoundationalist basis for grounding historically pertinent activities – paradigmatically, speech acts and the inquiries of science. In fact, one might reasonably say that, in his own account of speech acts, J. L. Austin uses the notion of "normal" use or "accepted conventional practice" very much in accord with what Wittgenstein was pressing home when he spoke of forms of life – even though Austin never really elaborated the full import of his own notion and probably had another view altogether of human society (and even though he was profoundly doubtful of Wittgenstein's entire enterprise).[75] In any case, his deconstructive critics (notably, Derrida – and in an expository way, Jonathan Culler) as well as his defenders (notably, John Searle) have oddly confused matters.

Derrida correctly noted Austin's contrast between the "serious" and "parasitic" uses of certain performative utterances.[76] But he then went on to claim, challenging Austin (and Searle, who comes to Austin's defense[77]), that "what is at stake above all is the *structural impossibility and illegitimacy* of [the given] 'idealization,' even one [*that* between 'serious' and 'parasitic'] which is methodological and provisional."[78] This rightly captures Austin's characteristic tentativeness and provisionality – which Culler really misdescribes. For, as Culler sees the matter,

> What the indissociability of performative and performance puts in question is not the determination of illocutionary force by context but the possibility of mastering the domain of speech acts by *exhaustively* specifying the contextual determinants of illocutionary force. A theory of speech acts *must* in principle be able to specify *every* feature of context that might affect the success or failure of a given speech act or that might affect what particular speech act an utterance effectively performed. This would require, as Austin recognizes, *a mastery of the total context. ...*[79]

But Austin's formula (which Culler cites) is simply treated by Austin himself as a mere "moral" – that is, as *not* an actual project to be "seriously" undertaken or even conceivably completed.[80] It is certainly Searle's view that the project is possible in a straightforward first-order sense and should be undertaken; but, then, that merely indicates how easily Austin could be misunderstood by supporters and opponents alike.[81] *Part* of the point at stake – the lesser part – is that Culler's objection is not actually pertinent to Austin's own project, since Austin never was confident about identifying "every" feature "exhaustively" with regard to anything that could be called "the total speech situation." Austin meant – as he quite explicitly says – to contrast the "serious" and the "parasitic" in a series of contexts in which distinctions can hardly fail to be incompletely determinate and systematically incomplete and incompletable.

Derrida (rightly) criticizes "the *structural* impossibility and illegitimacy" of totalizing, of specifying the function or import of a *particular* speech act *in context*, by way of locating it within a system of totalized possibilities. Still, there are two distinct challenges we may level against Derrida's account: first, Derrida wrongly charges Austin with totalizing, whereas the (ironic) truth is that it is Searle (with whom he actually debates) who believes he can provide both the necessary and sufficient conditions of any speech-act type and a complete table of the defining properties of any given speech-act type that distinguish it from any other[82] – Austin never suggests that he can do that, always resists the temptation; second, Derrida insinuates (if he does not actually charge) that, as a result of disqualifying totalizing, particular speech acts cannot, in the contexts in which they apparently obtain, ever be, there, determinately specified or identified, whereas (*per* Austin and Wittgenstein, if one may join thinkers who would not have cared to be thus joined) the discrimination of speech acts (*and of anything else*) *neither presupposes nor entails totalizing* – there's the Nietzschean *aporia* to which Derrida (also Foucault, in a different way) subjects himself. For all their acrobatics, what Nietzsche and Derrida say does make determinate sense, exposes totalizing or essentialist or universalist or logocentric presumptions in the arguments of others – in Derrida's *jeux*, structuralist and structuralist-like maneuvers.[83] If Derrida were a radical skeptic, then of course we should have to construe the rhetoric of his language as an insinuation of doubt embedded in the cognitively pertinent claims of his own would-be targets. But *he* means to insist quite straightforwardly that totalizing – in particular, the totalizing favored by Rousseau and Saussure and Lévi-Strauss (and Husserl and a hundred others) – *is* "impossible" and "illegitimate"; it is only the precise "relation" at the juncture of language and reality (a relation that disallows totalizing and enables us to affirm that impossibility) that cannot be *specified*: for, to be able to specify it would commit us to the very claim opposed. That is the point of Derrida's "*différance*" – the Medusa of correspondence that can only be glimpsed through the indirection of an expression antecedently known to be incapable of actual reference.[84] Deconstruction invites us – exuberantly and disastrously – to *apply* everywhere the principle of the infinite attenuation of determining context. But it is disingenuous of Derrida (and Culler) thus to chide those naturalists and phenomenologists who actually embrace the notion of context in the anti-foundationalist and anti-totalizing spirit. The critical point to grasp is that *doing that* affects (1) the whole of science and (2) the human sciences most particularly. In quite different ways, this is what Grünbaum and Derrida jointly fail to address.

Culler goes on to say something quite worthwhile about context, however, which catches up the more important part of the issue – the one in fact that Derrida originally pressed. Culler says,

> Context is boundless in two senses. First, any given context is open to further description. There is no limit in principle to what might be included in a given context, to what might be shown to be relevant to the performance of a particular speech act. This structural openness of context is essential to all disciplines. ... Context is also unmasterable in a second sense: any attempt to codify context can always be grafted onto the context it sought to describe, yielding a new context which escapes the previous formulation.[85]

Now, this is true but *misses the essential point* that Austin shares with Wittgenstein.

The "boundlessness" of context is and cannot but remain an insurmountable obstacle for any cognitivist strategy for fixing actual contexts: every such effort would have to confront – and would fail to overtake – the dual threat of transparency and totalized system. The determinate structure of context cannot be fixed with certainty because nothing can; nor can it be fixed, with whatever precision we can command, relative to an exhaustive and organized table of all possible lines of significance, because we can never assure ourselves of so much closure, or of any measurable approximation to so much closure.[86]

But context *can* be made determinate with as much precision as any distinction that is context-bound. We succeed in doing so everywhere. Notice that Culler's first objection affirms only that "any *given* context is open to *further* description." This is not to deny a functional measure of determinateness in fixing context; it is only to remind us of the holism of all discourse about contexts and of the impossibility thereby circumscribed of exhaustively or independently codifying (intensionally qualified) intentional phenomena.[87] So the "boundlessness" of context is a feature *of* effectively determined contexts, not a mark of the radical pointlessness of claiming relevance and confirmability for any discourse at all. By the same token, the deconstructive complaint *has no distributive application* to any particular conceptual scheme. Culler adds (quite rightly) that any improvement, enlargement, codification of a would-be system of categories can be "grafted onto" the context of that very system, generating the puzzle of context again. But *that* is not a barrier against determinate discourse: it is a feature of the indefeasible open-endedness of such discourse – just where it is successful. Deconstruction may warn us of our global incapacity to overtake this limit (if limit it be), but it cannot *improve, correct, assess,* or *confirm* the determinate details of particular conceptual schemes. It cannot descend to determinate details, for in doing that it ceases to be (or to be merely) deconstructive – ceases to deny the viability of fixing context or of fixing distinctions within context. This, for example, marks the precise point at which Foucault, identifying the determinate shift in the dawning orientation of seventeenth- and eighteenth-century conceptions of discipline and punishment, ceases to be (or, aporetically continues to be) a Nietzschean historian.[88]

Deconstruction is not irrelevant to science; its complaint is already absorbed *in* any inquiry that rejects transparency and totalizing. On the argument, in doing that, it inexorably favors Wittgenstein over Nietzsche, which is to say, it inexorably favors the theme of *praxis*, of grounding cognition itself in the habituated and habituating practices of an actual society – viable in sub-cognitive, biological terms but ensuring thereby a (holist) realism within which the intentional concerns of human investigators advance their own fledgling systems. Wittgenstein himself has no theory of history or of *praxis*, of course. Nevertheless he offers (however thinly) the master clue: the priority of forms of life.

What Culler – in effect, Derrida – objects to is the untenable *structuralist* presumption to *totalize* with regard to any sector of the human world. That, of course, is the point of Culler's term "boundlessness" as well as of Derrida's explicit insistence on "the structural impossibility and illegitimacy of such an 'idealization.'" But Austin, rather like Wittgenstein, was (implicitly at least) considering a conceptual alternative to structuralism – one in which "normal" or "serious" use neither presupposes nor entails reference to *any totalized structure of possibilities*. The irony is that Searle could not defend Austin properly, because he *was* a proper target for Derrida's attack. But the upshot is that the recent deconstructive followers of Nietzsche (very much as we saw risked in de Man) have utterly undermined any possibility of accounting for the actual success of communicative exchange. They simply oppose totalizing (and its allied doctrines); but they offer nothing in its place. They muddle the difference between internalist and externalist concerns.

The proper answer to the deconstructive formula that Culler offers – namely, that "meaning is context-bound but context is boundless" – is to propose a better one: "meaning is context-bound but context is not groundless." For that is a formula, rightly understood, that escapes the *aporia* of Nietzsche's extreme view and saves what is required to make sense of the practices of natural discourse within the changing contexts of human history.[89] To concede the point is to imply that Wittgenstein's notion of "forms of life" offers the only promising strategy for restoring a sense of cognitively exploitable context; for, against the cognitivists of naturalism and phenomenology, against radical skeptics and deconstructive anarchists, it obliges us to concede that context is the human setting of *praxis*, of habits of life that possess a stability grounded in and grounding the survival of actual societies – within which cognitive inquiry tacitly enlists that same stability in pursuing its own work. It simply never appeals to the other in a criterial or principled sense.[90] The anchoring of context is holistic, forestructured and forestructuring below the level of reflexive analysis. Deliberate inquiry *cannot*, in general, fail to accord with these subterranean processes. It is an organic part of them. It puts the deconstructive threat to rest *by surviving*, societally; and it determines the acceptable boundaries of context in every instance by referring its disciplined claims to consensual tolerance. And, yet, consensus and societal tolerance are never tests or criteria of any kind.

What Nietzsche (and subsequent deconstruction) shows is that totalized systems are impossible. What Wittgenstein shows is that totalized systems are unnecessary. What naturalist or empirical inquirers show is that system-like structures are, within limits, able to be *fitted to segments* of natural and cultural phenomena, without presupposing or confirming totalized systems. A strong version of the unity of science claims an (effectively totalized) system of universal law that obtains in a radically context-less sense in the whole of nature. A strong structuralism claims an (effectively totalized) system of universal transformational relations of a rule-like sort that obtains in a radically context-less sense in the whole of human culture.[91] But there is no known demonstration that, for cultural phenomena, there must be a system underlying the phenomena of any examined sector;[92] and the difficulty of fixing invariant physical laws raises reasonable doubts about the ubiquity of context-less systems in nature.[93] Either, therefore, the admission of context (or human history) is heuristically, functionally, or fictively introduced or it precludes real systems (redundantly: totalized systems) wherever it is itself realistically construed.

We must begin with saliences, then, wherever transparency is denied. In a praxical setting, even the context of what is salient is itself recoverably salient. History is the diachronic record projected from what is currently salient, regarding an Intentionalized past and an Intentionalized future – open-ended to be sure, in the double sense that the focus of the salient is bound to change (apart from its being inevitably pluralized in any complex society) and that such change is already anticipated in formulating a history in any present present. Context (and, therefore, history) conceptually excludes system; *or* obliges us to construe systems, whether in the natural or human sciences, as artifacts contingently dependent on the horizon and saliencies of our own present present. This is the essential source of the perceived "weakness" of the human sciences. But it is easily converted into a mark of their extraordinary strength. For, the puzzle remains: how to achieve any measure of explanatory regularity under conditions (either not present or not matched in the physical sciences) of

1 ranging over Intentional phenomena;
2 merely establishing empirical similarities and reidentifiable types among Intentional phenomena;
3 formulating causal regularities among Intentional phenomena;
4 entertaining causal laws among Intentional phenomena;

5 confirming any such regularities across the shifting horizons and shifting saliences of our own (Intentional) history; and
6 confirming any such regularities reflexively.

Any serious attempt to deny the bifurcation of the sciences must pause to answer the challenges implicit in this tally.

Notes

1 Friedrich Nietzsche, "On the Truth and Lie in an Extra-Moral Sense," in *The Portable Nietzsche*, tr. and ed. Walter Kaufmann (New York: Viking, 1954), pp. 45–7.
2 Jacques Derrida, "Différance," *Margins of Philosophy*, tr. Alan Bass (Chicago: University of Chicago Press, 1982), pp. 3, 11 (italics added).
3 Ludwig Wittgenstein, *Tractatus Logico-Philosophicus*, new edn of English tr. of German 2nd edn, corr. D. F. Pears and B. F. McGuiness (London: Routledge and Kegan Paul, 1972), p. 3.
4 Ibid., p. 5.
5 Cf. Georg Henrik von Wright, *Wittgenstein* (Minneapolis: University of Minnesota Press, 1983).
6 Wittgenstein, *Tractatus*, p. 5.
7 Rudolf Carnap, "Intellectual Autobiography," in Paul Arthur Schilpp (ed.), *The Philosophy of Rudolf Carnap* (LaSalle, Ill.: Open Court, 1963), pp. 25–9.
8 Cf. Ludwig Wittgenstein, *On Certainty*, tr. Denis Paul and G. E. M. Anscombe, ed. G. E. M. Anscombe and G. H. von Wright (Oxford: Basil Blackwell, 1969), section 215.
9 Wittgenstein, *Tractatus*, 6.4311.
10 Ludwig Wittgenstein, Preface to *Philosophical Investigations*, tr. G. E. M. Anscombe (New York: Macmillan, 1953), p. ix. Wittgenstein actually remarks, "It suddenly seemed to me that I should publish those old thoughts [the *Tractatus*] and the new ones [the *Investigations*] together; that the latter could be seen in the right light only by contrast with and against the background of my old way of thinking" (p. x).
11 Wittgenstein, *Philosophical Investigations*, pt II, p. 126.
12 See D. F. Pears, "Universals," *Philosophical Quarterly*, I (1951).
13 Wittgenstein, *On Certainty*, section 559.
14 Ludwig Wittgenstein, *Zettel*, tr. G. E. M. Anscombe, ed. G. E. M. Anscombe and G. H. von Wright (Oxford: Basil Blackwell, 1967), sections 55, 56, 60, 63. (The citation is abbreviated for convenience.)
15 See for instance Wittgenstein, *Philosophical Investigations*, sections 98–108.
16 Ibid., sections 201–2.
17 See for example Jacques Derrida, *Of Grammatology*, tr. Gayatri Chakravorty Spivak (Baltimore: Johns Hopkins University Press, 1976); Hans-

Georg Gadamer, *Truth and Method*, tr. Garrett Barden and John Cumming from 2nd German edn (New York: Seabury Press, 1975); Paul de Man, *Blindness and Insight*, 2nd rev. edn (Minneapolis: University of Minnesota Press, 1983); Michel Foucault, *The Order of Things*, tr. from the French (New York: Random House, 1970).

18 Wittgenstein, *Philosophical Investigations*, section 219.

19 *Wittgenstein and the Vienna Circle: Conversations Recorded by Friedrich Waismann*, tr. Joachim Schulte and Brian McGuiness, ed. Brian McGuiness (Oxford: Basil Blackwell, 1979), p. 154.

20 See William James, *Essays in Radical Empiricism* (New York: Longmans, Green, 1912); and Aron Gurwitsch, "On the Intentionality of Consciousness," in Marvin Farber (ed.), *Philosophical Essays in Memory of Edmund Husserl* (Cambridge, Mass.: Harvard University Prses, 1940). The connection with Gurwitsch I owe to conversations with J. N. Mohanty.

21 See Joseph Margolis, *Pragmatism without Foundations: Reconciling Realism and Relativism* (Oxford: Basil Blackwell, 1986), ch. 8.

22 A particularly glaring example of the confusion of the two foci appears in Nelson Goodman's *Ways of Worldmaking* (Indianpolis: Hackett, 1976). Goodman's thesis (not unlike James's) *should* be construed in an externalist sense; but Goodman means it to accord with his earlier account (in *Fact, Fiction, and Forecast*) of the puzzle of induction – which is an internalist issue. A similar objection may be put to Quine's extensionalism, the basis of the "Two Dogmas" thesis – which is not merely metalogical but an externalist issue as well.

23 Cf. Henry Staten, *Wittgenstein and Derrida* (Lincoln, Nebr.: University of Nebraska Press, 1984).

24 Wittgenstein, *Zettel*, sections 305, 306, 308.

25 Ibid., section 326. Cf. Wittgenstein, *On Certainty*, section 28.

26 Cf. Robert J. Fogelin, *Wittgenstein* (London: Routledge and Kegan Paul, 1976), pp. 204–5.

27 See Saul A. Kripke, *Wittgenstein on Rules and Private Language* (Cambridge, Mass.: Harvard University Press, 1982).

28 Paul Ricoeur, *Freud and Philosophy*, tr. Denis Savage (New Haven, Conn.: Yale University Press, 1970), p. 27. The expression "hermeneutics of suspicion" may well have been coined by Ricoeur – I'm not at all sure.

29 Ibid., p. 8.

30 Ibid., pp. 26–7.

31 See Paul Ricoeur, "What is a Text? Explanation and Understanding," *Hermeneutics and the Human Sciences*, tr. and ed. John B. Thompson (Cambridge: Cambridge University Press, 1981).

32 Friedrich Nietzsche, *The Will to Power*, tr. Walter Kaufmann and R. J. Hollingdale, ed. Walter Kaufmann (New York: Random House, 1967), bk3, section 486; cf. also sections 470, 474, 480.

33 This goes rather against the interesting interpretation of Nietzsche advanced by Ofelia Schutte, *Beyond Nihilism* (Chicago: University of Chicago Press, 1984), ch. 2.

34 Wittgenstein, *On Certainty*, sections 58–9.
35 Ibid., section 105.
36 Ibid., sections 109–10.
37 Ibid., sections 178, 194, 215.
38 Ibid., section 308.
39 Ibid., sections 155–6.
40 See Joseph Margolis, *Persons and Minds* (Dordrecht: D. Reidel, 1978) and *Art and Philosophy* (Atlantic Highlands, NJ: Humanities Press, 1980).
41 Paul Ricoeur, *Interpretation Theory: Discourse and the Surplus of Meaning* (Fort Worth: Texas Christian University Press, 1976), p. 23.
42 In Gadamer, *Truth and Method*.
43 Roland Barthes, "From Work to Text," in Josué V. Harari (ed.), *Textual Strategies* (Ithaca, NY: Cornell University Press, 1979), p. 77.
44 See Michael Riffaterre, "Sémiotique intertextuelle: l'interprétant," *Révue d'esthétique*, nos 1–2 (1979); Harold Bloom, *The Anxiety of Influence* (London: Oxford University Press, 1973) and Stanley Fish, *Is there a Text in this Class?* (Cambridge, Mass.: Harvard University Press, 1980).
45 See Thomas S. Kuhn, *The Essential Tension* (Chicago: University of Chicago Press, 1977), Preface and chs 12–13.
46 Nietzsche, *The Will to Power*, section 1067.
47 Ibid., section 1052.
48 Signs of developing concern for accounts couched in these terms – among psychologists and sociologists, theorists of history, literary commentators and linguistic analysts – may be found (rather primitive and tentative signs, to be frank) in such recent studies as Jonathan Potter et al., *Social Texts and Context* (London: Routledge and Kegan Paul, 1984), for instance pp. 76–7; Jon Barwise and John Perry, *Situations and Attitudes* (Cambridge, Mass.: MIT Press, 1983), for instance pp. x–xii; Hayden White, *Tropics of Discourse* (Baltimore: Johns Hopkins University Press, 1978), for instance pp. 64–5; Rex Martin, *Historical Explanation* (Ithaca, NY: Cornell University Press, 1977), for instance ch. 4.
49 See Michel Foucault, "Two Lectures," *Power/Knowledge: Selected Interviews and Other Writings 1972–1977*, tr. Colin Gordon et al., ed. Colin Gordon (New York: Pantheon, 1980).
50 De Man, "Criticism and Crisis," *Blindness and Insight*, pp. 17, 19.
51 Ibid.
52 De Man, "The Rhetoric of Blindness: Jacques Derrida's Reading of Rousseau," ibid., p. 107 (italics added).
53 De Man, "Criticism and Crisis," ibid., p. 18.
54 Quine may perhaps stand proxy for the entire clan; see W. V. Quine, *Word and Object* (Cambridge, Mass.: MIT Press, 1960).
55 See ch. 9, above.
56 Adolf Grünbaum, *The Foundations of Psychoanalysis* (Berkeley, Calif.: University of California Press, 1984). There is a convenient summary of the argument of the book in Adolf Grünbaum, "Précis of *The Foundations*

of Psychoanalysis: A Philosophical Critique," *Behavioral and Brain Sciences*, IX (1986), accompanied by an extended Open Peer Commentary.

57 Grünbaum, *The Foundations of Psychoanalysis*, p. xi (italics added).

58 Ibid., p. xii.

59 This much at least accords with Davidson's concessions; see Donald Davidson, "Mental Events," *Essays on Actions and Events* (Oxford: Clarendon Press, 1980).

60 This much at least accords with Dennett's sense of the constraints on intentional ascriptions; see Daniel C. Dennett, *Content and Consciousness* (London: Routledge and Kegan Paul, 1969). It also catches up Feigl's worries about the so-called "many–many" problem – that is, the problem that any determinate physical or biological state may be associated (I should say, may "incarnate") indefinitely many alternative, unsystematizably linked intentional (or Intentional) states; and, alternatively, that any intentional (or Intentional) states may be associated with ("incarnated" in) indefinitely many alternative, unsystematizably linked physical or biological states; see Herbert Feigl, *The "Mental" and the "Physical": The Essay and a Postscript* (Minneapolis: University of Minnesota Press, 1967). See also ch. 9, above.

61 In a very real sense, this is the critical theme favored in the work of Paul Ricoeur, Hans-Georg Gadamer, and Jürgen Habermas – whom Grünbaum rather unceremoniously attacks for their failure to appreciate the complexities of the *physical* sciences; see Grünbaum, Introduction to *The Foundations of Psychoanalysis*. What Grünbaum says of them has some validity; but it hardly bears on what *he* himself fails to consider regarding the complication mentioned.

62 See ch. 8, above.

63 See Ricoeur, *Freud and Philosophy*; Roy Schafer, *A New Language for Psychoanalysis* (New Haven, Conn.: Yale University Press, 1976); George S. Klein, *Psychoanalytic Theory* (New York: International Universities Press, 1976). See also Robert K. Shope, "Freud's Concepts of Meaning," *Psychoanalysis and Contemporary Science*, II (1973), which Grünbaum cites. In analytic philosophy circles, of course, a decisive case had already been made out by Donald Davidson, "Actions, Reasons, and Causes," *Essays on Actions and Events*, although Davidson drew the illicit conclusion that explanations by reasons must be a species of explanations by causes (because having reasons is a species of causes). See further Joseph Margolis, "Actions and Causality," *Culture and Cultural Entities* (Dordrecht: D. Reidel, 1984).

64 Grünbaum, *The Foundations of Psychoanalysis*, p. 84.

65 Ibid., p. 75.

66 Ibid., p. 85. See also, p. 93f; pt III, ch. 10.

67 Ibid., p. 80, in the context of pp. 75–83.

68 Ibid., p. 17.

69 Ibid., p. 16. Grünbaum specifically cites here the views of Habermas, in *Knowledge and Human Interests*, tr. J. J. Shapiro (Boston, Mass.: Beacon Press, 1971), ch. 11; and Gadamer, in *Truth and Method*, p. 311.

70 Grünbaum, *The Foundations of Psychoanalysis*, p. 18.

71 Ibid., p. 17; see also pp. 19–20.

72 I take this to be rather close to the point of Lawrence Sklar's reflections on deciding whether time-reversal variant or time-reversal invariant laws accord with reality: "we know which set of generalizations truly describes the world because we know, independently of our knowledge of the lawlike behavior of physical processes in time, what the actual time order of events really is. Only this 'independent' knowledge of temporal order would allow us to decide which of the lawlike descriptions is, in fact, the true lawlike description of the world" – *Space, Time, and Spacetime* (Berkeley, Calif.: University of California Press, 1976), p. 402 (italics added). Ironically, a related conclusion can be drawn from Grünbaum's own well-known discussion of time, although Grünbaum, unlike Sklar, wishes to disjoin the anisotropy of physical time from the directionality of the "transient now" within "the psychological (common sense) context" of human existence – Adolf Grünbaum, *Philosophical Problems of Space and Time*, 2nd enlarged edn (Dordrecht: D. Reidel, 1973), p. 329. Grünbaum explicitly declares, "I maintain with [Hugo] Bergmann that the transient now with respect to which the distinction between the past and the future of common sense and psychological time acquires meaning *has no relevance at all to the time of physical events*, because it has no significance at all apart from the egocentric perspective of a *conscious* (human) organism and from the immediate experience of that organism. If this contention is correct, then both in an indeterministic *and* in a deterministic world, *the coming into being or becoming of an event, as distinct from its merely being, is thus no more than the entry of its effect(s) into the immediate awareness of a sentient organism (man)*" (p. 324). Grünbaum provides (pp. 322–3) a convenient translation from Bergmann's *Der Kampf um das Kausalgesetz in der jüngsten Physik* (Braunschweig: Vieweg, 1929), pp. 27–8. The following is also instructive: "The distinction between *de facto* and nomological irreversibility is dubious; the idea of law-like irreversibility seems to be internally inconsistent. In the absence of any clear demonstration of a physical law which is temporally asymmetric, we must take law-like to *imply* reversibility. All fundamental equations seem to be invariant under time reversal. ... The distinction between nomological and *de facto* irreversibility ... seems to collapse; the irreversibility and asymmetry ... can only come from contingencies of initial or boundary conditions. ... Entropic irreversibility cannot be explained at the present time in terms of time anisotropy, nor can it be presented as the sole indicator of time anisotropy" – Henry B. Hollinger and Michael John Zensen, *The Nature of Irreversibility* (Dordrecht: D. Reidel, 1985), pp. 70–1. Nevertheless, there appears to be a developing interest in theories that insist on the law-like directionality of entropic processes. Ilya Prigogine, for instance, states matter-of-factly: "Now that we are at the end of this century, more and more physicists think that the fundamental laws of nature are irreversible and stochastic, that deterministic and reversible laws

are applicable only in limiting situations," "Irreversibility and Space-Time Structure," in David Ray Griffin (ed.), *Physics and the Ultimate Significance of Time* (Albany: SUNY Press, 1986), p. 232. The entire volume repays attention.

73 For a sense of some efforts to develop a formal account of the tenses (the problem of anisotropic time) without full attention to the intentionality of what Grünbaum dubs "psychological time" (human time, historical time), see Robert P. McArthur, *Tense Logic* (Dordrecht: D. Reidel, 1976); also, A. N. Prior, *Past, Present, and Future* (Oxford: Clarendon Press, 1961).

74 This of course is one of J. J. C. Smart's concerns regarding the prospect of genuine natural laws – that is, laws that make no implicit reference to particular individuals (times, places, parts of the actual world). See J. J. C. Smart, *Philosophy and Scientific Realism* (London: Routledge and Kegan Paul, 1965). See also P. F. Strawson, *Individuals* (London: Methuen, 1959). Quine would eliminate the problem atemporally, by replacing all proper names, indexicals and the like by general predicates or indefinite descriptions uniquely singling out whatever we would otherwise refer to. But there seems to be no real-time or cognitively pertinent sense in which the program can be fleshed out; see Quine, *Word and Object*, section 37.

75 Cf. for instance J. L. Austin, *How to Do Things with Words*, ed. J. O. Urmson (Oxford: Clarendon Press, 1962), pp. 22, 26. Austin actually shared with me, in conversation, his own serious doubts about Wittgenstein's work.

76 For instance ibid., p. 22.

77 John R. Searle, "Reiterating the Differences: A Reply to Derrida," *Glyph*, I (1977).

78 Jacques Derrida, "Limited Inc a b c ...," tr. Samuel Weber, *Glyph*, II (1977), 206 (italics added).

79 Jonathan Culler, *On Deconstruction* (Ithaca, NY: Cornell University Press, 1982), p. 123 (italics added).

80 Austin, *How to Do Things with Words*, p. 147.

81 See John R. Searle, *Speech Acts* (Cambridge: Cambridge University Press, 1969), chs 2–3.

82 Ibid.

83 See Derrida, *Of Grammatology*.

84 See Derrida, "Différance," *Margins of Philosophy*.

85 Culler, *On Deconstruction*, pp. 123–4.

86 See Joseph Margolis, "'The savage mind totalizes,'" *Man and World*, XVII (1984).

87 Compare the holism of discourse about rationality and mind; see Davidson, "Mental Events," *Essays on Actions and Events*.

88 See Michel Foucault, *Discipline and Punish*, tr. Alan Sheridan (New York: Random House, 1979).

89 Culler, *On Deconstruction*, p. 128.

90 This of course is what we have taken as the defining feature of pragmatism.

91 Cf. Margolis, "'The savage mind totalizes,'" *Man and World*, XVII.
92 The thesis is strenuously favored by Hjelmslev: see Louis Hjelmslev, *Prolegomena to a Theory of Language*, rev. edn, tr. Francis J. Whitfield (Madison: University of Wisconsin Press, 1961), p. 9. See also the recent discussion along similar lines (however supple) in Dan Sperber and Deirdre Wilson, *Relevance* (Cambridge, Mass.: Harvard University Press, 1986), particularly chs 1–3. Though an improvement on H. P. Grice's views of implicature, Sperber and Wilson's account oscillates between the claims of system and the claims of incompletely systematizable context: see particularly ch. 3, section 7. Cf. ch. 10, above.
93 See ch. 8, above.

12

The Society of Man

One of the most disconcerting questions, at once naive and searching, asks whether a society is merely an aggregate of individual creatures or a distinct creature or entity of its own. There is no need to press the answer: the question is hardly straightforward. It is meant to be a reminder of the radically opposed directions in which theory attempts to address the profound complexity of the relationship between man and society. Part of the underlying puzzle concerns whether, within the scope of some postulated *sub*-societal domain, the phenomena of social existence can somehow be derived and explained; or whether, though they are genuine but not thus derivable, they constitute more or less distinct structures effecting (and facilitating the development of) individual lives. Also (hardly negligibly), if, in spite of the fact that social phenomena are real, societies are fictions of some sort, then, against conversational habit, we must reinterpret what we mean in speaking of the "relationship" between man and society.

Human capabilities are formed by an intersection of genetic endowments subject to natural and contingent development and the culturally specialized grooming of such developing endowments. The process is construed largely in terms of differences between generations: "prevailing" practices are taken as biologically incarnate in the behavior and mental life of a parental generation; and the lag between generations and the cultural innocence of human infants are linked, by theory, both with the conceptual distinction of cultural processes *vis-à-vis* the biological and the contingency and diversity of actual historical cultures. This is an intuition disputed and drawn out in indefinitely many different ways. What is most sane about it is that it poses in a relatively untendentious way the master questions of the human sciences: how to understand the relationship between (1) the biological and the cultural, (2) the genetic and the societal, and (3) diachronic natural processes and human history.

But of course it is also an intuition that favors culturally entrenched habits of description and analysis.

In an early phase of his career, Paul Feyerabend had advocated rather sharply that one should develop a theory about what purport to be mental events, "without any recourse to existent terminology."[1] Part of his argument held that statements in "common idioms" (in effect, natural languages) "are adapted not to *facts*, but to *beliefs*," and hence are inimical to, and trivially capable of refuting, scientific inquiry about our mental life where it radically departs from the orientation organized and controlled by those "common idioms."[2] The notion is not altogether distant from that of Sellars's well-known contrast between the "manifest image" and the "scientific image"; and Feyerabend's (and Sellars's) instruction has proved remarkably vigorous and influential – no longer perhaps in terms of a theory of meaning, more in terms of a perceived need for scientific rigor (which, ironically, Feyerabend would probably not share).[3]

Now, the line of demarcation between the utter elimination of the human and the mental and their gradual replacement, and between replacement and analysis, is notoriously difficult to draw or defend. One can hardly fail to proceed piecemeal or to insist on supportive arguments favoring either strategy. Furthermore, there is a compelling sense in which the human agency exercised in any science, in the use of any natural language, in the invention of specialized idioms as remote as you please from our "common idioms," binds us (at least provisionally, in a way we hardly understand how to escape) to the effective reality of our own cultural world. When, therefore, it is affirmed that description and explanation in terms of such a idiom "suffers explanatory failures on an epic scale, . . . has been stagnant for at least twenty-five centuries, and . . . its categories appear (so far) to be incommensurable with or orthogonal to the categories of the background physical science whose long-term claim to explain human behavior seems undeniable" – so much so that "any theory that meets this description must be allowed a serious candidate for outright elimination" – one can only wonder whether we could ever segregate the human and physical sciences or invent or apply an explanatory schema to either that was not fatally infected with the conceptual disease of our common idiom.[4]

I

Having said that, we may conveniently fix our bearings by reminding ourselves of certain extreme but well-known opposing convictions about the study of the human world. Noam Chomsky and Emile Durkheim will serve us well here. Chomsky declares, quite characteristically, "I see no reasonable alternative to the position that grammars are internally represented in the mind, and that the basic reason why knowledge of languages comes to be shared in a suitably idealized population (and partially shared in actual populations) is that its members share a rich initial state, hence develop similar steady states of knowledge."[5] This is disarmingly straightforward. Chomsky says that there are no promising alternatives to a strongly nativist theory of language – "linguistic theory (or 'universal grammar') is what we may suppose to be biologically given, a genetically determined property of the species": the variety of natural languages (he claims) is somehow generated from underlying structures that are themselves internalized infra-organismically; and yet, at the same time, he admits that the theory holds only for a suitably "idealized" population and that there is some (benign) empirical discrepancy between the performance of actual societies and what the idealization posits. He also says that, although "the elements of syntax are not established on a semantic basis, and ... the mechanisms of syntax, once they have been constructed, function independently of the other components of the grammar, which are interpretive components" ("the thesis of the autonomy of syntax ... is probably correct"),[6] the possibility that grammar is *not* independent of semantic and "nonlinguistic factors" (beliefs and attitudes, for instance) – regardless of however weak the early work of the so-called generative semanticists might have been – must be acknowledged, though it "must [also] be empirically motivated."[7]

So, in the first place, Chomsky concedes that there *is* a conceivable, empirically available alternative that (for all we know) might actually *be* empirically favored over his own nativism.[8] Secondly, Chomsky's thesis about the modular autonomy of the grammatical is a thesis said to be empirically defensible *with respect to* actual natural languages – that is, fitted by some idealization *to* the linguistic practices of a society in which semantic, pragmatic, nonlinguistic factors of all sorts are inextricably intermingled at the "surface" or phenomenonological level of social behavior and reflexive reporting. Thirdly, *if* the counterthesis were empirically favored, then Chomsky's concession, that linguistically apt populations only "partially share" a knowledge of the underlying generative grammar of their own language, would *no longer mean* what it apparently now means; for Chomsky could then no longer claim that "grammars are [entirely or basically or essentially] internally represented in the *mind*," in any sense congruent with his own genetic claims. Fourthly, in the absence of a reasonably complete theory of linguistic behavior (which Chomsky has never provided and which bears on the plausibility of his modular grammatical thesis), and in the face of the obviously social and historical nature of other large practices – for example, in the arts, in political and economic life, and in religion – in which a biologically modular (and therefore non-socialized) practice seems most unlikely, it is insufficiently convincing to claim to account for any natural social practice without providing a central formative role for socialized instruction *or*, alternatively, without explaining in a fuller sense why such instruction should not be critically invoked *at* the originating point at which linguistic structures are pertinently ascribed to the behavior of the members of a new generation. In short, given the obvious independence of the empirical regularities Chomsky has discovered with respect to natural languages and the biological (nativist) thesis he offers in explanation, the best one can now say is that it has *not yet* been shown that, for given natural languages, the syntactic *is* independent of the semantic and nonlinguistic.

Chomsky's theory manages to combine a remarkable variety of doctrines that need not be held together, that individually require quite different sorts of empirical support and that, joined as a single thesis, can hardly be supposed to be empirically supportable merely by summing over the strengths of its various parts. It is a theory that is at once

1   functionalist, because it opposes the elimination of the cognitive and linguistic powers of human agents identified at the level of our "common idioms," the "manifest image," "folk psychology" and the like, and because it construes the mental (token-wise but not type-wise or by way of reductive identities or reductive analysis) as physical;[9]

2   nativist, in that it treats the language "faculty" as biologically determined and fixed in a species-specific way;

3   modular, in the sense that the putatively innate language faculty (by analogy, perhaps the faculties of perception and thought as well) is really a functionally ordered network of specific, limited, relatively autonomous, plural faculties that, ideally, cooperate to facilitate normal (linguistic) performance at the molar level of our "common idiom"; and

4   opposes, as such, certain empirical options themselves compatible with weaker variants of doctrines 1–3 – first, the option that the underlying syntactic structures of natural languages may be artifactually and variably generated by, may be subject to change because of, and may be internalizable in accord with, the changing experience and changing ways of handling the experience of the members of actual societies,[10] and, second, the option that the

underlying structures of natural languages are causally due (at least in part) to interpretive and consensual societal processes that are collective and not reducible to the terms of the merely mental, that infect in a structurally pertinent way whatever apt individuals do produce in performing in the socially acquired ways they do.

Doctrine 1 is open to empirical challenge in that the incarnatist option has been shown to be a viable and pertinent alternative to functionalism.[11] Doctrine 2 is open to empirical challenge in the sense that the language "faculty" may prove to be genetically much less determinate, or innately much more incipient than Chomsky admits, may even depend on deeper modular faculties of a prelinguistic sort, and may even entail more general *non*modular (or "horizontal") faculties at the innate level[12] – or may depend on modular but *not* innate habits or the like that themselves depend on "horizontal" uses of innate (but "shallow") capacities. And doctrine 3 has been regularly challenged on the grounds (which Chomsky himself acknowledges to be empirically eligible) that the syntactic (taken in the strong sense Chomsky favors) is not or is not demonstrably modular or is not independent of semantic and nonlinguistic determinants (that themselves are not innate). These considerations raise very complex questions that would require more patience and more labor than we can afford here. But they cannot be altogether ignored, and it may be possible to meet our needs here without affording a full account of the language "faculty."

In particular, the quarrel with Chomsky's nativism need not be construed as a disjunctive choice between affirming or denying an essential role to nativist competences. Admitting the unavoidability of the general nativist thesis, one may still dispute whether it must be modular or "horizontal" (broad-purpose) or both, whether it may be "shallow" or must be "deep," whether it must include the linguistic or at least the syntactic with respect to natural languages. But, apart from these relatively specialized, determinate contests *within* the nativist's own field, it is entirely open to us to insist, for reasons of methodological scruple, that *all* proposals of innate *linguistic* structure (as opposed, say, to innate sensory structure) must be dependent on the vagaries of fitting putatively deep structures *to* what we take ourselves (reflexively) to discern at the functional "surface" of actual linguistic behavior. We cannot (as we can and do in theorizing about sensory perception) appeal to known biological structures of specially designed organs. It is the presence of specific such organs that ensures the reasonableness of postulating fixed innate modules of sensory (or sub-sensory) processing essential to molar perception. In the analysis of language, we lack

independent evidence of the requisite "organs"; hence, we are obliged
to base our projections of innate syntactic structure on the skill of our
theorizing about probable deep grammars glimpsed through reflexive
analyses of linguistic behavior.

Clearly, this double empirical weakness has not dissuaded Chomsky.
But reflecting on how it bears on the nativist claim, we may insist that,
if many competing (but always idealized) grammars may be fitted to
distributed sentences that are themselves "selected" or idealized from,
or invented as reasonably conformable with, the complex play of actual
discourse – if these may be fitted by way of independently *parsing* such
sentences, *not* by way of analyzing how we actually first *access*
(cognitively) the torrent of moving language at the molar level at which
we encounter it – then the very "surface" of language may not really
yield a sufficient fixity of structure to justify positing, in the realist's
sense, a fixed innate structure by which it "must" have been generated.
There may be a profoundly heuristic feature in theorizing about deep
grammars that the nativists fail to appreciate: first, because they fail to
see that *their parsing* depends methodologically and epistemically on the
precision with which we can analyze *our accessing* of language; second,
because they fail to see that *what* we access may be extremely complex,
not entirely or even primarily captured by the uttering of sentences,
perhaps not even primarily focused on the grammar of isolable sentences,
certainly not analyzable in the flux of speech as yielding regularities
stable enough *there* (not among the hothouse sentences culled from it)
to confirm an innatism as fierce as Chomsky's. The fact is that Chomsky's
argument presupposes the necessity that the grammar of natural language
is a natural system – that is, that deep grammar (not language) must be
a system and that it must be a natural system.

The empirical regularities Chomsky is rightly admired for having
systematized are hardly incompatible with, or even (against his own
insistence) unfavorably disposed toward, the adjusted alternative options
broached just above regarding doctrines 1–4. Again, the famous thesis
of "the poverty of the stimulus"[13] (that is, the absence of empirical
stimuli sufficient to account for the rapid and apt mastery of language
by infants) is extremely difficult to assess in general, and even more
difficult to assess in terms favoring or disfavoring one or another option
with regard to Chomsky's reading of doctrine 3 – the most strenuous
of his doctrines, the one most explicitly opposed to the two options
mentioned with respect to doctrine 4. The trouble, as we have implicitly
remarked, is that Chomsky has never directly explored one absolutely
decisive range of linguistic phenomena – on which the empirical standing
of his own reading of doctrines 3–4 depends: that is, the actual behavior

by which apt speakers may be said to *access* the utterances of others *as* linguistic, *as* spontaneously (putatively) activating (internal) linguistic analysis or parsing. Broadly speaking, Chomsky's theory is addressed to activities of *parsing* sentences already accessed – sentences the meanings of which we are already more or less reasonably clear about, as apt speakers.[14]

If we insist, however, that, in order to validate Chomsky's own thesis, we need an account of human *cognizing* functioning at the surface of linguistic communication – so that the cues there accessed may be taken to be processed at some deeper "cognizing" level, innately perhaps – then the peculiar restrictedness of Chomsky's otherwise impressive work is clear at once. The point is neatly put by Robert Cummins, however allusively and however much Cummins intends to service a theory of an altogether different sort. As Cummins remarks, "whereas inferential characterization requires sentential interpretation of inputs and outputs [read this to mean that the 'inputs and outputs' are keyed to a run of linguistic behavior not antecedently known to have a specific sentential structure], the converse is not true: a device instantiating a transformational grammar transforms sentential inputs into sentential outputs, but the inputs and outputs are not related via any pattern of inference."[15] At the *accessing* level, we may say, our molar cognizing abilities address the *motley* of language inputs and outputs; at Chomsky's *parsing* level, our "cognizing" abilities address the fixed grammar of carpentered sentences. The question remains: how do we get from here to there? *What* is the "inferential" pattern that justifies "transforming" the "inputs and outputs" of the first into the "inputs and outputs" of the second? Chomsky never says. If we view matters in these terms, then it is even possible to challenge the full pertinence of the "poverty" thesis: for, it may be claimed that there *is* no distinct "poverty of the stimulus" at the level at which we cognitively access actual linguistic behavior; and the poverty alleged at the parsing level is itself an artifact of having implicitly supposed that, at that level, we discriminate cues linked to the idealized sentences we *must* be parsing. So one part of the theory ensures the plausibility of the other, while the essential condition for both is ignored.

Chomsky entertains, it is true, questions about the *acquisition* of a first language – but essentially *only* insofar as the performance of growing children permits us to map the developing congruity between their own developing mastery *and* the exemplary performance of apt adults. In doing that, Chomsky does *not* directly explore the problem of accessing language – which would oblige us to address empirically the modularity/nonmodularity issue, the issue of relatively "shallow" and "deep" modules of an innate sort, the possibility of non-innate modules, and the two complications mentioned under doctrine 4. Understandably, Chomsky interprets the congruity between infant development and adult performance in accord with his doctrine of innate competence; but, in whatever way the alternative options suggested could be fitted to adult performance (in effect, Chomsky's parsings), it is very likely that they would not encounter any decisive further difficulty in accommodating infant development. For, there, Chomsky is theorizing in a way quite analogous to that of Jean Piaget's "genetic epistemology"[16] – though of course with entirely different (in fact, opposed) substantive goals.[17] That is, Chomsky is theorizing, *from* the vantage of parsing *sentences* generated at the level of adult performance, *about whatever*, at the infant level (otherwise inaccessible), might account for the developing congruity already anchored (interpreted) *at* the adult level – in terms of the innatist theory. It is true that Piaget's theory lacks an empirically explicit, satisfactory account of the actual dynamics of cognitive development. Piaget offers only a formal succession of stages. Chomsky's theory betrays a similar limitation. But the strength of Piaget's proposal lies elsewhere. For, although he has curious structuralist pretensions (Chomsky does not) Piaget theorizes that the developing cognitive capacities of a human infant are a function of its active intervention in the world – so that, chiefly by what he calls "assimilation" and "accommodation," the child's developing experience and activity yields internalized, novel cognitive aptitudes that cannot be analyzed entirely in terms of a fixed set of innate "cognizing" capacities of any sort.[18] It is Chomsky's (and Fodor's) contention that this is "impossible" or "unbelievable." But Chomsky fails to demonstrate why this must be so (quite apart from the formalism of Piaget's account); and, as we may suspect, the force of Chomsky's contention rests entirely on the prior entrenchment of his own sort of innatism. Furthermore, it is just this sort of thinking that suggests the relative inflexibility (and implausibility) of Fodor's and Chomsky's conceptions of science.

To be sure, there are bound to be empirical considerations (the "poverty of the stimulus," for instance) that do bear on the processes of infant first-language acquisition. But there is no satisfactory specific way to assess "poverty" *independently of an empirically ramified account of how an adult speaker accesses linguistic utterances as linguistic utterances*: in the infant case, the "poverty" thesis is hardly more than a black-box confession – but in Chomsky's hands it seems to do much more empirical work than it should be permitted to do.

There is, therefore, an undeniable lacuna in Chomsky's important and influential argument. For instance, when, parsing a number of his usual splendid specimen sentences, he emphasizes (as he invariably does) the tacit, spontaneously operative, consciously unmonitored grammatical knowledge that we must call into play in using such sentences, Chomsky is likely to insist,

> We know these facts without instruction or even direct evidence, surely without correction of error by the speech community. ... This is knowledge without grounds, without reasons or support by reliable procedures in any general or otherwise useful sense of these notions. ... It seems there is little hope in accounting for our knowledge in terms of such ideas as analogy, induction, association, reliable procedures, good reasons, and justification in any generally useful sense, or in terms of "generalized learning mechanisms" (if such exist ...). We should, so it appears, think of knowledge of language as a general state of the mind/brain, a relatively stable element in transitory mental states once it is attained; furthermore, as a state of some distinguishable faculty of the mind – the language faculty – with its specific properties, structure, and organization, one "module" in the mind.[19]

But the inadequacy of psychologically internal general mechanisms of the largely inductive sorts Chomsky mentions (which we may concede) is hardly decisive regarding Chomsky's own modular innatism. There simply are other sensible possibilities. One might almost say there is something deliberately obtuse about Chomsky's way of obscuring the lacuna:

> Surely there is some property of mind P that would enable a person to acquire language under conditions of pure and uniform experience, and surely P (characterized by UG [universal grammar]) is put to use under the real conditions of language acquisition. To deny these assumptions would be bizarre indeed. It would be to claim either that language can be learned only under conditions of diversity and conflicting evidence, which is absurd, or that the property P exists – there exists a capacity to learn language in the pure and uniform case – but the actual learning of language does not involve this capacity.[20]

There *is* "some property of mind P" that enables a person to acquire language; but, under the circumstances Chomsky emphasizes (the "poverty" matter), that hardly leads us to UG as the uniquely possible or uniquely plausible option. (UG, Chomsky says, is "a characterization of the genetically determined language faculty," "'a language acquisition device,'" "an innate component of the human mind that yields a particular language through interaction with presented experience."[21]) Chomsky never really acknowledges that the "presented experience" thus accessed *may* be socially structured *and* socially decoded; he never acknowledges that humans *may* be innately preformed to access (at least in part) *whatever* is *socially* thus prepared and presented; he never acknowledges that *what* we (allegedly) ideally parse as instances of grammatical invariants *may* be so parsed as an artifact of cultural grooming at the very level of acquisition (or at the level of accessing the utterances of others – *by* consensualizing forces that operate at that level only or through psychologically internalized (*not innate*) "modules" dependent on the relatively stable habits of interaction functioning in one's own society.

"Generative grammar," Chomsky declares, "limits itself to certain elements. ... Its standpoint is that of individual psychology. It is concerned with those aspects of form and meaning that are determined in the 'language faculty,' which is understood to be a particular component of the human mind."[22] But how does this admittedly powerful (narrow) possibility lead to the conclusion that "there is no known alternative that even begins to deal with the actual facts of language"?[23] The finding is distinctly arbitrary and insensitive to the possibility that the *grammar* of natural languages is, at least in part, artifactually idealized, societally formed, formed in accord with whatever *sub*-linguistic innate competences we may possess by which we access *whatever* is thus societally formed and consensually reinforced. We may not even parse what we thus access – in the psychologically internal and idealized way Chomsky details; and what we do parse internally *we may* (stretching terms) *be said to parse "modularly" even if not by the use of innate grammatical competence*. We may do so (by internalizing socially acquired skills) in a way that is essentially linked to looser "inductive" processes favoring not much more than a salient conformity with consensually reinforced behavior. (This is meant only as a sketch of a counterstrategy, not as a full-blown alternative.) To be sure, the strongest empirical evidence in favor of such a thesis rests with analogies that may be drawn from other complex societal practices that no one really believes can be accounted for in terms of a strong modular innatism – the species-wide variations of painting and music and clothing and food preparation and religion for instance. But, without begging the question, why shouldn't we admit that it is an eligible, coherent, entirely plausible, and pertinent speculation? Chomsky never says.

We are touching, here, on an extraordinarily difficult question. It is also largely an invisible question. More accurately put, it is a question often pursued without genuinely grasping what is at stake. Chomsky's thesis is an account of linguistic aptitudes among humans, *within the following constraints*:

(i)   that language is species-specific;
(ii) that it is naturally acquired;
(iii) that it has determinate universal structures (species-wide) suitable for accounting for linguistic performance and acquisition;
(iv) that language (ideally) forms a system, in the sense (a) that the underlying structures postulated are incarnate in every natural language and (b) that the variant forms of the different natural languages are due to causal (law-like) interactions between those structures and the environing world and count (ideally) as alternative instantiations of an invariant set of rules adequate for generating the essential structures of those languages; and
(v)   that the governing mechanism for ensuring the species-wide distribution of the underlying structures is genetic, infra-organismic, confined (in a generous sense) within the terms of individual psychology.

II

The invisible theme of Chomsky's account is that language forms a *system*. Every resistance on Chomsky's part to an alternative account opposes any challenge to the thesis that language is a system. Hence, if language were not an autonomous mode of functioning, not a "module" within the range of human aptitudes, not a system of interfacing innate modules, Chomsky's claim would be threatened. This is the reason why Chomsky segregates the syntactic from the semantic and the nonlinguistic; segregates the psychologically generated from the socially interpreted; insists on construing UG as fixed and innate; denies (notably, against Piaget) the possibility of acquiring emergent cognitive aptitudes as a result of contingent interactions between organism and environment.[24]

This is also why Jerry Fodor, also adhering to the "system" thesis and borrowing Zenon Pylyshyn's trichotomous model of information-processing (transducers, input systems, and central cognitive processes[25]) maintains,

1   "the character of transducer outputs [say, an acoustical array enabling linguistic communication] is determined in some *lawful* way, by the character of impinging energy at the transducer

surface; and the character of the energy at the transducer surface is itself *lawfully* determined by the character of the distal layout [hence, is itself a modular system]";[26]

2 "*input systems* [say, Chomsky's UG applied to such an acoustical array, or mechanisms for processing of visual perception relative to 'the class of possible retinal (transduced) outputs'] *are modules*" and (such) "modular cognitive systems are domain specific, innately specified, hardwired, autonomous, and not assembled [that is, not assembled contingently as a result of interactions between organism and environment]";[27] and

3 "central systems" (the "systems" that permit inductive uses of the output of input systems, that bring background information to bear, that fix the cognitive beliefs of molar persons, that facilitate linguistic utterance and the like) are "relatively nondenominational (i.e., domain-*in*specific) psychological systems which operate, inter alia, to exploit the information that input systems provide ... [are also] *un*encapsulated, and (hence) not plausibly viewed as modular."[28]

"Central systems" remain systems, presumably, because they are parasitic on the information provided by input systems, because they behave in law-like ways, because they do not involve cognizing abilities not explicable in terms of the innate cognizing capacities of the species, and because the output of input systems is not "cognitively penetrable" to their operations – a point Pylyshyn was at some pains to ensure[29] – that is, because input systems function only and always in a "mandatory" way, are "information encapsulated," are immune as such to "feedback."[30] There is a great glut of conceptual apparatus here, certainly empirically unconfirmed as far as *language* (as opposed to sensory perception) is concerned. The empirical argument for applying it to language rests chiefly with tenuous analogies to what do appear to be modular sub-systems of sensory processing. It is alleged, however, that it rests on the "poverty of the stimulus" and the remarkable fit between UG and the idealized sentences we parse. But the truth is that the latter maneuver is relatively suspect – because, after all, it is itself part of the conclusion wanted. Also, behind the scenes, the ulterior quarrel concerns whether language is or is not a system – which, if not favorably settled, would effectively rule out Fodor's wholesale use of the cognitive scientist's contraptions.

The important pressure points – against Chomsky (and Fodor) – are easily collected. First of all, it is a deep-seated conceptual prejudice (an entirely understandable one) that every "domain" saliently so identified for sustained scientific inquiry forms or must form a *system*. The presumption has been classically formulated by Louis Hjelmslev, admittedly in order to service an adjusted Saussurean structuralism. Nevertheless, in following this line, Hjelmslev features just the maneuver all *functionalist* theories must adhere to if their sponsors are to bring the relevant sciences (biological, psychological, and social) into minimal accord with the unity model, however loosely or generously conceived. The notion of "system" among the physical sciences is (as we have already briefly noted) open to challenge if causality does not entail nomologicality and if the "domain" of natural phenomena does not behave homonomically. Among the *functionally* defined disciplines, Chomsky's and Fodor's in particular, the extension of the unity model cannot but be baffled if the functional or informational phenomena at stake are not anchored in nomological regularities *and*, *qua* functional (*qua* abstractly specified), such that they do not form a rule-like, generative, transformational, structurally ordered *system* (apart from not violating any fundamental causal laws). The matter is quite complicated. Chomsky explicitly says, "The rules [of grammar] are not laws of nature, nor, of course, are they legislated or laid down by any authority. They are . . . rules that are constructed by the mind in the course of acquisition of knowledge. *They can be violated*, and in fact, departure from the rules can often be an effective literary device."[31] Certainly, *if*, among the physical sciences, natural laws are not invariant or are invariant only by an idealizing effort on our part, in accord with the requirements of canonical explanation,[32] or if the causal agency of humans does not require that causality entail nomologicality,[33] or if the admission of the causal efficacy of the psychological cannot be reductively managed in accord with a homonomic model of the physical sciences,[34] then it may well be impossible to treat the natural world as a *system*. And certainly, *if* the physical world is not a system, then there can be no compelling reason for supposing, as Chomsky and Fodor do, that "natural" language (and "natural" thought) similarly form systems – now, not of a law-like causal sort but of a rule-like, syntactic sort.

Hjelmslev maintains, "*A priori* it would seem to be a generally valid thesis that for every *process* there is a corresponding *system*, by which the process can be analyzed and described by means of a limited number of premisses. It must be assumed that any process can be analyzed into a limited number of elements recurring in various combinations."[35] Hjelmslev intends his thesis in a realist spirit, though he concedes that the natural difficulties of analysis may require a certain tolerance for the heuristic. But there is no *conceptual* barrier against denying the ubiquity of system. One supposes the matter to be empirical. In particular, if the strongly innatist and modular thesis Chomsky and Fodor support proved to be false – *or* proved not to be sufficiently strong to preclude a non-innatist theory of language – then, given *their* general orientation (opposed, say, to that of the structuralist proper), *we would be unable to preclude functionalist (or incarnatist) conceptions of language that were not characterizable as systems.* In fact, given the relatively strenuous nature of the innatist thesis, such alternatives might well be preferred. Of course, a Wittgensteinian model of language would be just such a model; but, to motivate its adoption, one would have to show that certain pertinent challenges could be made to Chomsky's theoretical account – without at all disturbing the empirical regularities Chomsky has collected.

It may be useful to remark, by way of a small detour, that Robert Cummins, who, as we saw a moment before, very tellingly challenges Chomsky's formalism – the absence of attention, on Chomsky's part, to the cognizing powers involved in empirically accessing a language – himself sketches the nature of *functional properties* in a most unfortunate way. Cummins rightly refuses to restrict science to explanations concerned only with "subsumption" under causal laws: both because covering laws need not invariably be causal laws and because science may proceed by way of "analyzing" functional systems and functional and other properties.[36] Quite so. But Cummins goes on to insist that "psychological phenomena are typically not explained by subsuming them under causal laws, but by treating them as manifestations of capacities that are explained by analysis."[37] This is misleading because (on grounds supplied earlier on), it may be that psychological phenomena are open to explanation under causal laws (if reductionism be viable), are open to causal explanation even if there prove to be no invariant laws to be had, and enter in any case into causal explanations of other psychological phenomena.

But it is Cummins's notion of "analysis" that is particularly pertinent to our present concern. His intuitive account runs as follows:

> Many scientific theories are not designed to explain changes [that would call for causal subsumption] but are rather designed to explain properties. The point of what I call a property theory is to explain the properties of a system not in the sense in which this means "Why did S acquire P?" or "What caused S to acquire P?" but, rather, "What is it for S to instantiate P?", or, "In virtue of what does S have P?"[38]

On Cummins's view, "Analysis of a *system* [may] be called compositional analysis, to distinguish it from analysis of a *property*, which [may] be called functional analysis when the property is dispositional [that is, a property 'specified' but not 'analyzed' in 'transition theories' – theories that 'explain changes of state in a system as effects of previous causes' – hence a property specified (or indexed in some way) in 'laws of causal subsumption'], and property analysis when the property is not dispositional."[39] It is clear that Cummins takes it that the explanation of all functional properties (effectively, dispositional properties – including psychological and language-processing properties) *are* concerned with law-like changes – hence, are involved with systems. The "functional analysis" of a (dispositional) property need not actually invoke the laws of the system within which particular capacities obtain; but it may always be brought into accord with a "system analysis."[40] Cummins insists that

> Ultimately, of course, a complete property theory for a dispositional property must exhibit the details of the target property's instantiation in the system (or system type) that has it. *Analysis* of the disposition (or any other property) is only a first step; *instantiation* is the second. Functional analysis of a capacity C of a system S must eventually terminate in dispositions whose instantiations are explicable via analysis of S.[41]

Also, "*Any* functional analysis can be expressed in a program of flow-chart form [and] *any* problem of interpretive analysis [say, ascribing symbolic or semantic function to some physically specified functional property – assigning sentential meaning to an initial acoustical array] can be thought of as a symbol programming problem, since every functional analysis can be expressed in program or flow-chart form, and interpretive analysis specifies its capacities as symbolic transformations."[42]

Cummins's account serves as an example of a widespread prejudice. *If* natural, psychological, linguistic, and cultural domains need not be systems at all, then it is both futile and hardly cricket to define functional properties *tout court* in terms of systems – in fact, in the double sense of "system," in virtue of which

(i)   functional properties are invariably linked with systems exhibiting changes of state under invariant causal laws; and

(ii)   functional properties are themselves invariably capable of being analyzed without remainder in terms of systems of discrete states conforming to an orderly flow-chart program – that is,

clothing, kinship arrangements, music, painting, religion, counting); and extensionally, by way of ever simpler compositional sub-states and sub-sub-states.

The issues raised by (i) and (ii) are quite distinct, even independent at times; but the claims they make are certainly empirically dubious.[43] For example, that a programmed, physically instantiated computer accords with (i) and (ii) and with the priority of the physically specified "description" of its functional properties over the "interpretive" analysis of *those* properties does not show at all that every "natural" phenomenon – the activity of people speaking a natural language, for instance – *can* be analyzed convincingly in accord with (i) and (ii) *or* in accord with the priority of the descriptive over the interpretive. On any reasonable view, the "description" of the acoustic array of any language – the mere fixing of the admissible array itself – is indissolubly dependent on the "interpretation" of that language.[44] If, in addition, the acoustical patterns are embedded in contingencies linked to the changing historical experience of a people (to the multiple forms of the so-called "graduality" of natural language), and if the grammar of a language is similarly so embedded, then it is a foregone conclusion (as Chomsky himself admits is possible) that natural language cannot be a system in the sense required. Computers are systems (in the double sense given): both by definition and by effective invention. But whether "natural" cultural networks are systems (in the same double sense) – in particular, natural languages – is an empirical claim to which we are hardly driven for conceptual or empirical reasons. If, despite their ubiquity, music, painting, cooking, religion and the like do not form systems, then language need not form a system either, though of course it may.

We were considering how to challenge Chomsky and Fodor, at the point of making this small detour; and we had noticed how both appeal to the notion of a system and the notion of modular sub-systems within some encompassing system. Generally speaking,[45] the notion of innate modules (of a functional sort) is reasonably invoked at the explanatory level if

(a) there are distinct organs specialized for the mode of functioning in question (for instance, highly structured eyes, for modular visual discrimination);

(b) the phenomena to be analyzed obtain generically and naturally in a species-wide way with wide cultural variations (certainly, for instance, the phenomena of language, food preparation, clothing, kinship arrangements, music, painting, religion, counting); and

(c) there is a noticeable rapidity and aptitude in mastering the relevant practices in ways that accommodate "the povery of the stimulus" and exhibit a measure of cross-cultural convergence (as of course Chomsky's grammatical thesis affirms).

Now, the obvious fact is that, despite Chomsky's odd usage (in fact, more in accord with Fodor's terminological preference),[46] there is no known linguistic "organ" in the human body.[47]. Chomsky's theory, therefore, must rest on evidence in accord with (b) and (c). We must bear in mind, for example, that there does not appear to be any empirically promising system collecting the "syntax" or "grammar" of clothing, painting, music, social organization or the like – in spite of the double fact that all such practices are species-specific and species-wide and that all depend (as does language) on innate modular capacities (or competences) *of the sensory sort.* So it is hardly unreasonable to challenge the generalized modularity thesis or the modularity thesis as it may be specifically formulated for human language. (Needless to say, the structuralist's alternative focuses exclusively on feature (b) of the above tally and is fatally content to supply a "totalized" scheme of possibilities with respect to (b) – as in Lévi-Strauss's kinship system.[48] Also, on the usage favored by Fodor and Pylyshyn, "systems" such as those of clothing, music, and painting could be said to be "cognitively penetrable," reasonably orderly, naturally learnable, and naturally recognizable – that is, "Wittgensteinian." So the denial that language is organized as a system of innate modules is neither conceptually incoherent nor empirically outlandish.)

So much for the notion of system in general. If (to broach the second of our pressure points) natural domains – in particular, language – need not be systems, we need no longer construe modules as systems (or solely as systems). Innate modules may well be fixed systems; but we may also allow that, in *internalizing* social practices, individuals effectively *modularize* (in stable but contingent, diachronically changing ways) psychological aptitudes for responding smoothly, even improvisationally, to large ranges of socially formed and socially reinforced stimuli. This, for example, would be an entirely plausible way of accommodating Wittgenstein's "forms of life" and Bourdieu's "*habitus*" – to mention the most salient options. Clearly, we should, in so speaking, be departing from Chomsky and Fodor's usage. But why not? *Modules* would no longer need to be (only) autonomous systems, hardwired, "informationally encapsulated," or "cognitively impenetrable." They could, more or less as internalized habits, be a kind of provisional self-programming, something approaching a kind of software.[49] They would not actually *be* programs or software, because, once again, they would not be systems. But they could be open "networks," "schemata" of various kinds *for which* (in Wittgenstein's sense) apt agents (speakers in particular) would be able to formulate provisionally apt "rules."[50] On the other hand, if the terminological adjustment were refused, then we should simply have to introduce a *fourth* kind of mental structure (to be added to transducers, input systems, and general processes) – namely, the "internalizing" (habituating or "modularizing") structures just sketched. The adjustment would have the virtue of accommodating the coherent conceptual option Piaget, for instance, has regularly proposed (that Chomsky and Fodor regard as impossible[51]) and of accommodating the further possibility that, *if* such contingent powers may be formed *in* acquiring social experience, then (contrary to Fodor's strong objection) genuine, emergent forms of learning would have to be admitted.[52] The option is also hospitable to current opposition to essentialist and foundationalist conceptual strategies, and even to the various lines of theorizing about the social formation of the self.[53]

Furthermore, the admission of such a contingent and alterable "modularizing" capacity (and tendency) exposes the deep mistake embedded in Fodor's (and Pylyshyn's) worries about the "cognitive penetrability" of (their) modular input systems. For now we see that the input systems are not themselves *cognitive* systems at all (systems that actually produce or fix *beliefs*) but only the functional *sub*-systems (if modular systems they be) *of* whatever (molar) processes ("central processing systems," on Fodor's usage) are involved in the fixation of beliefs. In a word, the input systems process "information" (not beliefs) *assigned (as information) to modular sub-processes because those processes pertinently enter in a factorial way in the formation and fixation of beliefs.* But, if so, then of course input systems are *cognitively* impenetrable in the same sense as the movement of red blood corpuscles through one's body is "volitionally impenetrable" when one decides to raise one's arm. The "information" (assigned) at the retina, in the formation of visual beliefs, is not itself a belief but only a functional (informational) component thought to contribute to the explanation of visual belief: the information is "impenetrable" because there is no cognitive interface between molar and sub-molar "cognitive" functioning. There simply can't be – for logical reasons.

Once we see matters in this way, then, it is clear that the quarrel between Chomsky and Fodor, on the one hand, and ourselves, on the other, rests primarily with querying how "shallow" or how "deep" are the innate modules of language or of comparable practices. It is not a question of dismissing innate modules altogether. In admitting *non-*

innate (more-or-less) modularized thinking internalized in the process of living within a natural society, we effectively press Chomsky and Fodor's particular innatist claims to more "shallow" levels then they favor: conceivably, we might admit only incipiently linguistic or even sub-linguistic modules. The issue is an empirical one. But (to broach the third of our pressure points) the resolution of the quarrel depends essentially on *just how strong are the claims regarding the universality of the innate modular structures underlying the species-wide (central) practices in question.* If Chomsky is correct in holding that UG *is* strongly determinate and universal among natural languages, then the innatist claim is clearly vindicated: its scope would be proportional to the detail empirical confirmation would reveal. But, if UG (or large parts of UG) were artifactual indealization of some kind fitted, reasonably smoothly, to (molar) linguistic behavior, then some concession favoring our "fourth" (psychological) ingredient (against innatism) would have to be made – and its importance would similarly be proportional to the kind of detail we found ourselves unable to interpret in the innatist's favor. Contingently modularized linguistic habits *would* of course be "cognitively penetrable." Neither Chomsky nor Fodor offers the least argument against that possibility – either conceptual or empirical. Furthermore, the option very neatly accommodates cultural divergence, genuine learning, smoothly internalized habits of core performance, diachronic changes in such habits fitting changing experience, and whatever may prove to be the hardwired information-processing competences of well-confirmed biological modules that provide the sub-functional functions of our (molar) cognitive aptitudes. Certainly, Chomsky and Fodor fail to allow for the conceptual contingency that a strongly innatist account of language either fails or succeeds as the artifact of a carpentered idealization of actual language. The problem of accessing language in a particular cultural space – the "frame" problem in all its variety – *cannot* be accounted for in terms of *input systems*, simply because (1) input systems process "information" and not "beliefs," or are factorily assigned doxastic import because of their informational contribution to the formation of beliefs, and (2) the contextual features affecting the accessing of natural language depend to a significant extent on consensual practices of a societal sort that cannot be completely internalized psychologically.

We may now collect a very important strand of the argument by reminding ourselves that a theory of language must include a theory of accessing linguistic behavior – in addition to including a theory of first acquisition and parsing (which, as we have seen, Chomsky almost exclusively favors). Accession features the apt performance of a native speaker regarding both the analysis or interpretation of a stream of actual discourse and the production of punctuated contributions to it to which others may similarly respond. Now, this dual performance is complicated by two (interrelated) considerations at least. In the first place, if a significant part of the dominant surface structures of a given natural language – acoustical, semantic, syntactic, pragmatic – are subject to contingencies of partial change due to the collective experience of a given society (possibly not completely systematic at any moment, though to some extent regularized), then it is impossible to prise the theory of acquisition and parsing apart from the theory of accessing language. Secondly, if a significant part of the accessible surface structures of a natural language, including the contingencies just mentioned, are not and cannot have been internalized psychologically by every apt speaker, then it is impossible to explain the phenomenon of natural language in terms restricted to individual psychology – *a fortiori*, in terms restricted to an innatist psychology. Thus, even if a phonetic change occurs in a strongly regularized way over time – never all at once – the effective accessing of a given language will require an adjusting familiarity with the pattern as it changes. Henry Hoenigswald reports, for instance, that "Proto-Semitic *p*, reconstructed phonetically as a bilabial stop, is in Arabic replaced by a labiodental spirant, *f*. This alteration has left all contrasts intact, and it has not introduced new contrasts."[54] Hoenigswald also acknowledges "successive sound changes" that, only after a series of stages, preserve larger isomorphisms recognized in a given language.[55] All sorts of such "gradualities" are bound to obtain in a natural language.

If we imagine acoustical changes that are not so regular, that may be due to certain historical encounters between distinct peoples effecting small, numerous, somewhat discrete "islands" of influence still sufficiently regular to be managed by apt speakers, if we imagine deeper changes of such sorts affecting semantic *and* even syntactic features of a given language, if such changes are sufficiently characteristic of complex languages developing among shifting, interpenetrating, mutually influential but distinct populations, then it becomes perfectly clear that Chomsky's theory would be severely undermined on empirical grounds. There would be no way to segregate surface and deep structure *in the innatist direction* and there would be no way to segregate parsing from accessing, methodologically. The interesting thing is that Chomsky's idealizations do not register such historical developments or account for their dynamic role.

One can well imagine that the formation of pidgins before full creoles evolve might exhibit such patterns. Robert Hall, for instance, commenting on the spread of pidiginized English (and other European languages) in the South Pacific, notes the white man's characteristic lack of interest in acquiring native languages and his automatic assumption of intellectual superiority. He adds, with regard to managing the halting efforts of natives to speak English (or another such language),

> So the European would conclude that it was useless to use "good language" to the native, and would reply to him in a replica of the latter's incomplete speech, adding also some of the patterns of baby-talk commonly used by mothers and nurses in his own country. The aboriginal not knowing any better, would assume that this was the white man's real language, and would delight in using it. . . . He would also carry over into the new pidgin various habits of his own native tongue, not only in pronunciation, but also in grammatical forms and syntax, and of course in vocabulary. . . . The pidgin thus becomes institutionalized, with its grammatical structure crystallized at this initial stage of language-learning.[56]

Obviously, even if creolization tended to yield idealized regularities consistent with Chomsky's regularities, the accessing of language patterns changing in the manner indicated would go contrary to Chomsky's theory, in accounting for such regularities and for their detection.

Hoenigswald offers the further illustration of the following sort of phenomenon among mature languages: in the Balkans and Central Europe, where a rather incredible mix of different languages interpenetrate, one may find that the *grammar* of certain languages – say, Czech – has, over time, come to exhibit features that link it more closely with German than with the "family" of Slavic languages with which it is closely linked "genetically."[57] Here is the effect of a form of graduality that cannot be detected in actual use by the shifting generations of apt speakers of the language – that may even have changed at different rates in the life of the language (as a result of historical events) within the behavior of particular generations. Normally, the latter change will also go undetected. Here, then, the dynamics of actual language use and linguistic history confirms the reasonableness of linking the analysis of the grammar of a language to a careful account of molar accession. The gradualities of language use, the special phenomena of pidginized languages, the special phenomena of interpenetration and borrowing among mature language families of different kinds, all attest to the implausibility of segregating the deep and surface features of a language – or at least they attest to the methodological need to draw a responsible theory of the deep features from an empirically disciplined account of its surface features. All this, of course, raises substantial doubts about the validity of an extreme innatism.

Let it be said, then, that no one (well, almost no one – not even Locke) is *not* a nativist to some extent. To admit that, however, is not yet to admit Chomsky's or Fodor's "deep" version of innatism. The empirical evidence is peculiarly slippery: not because there is no evidence in favor of Chomsky's empirical regularities but because the evidence we have appears not to be decisive with regard to choosing between the "shallow" and "deep" readings that would affect the standing of Chomsky's particular doctrine. At the very least, it is hardly impossible to formulate a conceptual alternative to Chomsky's reading of the linguistic data.

More to the point, we must be quite clear that Chomsky never examines the question of *accessing* language as language: he restricts his speculation to the application of UG in the *parsing* of sentences already (assumed to be) accessed and to the *acquisition* of a first language congruent with his own innatist reading of how we parse language. This is particularly clear from the fact that Chomsky invariably takes it that the structure of a given *sentence* can be perspicuously fixed without any attention to intentional context. Fodor has insisted on the plausible finding that "the computational systems [the innate input systems he posits] that come into play in the perceptual analysis of speech are distinctive in that they operate *only* upon acoustic signals that are taken to be utterances."[58] That is, Fodor's thesis (effectively, Chomsky's) holds that "the structure of the sentence recognition system is responsive to *universal* properties of language and hence that the system works only in domains which exhibit these properties."[59] But this, of course, *is* to admit the distinction between the accessing problem and the parsing problem *and* to admit the dependence of our solving the parsing problem on the precision with which the accessing problem can be specified and solved. The trouble is that the accessing problem cannot be determinately specified *in linguistic terms* so that what is accessed may be suitably and uniquely restricted to the *sentences* Chomsky parses. *What* is accessed is not yet sufficiently analyzed[50] for the regularities assigned to sets of such sentences to be said to confirm that apt speakers (ideally) *do* parse *them* (in a psychologically real sense) in ways that approximate UG. It is entirely possible that speakers respond in ways that may be aptly simulated by such grammars and that, once actually introduced to such grammars or related approximations or fragments of a grammar, such speakers may be additionally disposed to favor such idealizations as capturing *their* actual patterns of linguistic accession.

It should be noticed that the counterthesis is considerably strengthened, if sensory modalities are treated as (or as involving) innate modules, general inductive and reasoning capacities are similarly modular, non-innate "modules" (of psychological internalizing) are conceded, actual societal practices need not form closed systems, practices may be grounded in (Wittgensteinian-like) "forms of life," and the human species exhibits convergent "prudential" (molar) concerns incorporating the output of innate modular perceptual and actional capacities. For then, even in the absence of a strictly universal grammar, it may well be the case that the contingent variations of grammar (within the species-specific and species-wide practices of language) would still yield a sense of strong convergence ranging over the *structure of sentences*; would do so in spite of the fact that sentences serve only as an approximation of what it is human agents do access in the context of full linguistic communication. *Humans may initially access the culturally contexted behavior of other agents.* Sentences may serve as an abstraction (in this respect) of the complex incarnate phenomena that are naturally accessed. On that view, we need not, or need not always, or need not in the innatist's sense, compute the structure of the idealized sentences we abstract from the accessing context, *in accord with an innate grammar.*

These considerations clearly demonstrate that there may be important disanalogies between language and the perceptual modalities. Fodor, therefore, may have been excessively sanguine when (reflecting on the apparent failure to "register" visual information about the numerals of one's watch – which "must have been registered" in looking at one's watch) he goes on to say, "There are quite similar phenomena in the case of language, where it is easy to show that details of syntax (or of the choice of vocabulary) are lost within moments of hearing an utterance, only the gist being retained"; and then concludes on the strength of that analogy: "Yet *it is inconceivable that such information is not registered somewhere in the comprehension process* and, within limits, it is possible to enhance its recovery by the manipulation of instructorial variables."[60] Either it is not inconceivable – on the argument that certain "information," for instance about idealized sentences, was never in "the comprehension process" to begin with; or it is inconceivable *but only because the* "information" has already been conceded to have been internalized in a non-innatist way. There you have the double challenge which the accessing problem imposes on the nativist.[61]

The challenge may be put most economically in the following way. Improvisatory and conversational thinking, though conveyed linguistically, appears to favor extraordinarily wide, even radical, differences among historical societies, from age to age within a society, and across any time slice of a single society's life. These differences may reasonably be viewed as accessible intra- and interlinguistically; but there is no promising or inclusive schema of the innate, core structures of species-wide human thought that could account generationally for all the varieties of human conceptualization. Nevertheless, if natural language is to be reconciled with the "poverty of the stimulus," then so, too, must natural thought. (Is it irrelevant to remark that scientific theories, literary fictions and the like are all distinctly underdetermined by whatever may be designated as the "data"?) If the parallel is favored, then there actually is a huge, essential phenomenon that *empirically* challenges the likelihood of a Chomskyan-like solution (in fact, just the one Fodor offers – in a space not obviously quite the same as language though inseparable from it). At the very least, then, we cannot afford to suppose that Chomsky's thesis is the only promising one we have – or even the most promising.

One last, hopelessly large qualification remains to be added. Either it is not true that such "basic categories" (or natural-kind terms) as "dog" (as opposed to "animal" or "poodle") are innately specified for visual perception, or it is reasonably open to us to argue that such categories represent workable, *non*-nativist idealizations (at the molar level) regarding a range of (postulated) innate visual processings. Quine's indeterminacy argument may well be incompatible with a strong nativism joined to a strong non-nativist sense of the salient prudential interests and needs of the species; but a moderate indeterminacy thesis at the level of belief and speech may be both realistic and quite feasible. The point to notice is simply that Chomsky's and Fodor's brand of nativism is neither the only nor the obviously most reasonable option to favor; *and* the force of their thesis presupposes an answer to the accessing problem. Merely concede the difficulty of *knowing* whether humans universally access dogs in relevant visual perception, and you will see at once the difficulty of confirming the theory that they do so innately (through visual input systems); and if the visual problem remains unsolved, it is a foregone conclusion that the linguistic problem will remain unsolved as well.[62]

On the strength of that stalemate (if stalemate it be), the innatist theory has utterly failed to dislodge the strong sense in which the psychological and the societal are symbiotic – *a fortiori*, the strong sense in which language and similar cultural practices cannot be satisfactorily analyzed without admitting the substantive, irreducible role of the grooming processes of natural human societies.

Also, apart from the empirical difficulty of formulating invariant grammars for the immense variety of natural languages and the potentially artifactual nature of at least some general uniformities favored as a result of idealizations fitted to the linguistic practices of actual societies, Chomsky's underlying essentialism is basically at variance with current thinking about the very status of science: *if*, therefore, Chomsky is searching for real linguistic universals on the strength of the thesis that "our minds are fixed biological systems with their intrinsic scope and limits,"[63] then there is good reason to believe that grammar is probably *not* able to be disengaged from semantic and nonlinguistic factors in the autonomous way he favors.

Nevertheless, Fodor's and Chomsky's programs are relatively moderate. The radical use of the initially modular notion of syntactic modeling obtains when the top–down or "folk" orientation is dismissed altogether. Thus, Stephen Stich, who opposes the solipsistic thesis and all reliance on "folk psychologies," nevertheleses advocates the autonomy (and adequacy) of the syntactic thesis. His own recommendation (the so-called "Syntactic Theory of the Mind") holds that "syntactic theories can do justice to all of the generalizations capturable by quantifying over content sentences while avoiding the limitations that the folk language of content impose[s]." Apparently, the formulas or sentences Stich has in mind are "no more than an infinite class of complex syntactic objects [that] *has no semantics*" – for which the (empirical) vocabulary to be supplied appears to be more or less restricted to "physical states of the brain" and "neurological properties."[64] What Stich means is that, in "mimicking content-based theories ['folk psychology' or psychology drawn from the 'common idioms' of natural language or 'the manifest image' or the like] as closely as possible," his (novel scientific) language will characterize mental states (drawn from folk psychology) "not ... by their content sentences but, rather [only] by the [new] syntactic objects to which are mapped."[65] When he says, therefore, that "syntactic theories can do justice to all of the generalizations capturable by quantifying over content sentences" (including causal, ideological, referential, and similar considerations), he merely means that his own idiom will (1) be *formally* adequate to permit the mapping of sentences involving infinitely many or all possible categories that *could be drawn from folk psychology*, and (2) would compare favorably in an *empirical* way with the success of *any theory like Fodor's*. But, as Stich very reasonably adds,

> an STM theorist [one committed to the Syntactic Theory of the Mind] will not have nearly enough [empirical] generalizations [formulated in terms of the vocabulary favored] to put his theory to the test, at least not in the early days of theory construction. And for the foreseeable future, all days will be early days. So if theory is to confront data, the syntactic theorist will have to make a significant number of ad hoc assumptions about causal links

between [postulated] states on the one hand, and stimuli and behavior on the other.[66]

Just so. And yet, Stich's original purpose was to *replace* an empirically inadequate folk psychology.

One telltale application of the thesis cuts through all the abstractions. Observing that "as subjects [become] increasingly ideologically distant from ourselves, we [lose] our folk psychological grip on how to characterize their beliefs," Stich says that that difficulty disappears for the syntactic theory – "since the characterization of a B-state [that is, a state roughly analogous to a belief, in the folk psychology] does not depend on the other B-states that the subject happens to have [or states of other types holistically linked in the folk psychology to belief states, for instance intentions, desires and the like]." The syntactic theory "ignores the subject's linguistic context," his ideological orientation, socialized experience, and the rest – in effect, rejects psychological holism. Nevertheless, Stich claims, "*if* the generalization is there" – that is, capturable by *some* folk-psychologically based account – "*it can be captured by a syntactic theory.*"[67] The point is that it *can* be captured, in principle, in the purely formal sense that it can be mapped by the new language (a claim which might still be false, if, say, the *new* language were thoroughly extensional and the relevant empirical ["folk"] aptitudes generating admissible sentences were incapable of being extensionally mapped); but, apart from that, we don't know, and Stich admits that he doesn't know, whether it can be captured in real-time terms in an empirical or cognitively relevant way. For example, he urges (for the new science of psychology) that "any differences between organisms which do not manifest themselves as differences in their current, internal, physical states ought to be ignored by a psychological theory."[68] Obviously, Stich is simply betting that an empirical psychology of his own making, formulated according to the scruples of his STM, will be empirically better or at least as good as a folk psychology. But that clearly depends, for one thing, on whether or not he has delimited his empirical vocabulary and empirical procedures beyond what the domain requires. (Perhaps he has impoverished them to the extent that he can no longer recognize what is being sacrificed – a matter that cannot be decided by purely formal means). Certainly, Stich has risked his own theory without clearly examining whether "folk psychology" is inherently "holistic" or intensionally complex (in Davidson's sense), whether a formal, syntax-driven system can accommodate all pertinent semantic and nonlinguistic aspects of actual human experience, whether any extensional simulation of a segment of complex human behavior can be expected to lead to incrementally more and more complete simulations (or analyses) of full capacities (in the manner of AI simulation), or whether the societal dimension of language can be effectively ignored. The essential weakness of Stich's strategy is simply that it is methodologically parasitic on theories such as Fodor's functionalism or "folk" psychology and that it remains empirically indifferent to whether it actually offers a promising hypothesis.

III

Speaking very broadly (and somewhat incautiously), we may conclude that theories of the nativist, infra-organismic, computational, syntax-driven sort, or of a sort entirely opposed to theories or vocabulary favoring the "manifest image" or the like, tend to eliminate the social dimension of human existence from playing any central or irreducible role in the description and explanation of the phenomena of the human world.

At the opposite pole of such speculations stands Emile Durkheim, who declares that "society is not a mere sum of individuals. Rather, the sytem formed by their association represents a specific reality which has its own characteristics. Of course, nothing collective can be produced if individual consciousnesses are not assumed; but this necessary condition is itself insufficient. These consciousnesses must be combined in a certain way; social life results from this combination and is, consequently, explained by it. Individual minds, forming groups by mingling and fusing, give birth to a being, psychological if you will, but constituting a psychic individuality of a new sort."[69] Certainly, the minimal key to Durkheim's entire theory – as well as to Auguste Comte's, apart from Durkheim's rather ineffectual pursuit of the principled autonomy of sociology and apart from his extreme ontological tastes – lies with the notion of the irreducibility of the social. Durkheim does say that "there are in societies only individual consciousnesses; in these, then, is found the source of all social evolution";[70] and in a footnote he adds that, in introducing the distinction between

a collective consciousness [*conscience collective*] and individual consciousnesses ... it is not necessary to posit for the former a separate personal existence [although] the states which constitute it differ specifically from those which constitute the individual consciousnesses. This specificity comes from the fact that they are not formed from the same elements. The latter result from the nature of the organicopsychological being taken in isolation, the former from the combination of a plurality of beings of this kind.[71]

But it is probably hopeless to achieve a fully consistent reading of Durkheim. At the very least, he produces a doctrine of *homo duplex* (here, adumbrated) – an uneasy union of biologically and socially focused sensibilities; at the worst, he invents a collective organism that apparently has its own psychology capable both of advancing and thwarting the lives of individuals; for, as he also remarks, "The group thinks, feels, and acts quite differently from the way in which its members would were they isolated."[72] There is no doubt that Durkheim's *duplex* notion is a conceptual disaster, though it is an attempt to hold together all the relevant considerations. The point to notice, however, is that it really fails in both of its most important conceptual functions: (1) it intrudes an ontological dualism (composition from different elements), where what is wanted is a realistic sense of conflicts of interest within the complex lives of biologically differentiated individuals – consistent with their socialization; and (2) it multiplies types of individual agents or states of agents (agents or states composed from different elements), where ordinary ontological economy would have been sufficient for Durkheim's purpose.

Two remarks fix the fatal mistake and its motivation: first, Durkheim says that "[the] essential characteristic [of social phenomena] is their power of exerting pressure on individual consciousnesses" – which (in Durkheim's mind) confirms the "different nature" of the two forms of consciousness *and* the autonomy of sociology; second, he says that the harmony and order of "social life [is due to] a correspondence between the internal [biopsychological] and the external [primarily social] milieu, [which] is only an approximation; however, it is in general true" – which (again, in Durkheim's mind) provides the dynamics for explaining such phenomena as *anomie* and for validating the moral instruction of individual men.[73] The fatal mistake rests with Durkheim's literal–minded emphasis on the "external" (the milieu of society). Durkheim (equivocally) construes the "external" as the *externally coercive*, whereas his own studies tend to show that, apart from the trivially obvious coercion that *other* (in this sense, "external") individuals in a society might exert on a particular agent, individuals normally *internalize* the coercive or constraining rules of the society to which they belong (in a sense not terribly distant from that of Freud's superego). In putting things in this way, Durkheim obscures what is defensible in methodological individualism – namely, that *only* individual human agents (or aggregates of agents) are effective *agents*; and he obscures the point of quarrel between a reductive and a nonreductive individualism – namely, that the *attributes* of socially organized aggregates may or may not be *analyzable into nonsocietal, psychological attributes ascribable to single individuals*.

The decisive case, once again, is surely that of language. It is certainly clear that no one can have internalized the meanings of all the words of, or the range of all acceptable ways in which words may be used in contexts of activity and experience with respect to, the whole of any complex natural language. Also, *if* (as we have speculated) syntax cannot be autonomous, then a natural language is the paradigm of what may be ascribed only to an entire society – that members must have internalized (doubtless, different "parts" for different individuals). Furthermore, the paradigm of language shows (*contra* Durkheim) the essential sense in which the practices, traditions, institutions, rules and the like of a *society* are smoothly internalized in – and thereupon constrain – the behavior and sensibilities of individual agents. In this sense, language is surely the generic phenomenon relative to which ideology (Durkheim's preference) is simply a special case. Language focuses the most inclusive form of the puzzle of social communication at the fully human level; whereas, however important, ideology must count as a restricted communicative network serving quite special functions. There is a fair sense in which Durkheim construed *conscience collective* primarily in terms of ideology rather than in terms of minimal linguistic communication – which perhaps explains his peculiar fixation on the primitive nature (hence, the essential function) of totemism and the moral instruction of the French.[74] He seems to have had remarkably little to say about language.

We, however, may usefully exploit the contrast between Chomsky and Durkheim, because *both* have somehow missed the import of the social dimension of language and human life: Chomsky, by effectively refusing to admit even the eligibility of pertinent speculation; Durkheim, by his inabiity to acknowlege the social or societal dimension without encumbering it immediately in the hopelessly organismic way he does. The conceptual problem stares us in the face: namely, *how to isolate the collective, societally structured features of human existence without postulating any distinct collective entities* such as societies, socioeconomic classes, nations or collective states of mind. In fact, without pursuing for the moment a most promising clue, it helps immensely merely to mention that Wittgenstein's notion of "forms of life" (and, within "forms of life," "language games") does provide a model in which

(i)   the societal dimension of language is causally efficacious and not reducible to the mental or intra- or infra-psychological;

(ii)   actual linguistic agents are restricted to individual speakers and aggregates of speakers; and

(iii) no collective entities are ever postulated.

So Wittgenstein's model is radically opposed to Chomsky's – which may be dubbed *psychologistic*, in the narrow sense that, although it affirms (ii) – *a fortiori*, affirms (iii) – it also (in effect) denies (i); but Wittgenstein's model is also radically opposed to a Durkheimian-like theory that would entail some complex adjustment of (i)–(iii) such that the societal dimension of language requires a symbiosis between individual and collective entities. The problem, in short, is precisely the issue of *methodological individualism*. But, since that itself is a most vexed matter, we do well to approach it as circumspectly as possible.

There are many promising models that facilitate understanding what is misleadingly described as the *relationship* between aggregated individuals and the society to which they belong, in terms that assist us at the same time to understand the psychologically internalized forms of acculturation. But one prominent model for analyzing human speech fails (in a most instructive way) to capture what is minimally needed – in a way that recalls the solipsistic or infra-organismic models we have already reviewed (though it is not one of them), and in a way that does not quite accommodate the social facts it obviously has in mind (that Durkheim had also been trying to feature in his own impossible theory). This is curious, because the model in question actually focuses on the very activity of speech used to achieve a certain communicative objective. Its failure, therefore, is particularly helpful in pointing us in the right direction.

Consider the extremely well-known formula offered by H. P. Grice (there are a number of variants that Grice has sketched regarding the meaning of a normal linguistic act or utterance):

"U meant something by uttering x" is true if, for some audience A, U uttered x intending

(1) A to produce a particular response r.

(2) A to think (recognize) that U intends (1).

(3) A to fulfill (1) on the basis of his fulfillment of (2).[75]

Apart from technical problems about whether provisions (1)–(3) could preclude unwanted counterinstances,[76] Grice's formula has the odd feature that it does not attempt at all to account for the conditions on which a speaker might reasonably expect an audience to recognize his intention for what it is, and does not attempt to account for the conditions on which an audience might reasonably be expected to recognize a speaker's intention in uttering what he utters. But there is no question that Grice is addressing, however obliquely, the issue of accessing language.

Grice was interested in the *sui generis* nature of linguistic intentions; but it is not clear that these really are different from cultural intentions in general, or that Grice captured linguistic intentions at all. For, suppose a speaker and his audience actually *share a language*. Then, when the speaker speaks, either he intends *what is normally intended in speaking* in accord with the language shared (whatever that may involve) *or*, dependently and in accord with other cultural practices that speakers and audiences share, his (second) intention to use what is normally intended in speaking is recognized by his would-be audience because they share the appropriate (second) practice or because they can infer his (second) intention from the practices they do share. The linguistic and cultural intentions of speakers and agents are, then, in a certain obvious sense, at least in part, tautologically embedded in the practices they share.[77] Individual speakers (and other cultural agents – hat-doffers, for example, in familiar contexts) may rightly be ascribed intentions that (assigned to individual agents) merely correspond to the somewhat rule-like (intentional) regularities of distinctly conventional practices. The practices don't come into existence because isolated individuals first intend what Grice says they intend: that would be the solipsist's way (or, perhaps, a linguistic version of Rousseau's paradoxical social contract) and would never produce *social practices*; on the other hand, in sharing a practice *by* acting conformably, agents need not and normally do not *in addition* intend *to* conform – though they may of course have second intentions regarding how (in some ulterior way) to use what (tautologously) they intend merely in sharing a given practice.

The great irony of Grice's model is that instantiating the intentional formula cannot of itself generate a social practice to be shared or the actual sharing of that practice; and, if the practice thus shared is presupposed (which is obviously what Grice has in mind), then the mere intentions of individual linguistic agents are already tautologically entailed by the very functioning of the practice. The mystery of how this obtains is not in the least clarified by the intentional model itself: on the contrary, it presupposes the answer. In fact, in a somewhat startling way, what Grice has attempted in miniature (in the analytic idiom) is remarkably close to what Husserl originally attempted – also quite unsuccessfully (in his phenomenological idiom) – on a monumental scale. For Husserl, too, begins as a methodological solipsist and tries to derive, by what he regards as his transcendental strategies, a vindication of the existence of a field of socially shared life involving *other* agents and historically *shared* practices. But, rather like Grice (if one may put things this way),

Husserl is obliged to draw out of his solipsism what he cannot but have presupposed in characterizing its powers as he does.[78]

The most important, most comprehensive puzzle that language poses for the human sciences (once they are permitted to claim their status as genuine sciences) concerns how to make sense of the apt behavior of aggregated individual agents who cannot have completely internalized (individually) the whole of, or quite the same "parts" of, the practices that their society "possesses." *If* the structuralists and the nativists are wrong about language and thought, there is and can be no totalized or totalizable *system* which apt social participants can actually share. On that condition, there are only two possible ways – each individually inadequate, each requiring the other – in which the spontaneous, natural, improvisational, smooth functioning of social practices can be accounted for

(a) by some biologically preestablished harmony; and
(b) by some process for achieving harmony at the level of culturally significant behavior.

There is, actually, some strong empirical evidence favoring (a), as well as quite reasonable second-order (pragmatist or transcendental) arguments. For instance, the genetic uniformity and viability of the species is clear enough. And, even at the level of behavior, the peculiar responsiveness of nurtured infant and nurturing mother and of sexual attraction and copulation indicate a kind of preharmonized specialization, a significant measure of biologically favored social cofunctioning – hardly, of course, as dramatic as what we find among the social insects or among the sponges and social amoebae, or for that matter in the ethology of the largely instinct-driven higher animals.[79] But that *is* precisely what we should expect if man is a creature that inhabits a contingent culture only partially explicable in terms of his biological endowment. Furthermore, there is considerable evidence of a biologically preestablished ecological harmony between percipient organisms (including man) and the environing world[80] – which, in being species-wide, services whatever preharmonized social (or cultural) cofunctioning may be taken to obtain. In a sense, evidence of this sort provides a first-order basis for such second-order speculations as that of Quine's notion of a favored "quality space" (that must be presupposed in successful linguistic communication) or of Peirce's conception of man the scientist as the cognitive organ of nature (which, of course, is merely a fanciful way of making sense of the evolutionary basis for the achievement of a cooperative science actually penetrating the structures of nature).[81] But there is *nothing*, either in first-order or in second-order terms, that compels or even plausibly persuades us that the harmonious functioning of cultural practices (b) is reducible to or explicable solely in terms of (a). In a sense, this is simply the valid and recoverable insight of Durkheim's fulminations.

If, then, we suppose that the cultural is not reducible to the biological, and if we wish to avoid Cartesian dualism at the same time as we hypothesize that the harmonious functioning of cultural practices – paradigmatically, linguistic practices – must depend on (and in a certain sense, complete) biological harmonies, we are bound to conclude (1) that cultural phenomena are *emergent* with respect to the biological (that is, not analyzable or explicable solely in terms suited to the physical and life sciences);[82] and (2) that cultural phenomena are *incarnate* in biological phenomena (in the sense that cultural attributes are indissoluble, whether analyzed as monadic or polyadic, though they exhibit both physical or biological and historical or intensional or rule-like aspects). For example, painting in a Cubist way and signing a deed are acts that are intelligible only in the complex space of an historical culture; but, in that space and as the particular acts they are, they are, indissolubly, the exertions of living organisms. Cultural attributes and phenomena *cannot* be treated merely as functionalist (abstract) phenomena. Interestingly, it is precisely the reductive or functionalist accounts that the syntax-driven models of explanation (Fodor's and Stich's for instance) are inclined to favor. Something like the joint distinction of the emergent and the incarnate is what is lacking, also, in P. F. Strawson's otherwise extremely suggestive metaphysics of persons[83] – which, by that defect, is neither explicitly insured against dualism nor clearly adequate as an account of the agents of a palpable human world.

The question remains, how can we account for (b), for harmonious social practices? Eschewing structuralism, Hegelianism (in the sense of a rational theodicy incorporating collective life), all talk of totalized systems, *and* favoring the profoundly historicized nature of human existence, there is really only one promising line of strategy available: that is, to construe the sharing of a practice as ineliminably *improvisational* and *interpretive*. Both the improvisational and the interpretive must be viewed *as social* or societal (but not Durkheimian) processes: this is precisely what is slighted or excessively delayed (if not, actually lacking) in, for instance, Piaget's constructivist notion of the development of intelligence in children – even with respect to language. Curious as it may seem, Piaget begins, effectively, from a standpoint very close to that of a methodological solipsist. This is especially clear in his strong emphasis on "egocentrism" – which has been convincingly criticized for that very reason by Lev Vygotsky and Jerome Bruner.[84] In a sense, it confirms the remarkably widespread tendency in psychological theory to avoid as much as possible admitting a robust sense of social phenomena.

In terms of distinctions introduced earlier, in airing Chomsky's thesis, we may now suggest that the process of *accessing* culturally freighted materials (*texts*, in the largest sense, including linguistic utterances) – a process groomed in first-language *acquisition* and presupposed in linguistic (or, by however loose an extension, lingual) *parsing* – *must incorporate a measure of consensually supported interpretation and improvisation.* The importance of this seemingly modest admission can hardly be overstated. For (1) there is *no* plausible innatist or modular reading of such a process; (2) such a process is, on the face of it, a societally specified phenomenon not entirely reducible psychologistically; and (3) its admission precludes or counts heavily against construing any sector of culture's functioning as a *system* of any sort.

Broadly speaking, would-be systems (within the space of the *human* sciences) are of three possible sorts. They could be Chomskyan. They could be Hegelian, as in supposing that every finite, historicized era had its own generating or totalizing *Geist*. (That thesis is easily defeated, however, for either it yields a system as a retrospective artifact fitted to a historical era or it fails to ensure a system for any historically open-ended world or world in "transition."[85]) Or else they could be structuralist, in the sense favored by the tradition of Saussure and Lévi-Strauss. But, again, such an account would be defeated by the impossibility of totalizing over *all* possible structural variants, by the paradoxes of the doctrine of internal relations, which the structuralists clearly did not grasp, and by the epistemic irresponsibility of the structuralist option itself.[86]

By *system*, here – in a sense suited to both the physical and the human sciences – we may understand an actual domain that is

1   determinately structured (well-formed) in terms of some finite set of basic alphabetic or relational ingredients governed by a finite set of rules or laws;
2   deterministic or normative, with regard to the possible variety of phenomena it may exhibit;
3   extensionally structured;
4   homonomic with respect to a match between its descriptive and explanatory vocabularies (whether causal or noncausal);
5   closed, with respect to its universal explanatory principles, laws, rules or the like;
6   totalized or universal, with respect to all possible pertinent phenomena or to all the pertinent phenomena of a given domain; and

7   ideally capable of being thus characterized synchronically.

Viewed in these terms, structuralism (what is often called analytic structuralism, the structuralism best illustrated by Lévi-Strauss,[87] is a deliberately abstract functionalism fitted to a *societal* order (just because the societal is not reducible psychologistically and because the societal is needed in the human sciences), "mythically" outfitted (as Lévi-Strauss is fond of saying) with homonomic universals (just in order to ensure suitable comparison with what is assumed to obtain in the physical sciences). Thereupon, structuralism is construed as affording a reasonably parsimonious enlargement of something like the unity-of-science canon.[88] Structuralism, therefore, rather glibly manages to eliminate both actual societies and actual human agents (while remaining functionalist, methodologically), by the device of generative systems that entirely replace actual human domains. Chomskyan innatism is, clearly, diametrically opposed to structuralism, in preserving individual molar agents (and sectors of the actual world). It also reduces or eliminates societies *vis-à-vis* actual individuals. Eliminationists of course eliminate the very need for functionalism – hence, they attack both Chomskyan and structuralist alternatives by way of construing the systems of human existence as indistinguishable in principle from the systems of physical nature.[89]

But the space of human life *is* (at least provisionally) different from the merely physical world: not just because of the intentional, significative, historical, interpretive features of the human world but also because those features have never been shown to be describable or explicable in terms of natural laws confined to the inanimate physical world explicable in *physical*$_2$ terms, in Feigl's sense[90] or in terms of laws holding, conformably with the unity program, among biological phenomena. This is why the Intentional or cultural properties of the human world *are not properties of (mere) physical phenomena:* they cannot, or cannot as yet, be ontically adequated (as we may say) with purely physical phenomena, because we have no idea how, in principle, in terms of causal regularities, to account for their emergence *as* the properties of physical systems. This is the point of insisting that culturally informed psychological properties are (indissolubly) *incarnate* in the physical or biological and are not merely local properties *of* the physical or biological.

The point is essentially missed in John Searle's recent account of the mind/body problem. Searle declares,

> the mind and the body interact, but they are not two different things, since mental phenomena just are features of the brain. One way to characterize this position is to see it as an assertion of both

physicalism and mentalism. Suppose we define "naive physicalism" to be the view that all that exists in the world are physical particles with their properties and relations. The power of the physical model of reality is so great that it is hard to see how we can seriously challenge naive physicalism. And let us define "naive mentalism" to be the view that the mental phenomena really exist. There really are mental states; some of them are conscious; many have intentionality; they all have subjectivity; and many of them function causally in determining physical events in the world. ... Naive mentalism and naive physicalism are perfectly consistent with each other. Indeed, as far as we know anything about how the world works, they are not only consistent, they are both true.[91]

The trouble is that Searle never addresses the question of the grounds on which we could show that mental properties just *are* the local ("macro-level") properties of physical brains – properties which may be causally explained by some suitably selected neurophysiological ("micro-level") elements of such brains.[92] Searle has simply not earned the right to claim that they can be thus explained. Perhaps they cannot – without yet violating any known physical processes and without invoking Cartesian dualism. The thesis that the mental – or, in particular, the culturally freighted mental life of humans – is *incarnate* in physical properties rather than *identifiable as* (as opposed to being *identical with*) physical properties is meant as a scrupulous accommodation of the admission, at once methodological and ontological, that we lack not only the account Searle pretends we may be certain of, but also any principled reason for believing that the cultural, the linguistic, the lingual, the Intentional (which qualify all the saliencies of human psychology) *can* be analyzed in physical terms. We capture that admission by acknowledging that the psychological or mental properties in question are the properties of *selves* and not mere bodies. Searle nowhere attempts to explain the conceptual relationship between selves and bodies. So it is quite insufficient to insist (as he may rightly do) that "our commonsense mentalistic conception of ourselves is perfectly consistent with our conception of nature as a physical system"[93] After all, the incarnatist thesis, like the identity thesis and the mere redefinition of the "physical" to include the mental, also precludes dualism. The theoretical difficulties of reducing the Intentional to the physical may well favor incarnatism. It may be, as we have had more than one occasion to remark, that the complexities of the Intentional world of human culture are able to be marked *only at that emergent level at which they appear*; and that, *as inquiring selves*, we may be quite unable to analyze or reduce them (admitting them to be real) in terms of any lower-level processes.[94] If so, then we may well be confined to the idiom of the incarnate. There is no incoherence in the proposal.

IV

To remind ourselves of an earlier clue, Wittgensteinian "forms of life" preserve both aggregated human agents and their societal or cultural space, without any felt need to intrude a Durkheimian collectivism or holism; they replace an abstract functionalism with an implicit incarnatism; and they preclude any and all forms of (cognitive) foundationalism and the perceived need for a system (in the technical sense supplied). They accomplish much more, in fact. For the Wittgensteinian account also provides what is probably the most convincing defense, intuitively, for construing the accessing issue we raised earlier on, in terms of a consensually tolerant, pluralizing practice of improvisation and interpretation.

The analysis of social process and social attribute may be collected at two foci: first, one needs a phenomenological account of what is distinctive about cultural features; second, one needs a conceptual sketch of what makes the cultural coherently and irreducibly *sui generis*. Perhaps the most famous phenomenological account (if we may use the term, without risk of being misunderstood) *is* Wittgenstein's. In any case, Wittgensteinian "forms of life" surely accommodate the following themes:

1   there are indefinitely many human societies that exhibit distinctive ways of organizing and sustaining life;

2   a recognizable community somehow does not thwart infants developing among its apt members from learning and internalizing its practices, so that in time they also become spontaneously apt practitioners;

3   acquiring such practices is natural, in the sense that it proceeds, as one grows up, largely without explicit or formal instruction and in a way in which such instruction itself presupposes the achievement in question;

4   the members of a viable society must be supposed to know, in general, the practices they share, in the sense that their spontaneous behavior, particularly with respect to whatever is relatively basic and common to most parts of that society (language, notably),

must in general count as correct or admissible instantiations of such practices;

5 their being able to perform in such shared ways entails sharing an indefinitely large body of beliefs as well (which surely contribute to the viability of their society);

6 although the apt members of a society may (variably) project rule-like generalizations that they take to fit particular runs or segments of a diachronic practice and that are more or less in accord with how that practice may be acceptably extended, a natural (or naturally acquired) practice is not in any sense actually governed by higher-order rules, and would-be rules are no more than abstractions of the tendency of a regular practice to favor this or that direction – reflexively formulated (possibly heuristically) by those who have actually mastered that practice in a natural way;

7 apt practitioners, behaving spontaneously in societal contexts, do not normally guide themselves (and need not guide themselves) by explicit rules, whether reflexively projected or projected in consultation with other apt agents;

8 at any point at which apt agents act in accord with a natural practice, and, acting thus, effectively extend the practice, there is no formal constraint on their behavior – of any sort that would determine whether or not they had (acting thus) actually conformed with that practice;

9 the practice may be extended in indefinitely many different ways, gradually and tolerantly, in accord with how the members of a society are at the time disposed to favor its continuation; and

10 the practices and what are identified as salient instances of given practices are bound to change diachronically as a result of a general consensual drift among the apt members of a given society.

There is of course no short text in Wittgensteinian's *Philosophical Investigations* (or elsewhere) from which this compendium can be directly drawn. It was not even Wittgenstein's habit to order matters in this way. Nevertheless, it is a reasonable approximation of the full sense of his notion of "forms of life." Perhaps two very brief citations will remind us of the thrust of his sense of social existence: "What has to be accepted, the given," he says, "is – so one could say – forms of life";[95] and

"How do I manage always to use a word correctly – i.e. significantly; do I keep on consulting a grammar? No"; ... understanding [what I mean in speaking] is [not] the activity by which we shew that we understand. ... [As in extending an arithmetic series by adding another number, one] must go on like this *without a reason*. Not, however, because he cannot yet grasp the reason but because – in *this* system – there is no reason. ... And the *like this* (in "go on like this") is signified by a number, a value. For at *this* level the expression of the rule is explained by the value, not the value by the rule. For just where one says "But don't you *see* ...?" the rule is no use, it is what is explained, not what does the explaining. ... To guess the meaning of a rule, to grasp it intuitively, could surely mean nothing but: to guess its *application*. And that can't now mean: to guess the kind of application; the rule for it. Nor does guessing come in here ... [it] presupposes that one has already mastered a technique.[96]

Certainly, it is clear that Wittgenstein is opposed to anything like a structuralist system in attempting to understand natural societies: there are no finite, adequate, formational and transformational rules by which to generate – even for a particular society – apt instances of given practices (even of seemingly formal practices). We are to suppose that the (natural) acquisition of a (natural) practice is largely tacit, attentive to the open-ended, indefinitely extendable nature of such practices. That is, a *practice* is itself some sort of abstraction (which therefore encourages the search for rules) drawn from various particular acts (notably, instructionally favored examples) that need not themselves be formulably identical in any way or aggregatively inhospitable to perceptible differences or diachronic drift among the accumulating, new instances of that same putative practice. Mastering a natural practice makes no sense, for Wittgenstein, without one's acquiring the tacit ability to *improvise* spontaneously: to *add* to the (variably) collectible series of instances that could be said to belong to a practice (and could be "shown" to do so, heuristically, by formulating would-be rules for such inclusions) – so that, despite any perceptible novelty in such behavior (perhaps more noticeable when collected with other additions), the community affected will not normally reject, challenge, be dubious about or troubled by, or similarly question, the "inclusion" of that behavior "within the scope of practice." (The use of scare quotes, here, signifies of course the largely mythical nature of such surveillance.)

The point is that natural practices entail an improvisational skill: pertinent improvisation invites, in some tacit sense at least, a consensual acceptance within the society affected; *and* both improvisation and consensus cannot but be *interpretive*. An agent must (tacitly) understand that, however distinctive his action may be, it will need to accord with the practice in place and will need to be consensually favored; and the members of the society affected (including the agent) will (tacitly) need to confirm the congruity between the act and practice in question. Humans, therefore, are entirely capable of maintaining, at the level of cultural practice, a distinct form of social harmony

(i)   in the absence of any totalized or universal system of rules; hence,

(ii) in spite of no one's having internalized any necessary or sufficient or totalized set of generative rules;

(iii) compatibly with plural and nonconverging interpretations of given practices;

(iv) compatibly with a fragmentary and imperfect (variably fragmentary and variably imperfect) familiarity, on the part of apt practitioners, of precisely how a putative practice has actually been, and has been tacitly conceded to be, instantiated; and

(v)   despite ineliminable indeterminacies of context in actually extending a practice through improvised behavior.

It may well be item (v) that the partisans of system mistrust most; and it is (v) that Wittgenstein most satisfactorily accommodates through the notion of "forms of life."[97] This answers in a way a persistent question of Chomsky's: for Chomsky regularly links the postulation of a *system* of (universal) linguistic rules with the "highly degenerate sample" of language, largely "irrelevant and incorrect," "ill-formed, inaccurate, and inappropriate," and decidedly "limited" as far as illuminating the structure of a natural language, on which sample a child is (falsely) supposed to base its developing "theory" of the language it eventually masters.[98] What, in effect, we have shown is that *we need not postulate a system* (in the sense both structuralists and nativists favor) *in order to account for the harmonious processing of cultural practices in accord with the assumption of methodological individualism – namely, that the only intelligent, culturally informed agents are human individuals.* It is entirely possible, as we see, that the required harmony – even under conditions of historical change, open-ended process, division of social labor and of social mastery with regard to given practices – can, in principle, be accounted for, *if* we concede the interpretive (or hermeneutic) nature of the relationship between practices and their instantiations (in the direction both of generating pertinent behavior and of consensually accepting improvisation). Also, of course, the rejection of a totalized or universal *system* of language or social practice, the rejection of total infrapsychological internalizing, the insistence on the social nature of language and cultural practice, the admission of improvisation and diachronic change within natural practices, *and* the provision of consensual interpretation as essential to the smooth functioning of cultural practices all conspire to undermine a merely computational or representational (or modular) theory of mind.

Viewed through these themes, Wittgenstein's notion of "forms of life" proves a most promising sketch of a theory of *accessing* societal or cultural "texts" – one that outflanks Chomsky's nativism, shows why we are hardly obliged to subscribe to it, and pinpoints its enormous and essential lacuna. The obvious irony is this: *if* Chomsky is right that a child does not rely primarily on the "degenerate sample," then the accessing problem is even more pressing than one might have supposed. In that case, only a theory such as the Wittgensteinian seems capable of providing the social precondition on which the solution of the "frame" problem Chomsky's own parsing thesis presupposes could be supplied. (We might of course speak here of human societies as "open systems," in the sense already sketched, except that the notion of an "open system" is not explicitly opposed to the formal constraints of "systems" in the technical sense here assumed. Our thesis means to insist, rather, on the peculiar asystematic nature of cultural life – within the terms of whatever regularities may obtain, at physical, biological, psychological, and societal levels, that may encourage the search for system-like domains.)

Nor need the muddles of methodological individualism concern us unduly. Even Popper, certainly a strong advocate of methodological individualism, rejects what he terms "methodological psychologism" – the doctrine (which he attributes to Comte and Mill) that all social phenomena may be explained entirely in terms of "psychological propensities alone" – and construes individual agency as pertinently functioning in a decidedly "institutional" context. As Popper has it, "we must try to understand all collective phenomena as due to the actions, interactions, aims, hopes, and thoughts of individual men, and *as due to traditions* created and preserved by individual men. But we can be individualists without accepting psychologism."[99] Popper is quite correct in his emphasis; although very probably he cannot be drawn into the analysis of the relevant process because of his basic opposition to what he takes historicism to be. (Also, Popper has remarkably little to say about the human sciences.)

There are rather few attempts at fixing the dynamic and dialectical process of mastering a practice and of altering it by the exercise of a given generation's mastery, which then presents a somewhat different "structure" to succeeding generations; but there are many promising accounts that resist reducing the social to the psychological and that feature the historical nature of human existence. Possibly one of the most perceptive summaries of the essential process is provided by the French anthropologist Pierre Bourdieu, who (conveniently for our purpose) opposes structuralism and nativism and manages to combine elements of Wittgensteinian, Marxist, and phenomenological themes.

First of all, Bourdieu rejects "the *realism of the structure*, which hypostatizes systems of objective relations by converting them into totalities already constituted outside of individual history and group history." Secondly, he insists on isolating the "*modus operandi*," the actual productive process of any "observed order," rather than any "algebraic" approximation of the putatively finished "*opus operatum*." Thus focused, he introduces the notion of *habitus*,

> systems of durable, transposable *dispositions*, structured structures predisposed to function as structuring structures, that is, as principles of the generation and structuring of practices and representations which can be objectively "regulated" and "regular" without in any way being the product of obedience to rules, objectively adapted to their goals without presupposing a conscious aiming at ends or an express mastery of the operations necessary to attain them and, being all this, collectively orchestrated without being the product of the orchestrating action of a conductor.

Habitus functions thus to enable "agents to cope with unforeseen and everchanging situations."[100] It is reasonably clear that Bourdieu's formulation takes into account the narrower (harmonizing) functions of ideology and the larger (harmonizing) functions of natural language as well as the linkage between the two. As Bourdieu rather trimly observes,

> the constitutive power [that is, the power to constitute recognizable social structures] which is granted to ordinary language lies not in the language itself but in the group which authorizes it and invests it with authority. Official language, particularly the system of concepts by means of which the members of a given group provide themselves with a representation of their social relations (e.g. the lineage model or the vocabulary of honor), sanctions and imposes what it states, tacitly laying down the dividing line between the thinkable and the unthinkable, thereby contributing towards the maintenance of the symbolic order from which it draws its authority.[101]

In effect, what Bourdieu confirms is the robust coherence of admitting

"regulated improvisations," improvizations responsive to what are reflexively perceived as the "objective conditions" of historicized life – so that
we can speak in a realist manner of culture as man's "second nature":
without invoking fixed or totalized rules or structures, without denying
order or harmony or a distinctive spirit to individual societies, and
without denying a measure of choice and alternative contingencies within
the real-time capacities of such societies.[102] Bourdieu's *habitus*, in short,
bridges the tacit mastery of a "form of life" (in Wittgenstein's sense)
and the relatively explicit forms of strategy, ideology, grammar, official
pronouncement and the like that serve at least as heuristic (as well as
effective) schemes in terms of which to coordinate and direct social
consensus, to facilitate favored internalizations of what a society's
practices are like, and (as we may now add) to guide (in both directions)
the interpretive harmony that a culture requires.

This is a substantial gain, therefore; for, without actually advocating
any particular ideological direction, we have offered a sketch of a
systematic answer to all those reductive theories of human intelligence
that have held it impossible to reconcile a strongly historicized sense of
human existence, a distinction between the psychological and the social,
and a complete absence of (totalized or universal) systems within which
(alone) apt behavior is said to be intelligibly generated and interpretively
grasped.

Two brief remarks may suggest where to find cognate notions in other
traditions. Lev Vygotsky, for instance, offers a broadly Marxist account
of the development of higher psychological processes – adumbrated in
the formula, "If one changes the tools of thinking available to a child,
his mind will have a radically different structure," which Vygotsky tried
to confirm by comparative studies of nonindustrialized and industrialized
societies and through which he takes issue with Piaget's egocentrism.
Vygotsky seems, however, to have bifurcated (in a manner that remotely
recalls Durkheim's thesis and less remotely Piaget's) a biologically
grounded and a culturally developed psychology – a disjunction his
Soviet colleagues rejected. But, even with respect to that issue, his
thought is at least a somewhat crude anticipation of Bourdieu's subtler
theme.[103]

Vygotsky was largely opposed to the non-historicized and undialectical
views of psychological development in Western behaviorism and in
Piaget. Another, possibly even more original Soviet thinker, Mikhail
Bakhtin, is more attracted to historicized communication in general than
to the orthodox (Soviet) Marxist account and is particularly opposed to
the static, totalized structures of Saussurean linguistics. Bakhtin's
remarkably rich conception of the multiple, interlocking strands of social

exchange – that meet and interact in all discourse (both within the utterance of any individual speaker and between speaker and audience, resonating through an entire society) – defies all system and organizes meaning in a strongly contextual way that (withal) does not lose a sense of linguistic and communicative order. Again, the notion is not really distant from Bourdieu's theme, although it suggests further, remarkably deep complications. The following summary is entirely representative:

> Every concrete utterance of a speaking subject serves as a point where centrifugal as well as centripetal forces are brought to bear. The processes of centralization and decentralization, of unification and disunification, intersect in the utterance; the utterance not only answers the requirements of its own language as an individualized embodiment of a speech act, but it answers the requirements of heteroglossia as well [that is, of the contextually pertinent different "voices" within and between speakers and respondents in their social and historical world]; it is in fact an active participant in such speech diversity. And this active participation of every utterance in living heteroglossia determines the linguistic profile and style of the utterance to no less a degree than its inclusion in any normative–centralizing system of a unitary language. Every utterance participates in the "unitary language" (in its centripetal forces and tendencies) and at the same time partakes of social and historical heteroglossia (the centrifugal, stratifying forces).[104]

The essential point has already been noted, of course: that is, the peculiarly resilient way in which one can link psychological and social processes so as to make human intelligence itself intelligible – in strongly historicized or praxicalized terms, without yielding to any totalized or nativist system. In doing that, what we must grasp is that the regularities that function in language and culturally cognate ways are to be drawn out of the *social existence* of biologically endowed persons – historicized, contingent, largely tacit, consensually disposed to accommodate improvisation – not, or at least not primarily, in accord with *particular abstract rules, conventions, codes of any sort.* This, ultimately, is the huge theme at stake in the disputed priority of syntactic or formal structures of any sort with regard to the intelligibility of the phenomena of the human sciences. Favoring the reliability of social existence over formulated rules entails favoring the open-ended, indefinitely pluralized, historically perspectivized, ecologically functioning cognitive capacities of entire societies over essentialist, foundationalist, and ahistorical claims.

To emphasize historicity is to emphasize how difficult it is to deny the strongly relativized (but hardly arbitrary) sense in which what will appear rational or reasonable in one historical setting may not appear so in another. In this sense, in being historicized, rationality is also relativized. A Gadamer, for instance, who would save "essential" humanity through the very transient contingencies of history – always opposing timeless, unchanging essences of course, always insisting on "the primordial communality that unites all the thought attempts of humanity, including that of our Western tradition"[105] – remains, for all his resistance, a closet essentialist. Gadamer cannot account for the "primordial communality" itself: from a historicized vantage, every regularity seemingly approximating the limit he favors is a fortunate contingency that both facilitates and depends upon our cognitive practice. There is no alternative. We cannot, in the short run, appeal to such a theme in adjudicating disputes about method or substantive claims *or* in dissolving the need to; and, since, in the historicized short run, there *is* no single, reliable principle of reason to pursue, we are bound to concede both the provisionality (possibly even the nonconvergence) of postulated alternative maxims or norms of rational inquiry (and conduct) *and* the significant difference between what may be plausibly so postulated for single or aggregated individuals at a given time and for a society taken collectively at that time.

Hence, just as a language is an idealization, over time, over the aggregated linguistic behavior of individual speakers, so, too, human rationality or (more specifically) the rationality of communities of scientific investigators is a (collective) idealization of some kind over the behavior of an aggregate of historically located individual investigators.[106] It is in this sense that the human sciences address themselves to collective human phenomena as well as to individual and aggregated phenomena – in spite of the fact that collective entities are fictional. Language, tradition, practice, institution, rule, rationality, method, era, genre, style, ideology, science, art, and related concepts are, pertinently, all addressed to the collective. But though collective entities are fictions, concepts of collective life are not; the need for a sense of historical continuity through change and the threat of change seems to ensure their nearly irresistible attraction for contingent aggregates of humans. Accordingly, they affect our lives and plans of action;[107] by a rather nice irony, they help to keep our sense of language and rationality from collapsing into no more than the aggregated small samples of behavior we ourselves exhibit in any instant or short interval of time.

V

We must catch up one loose end at least. The formal solution to the problem of methodological individualism requires that we distinguish collective *entities* (societies or collective states of mind) from collective or societal *attributes*; requires that we admit only individual human agents or aggregations of such agents as real; *but* requires that we also concede the ascribability (in realist terms) *of* societal attributes *to* such individuals. Broadly speaking, this accords with Popper's admittedly thin (perhaps otherwise questionable) advocacy of methodological individualism and rejection of methodological psychologism. (For Popper takes a very dim view of history – not very distant from Carl Hempel's. He has little to say that is promising about the human sciences; and he regularly intrudes his unsatisfactorily developed account of World 3 in speaking of human cultures.) In any case, *if* we admit that collective or societal attributes may be incarnate in biological phenomena *in the very same sense in which* distinctive psychological phenomena are incarnate, then *if* the societal or cultural cannot be reduced psychologistically (as Popper insists), the Intentional complexities of individual life must presuppose the distinctive Intentional dimension of a cultural space *within which* the other obtains and only obtains – even though (along the lines suggested by Wittgenstein, Bourdieu, Vygotsky, and Bakhtin) the relation between the individual and the societal is essentially symbiotic and subject, diachronically, to shifting causal influence in either direction (precluding "systems").

It may be shown[108] that the quarrel between Chomsky and Piaget turns on the possibility of grounding (cognitive) practices without admitting closed systems. But what may not be clear regarding their disagreement is that the full contest between them turns on the viability of a model developed more or less along the general lines of Wittgenstein's forms of life: that is, a model that

1 captures the internal coherence, plausibility, and explanatory power of irreducible societal or cultural structures;
2 concedes the symbiosis of the individual and the societal;
3 does not construe the collective entitatively;
4 accommodates the historicity of human existence;
5 precludes treating particular sectors of human space as systems; and
6 reconciles inquiries thus informed and the rejection of all forms of (cognitive) transparency.

Clearly, to admit that much is (effectively) to disallow any model of science adequate merely for the sub-Intentional phenomena of the physical sciences.

But our conclusion suggests the need for a final clarification. By a sort of serendipity, we may find what we need in scanning Jon Elster's reading of Marx's philosophy. In his paper "Marxism and Individualism" Elster characterizes "methodological individualism" (MI) thus:

> MI ... is the claim that all social phenomena – events, trends, behavioral patterns, institutions – can in principle [hence, "will sooner or later"] be explained in ways that refer to nothing but individuals – their properties, goals, beliefs and actions. In addition, MI claims that explanations in terms of individuals are superior to explanations which refer to aggregates. In a word reduction is both feasible (in principle) and desirable. The denial of MI is methodological holism [explanation in terms of irreducible "supra-individual entities," said to occur irreducibly "within intensional contexts"]."[109]

But this is a mistake. For it is entirely possible (indeed, it is quite reasonable) to reject Elster's version of MI (which Elster finds "trivially true" but which is *not* equivalent to Popper's formulation – also often taken to be "trivially true").

Elster's interpretation, which (he indicates) Marx on occasion "denies" or "violates"[110] but which he (Elster) believes could strengthen Marx's view in a "scientific" spirit, allows reference to collective entities only within the intentional content of individuals' beliefs and goals and the like. Certainly, the adjustment (read nonexclusively) is entirely reasonable and entirely in accord with all versions of MI. Elster's intended reductionism requires only that reference to collective entities *outside* the scope of the intentional operators ranging over individuals' beliefs and goals should and will eventually be reduced to ascriptions within their scope. The result is that Elster never provides for collective or societal *attributes* that neither entail supra-individual entities nor require reference to collective phenomena by way of making reference to collective entities within the intentional scope of the operators mentioned. In short, Elster never makes provision for the institutional or cultural space *within which* (and within which only) individuals exhibit the intentional life he concedes. To predicate that human beings speak a language, for instance, to admit that language has real structures that are irreducible psychologistically (in accord with Elster's MI), and to hold (nevertheless) that languages are not collective entities, is to deny

Elster's MI and not (yet) to subscribe to methodological holism. It is to admit only that the psychological states, behavior, and mode of existence of human beings possess properties that are "collective" but not reducible solely or adequately in terms of the intentional content of individual mental states or individual actional objectives or the like. Human existence is culturally structured; collective or societal attributes are ascribable first to aggregates of individuals and thereupon to the individual members of social aggregates; but the properties themselves are *not* expressible *in terms of aggregating the properties attributable in any way to the individual members of an aggregate* – unless (the very issue at stake) those properties already include properties not analyzable in terms of the intentional structure of individual belief states, actional objectives and the like.

To catch up what we have already explored: Elster's model would accommodate Chomsky's linguistics all right, but not Wittgenstein's view of language. Since the matter is plainly an empirical one, Chomsky clearly fails to entertain a distinct and viable alternative and Elster defines MI in an unacceptably narrow way. It is certainly reasonable to attempt to discern the "fine-grained" structures of Marx's notions of class struggle and class consciousness. But it is hardly a foregone conclusion that their proper analysis can be counted on to yield along lines restricted to Elster's MI: in moving from collective entities to aggregates of individuals to the individuals themselves, *nothing* is assured regarding the reducibility, or logical constraints on the use, of the collective predicates that Marx appears to require. It is not unreasonable to suppose that Marx may refer to "supra-individual entities," though that is not nearly as clear a matter as Elster apparently supposes; but it is more reasonable to suppose that, where he does not so refer but still speaks of collective phenomena, Marx does not (and could not) subscribe to Elster's MI and would (rightly) resist doing so. Elster offers no empirical reason for thinking that reduction in accord with his own version of MI is likely to be, or actually is, scientifically preferable to its denial (where, contrary to his claim) the denial does not entail methodological holism. He does not insist on the wholesale reducibility of all intentional discourse.[111] But if he does not, then he cannot but concede that the irreducibility of collective cultural attributes – salient at least in the human sciences – leads, in the absence of any successful reversal, to the bifurcation of the sciences themselves.

Furthermore, if the irreducibility of collective attributes is acknowledged, then it becomes entirely feasible to recover descriptions and explanations in accord with "sociological functionalism" (associated with Durkheim and, even more interestingly, with Ludwik Fleck)[112] – though without needing to adopt specific Durkheimian doctrines. Elster rejects such explanations: partly because he makes no provision for collective properties and partly because he requires "consequence laws" before admitting (in a Hempelian spirit) functional explanations. He offers the following summary (following Robert Merton's account) of what "a valid functional explanation in sociology would look like":

> An institution or a behavioral pattern X is explained by its function
> Y for group Z if and only if:
> (1) Y is an effect of X;
> (2) Y is beneficial for Z;
> (3) Y is unintended by the actors producing X;
> (4) Y – or at least the causal relation between X and Y – is
>   unrecognized by the actors in Z;
> (5) Y maintains X by a causal feedback loop passing through Z.[113]

The critical condition is (5) of course, which Elster would admit if something like a social analogue of biological or evolutionary feedback were available – which he thinks is not available.[114] But (5) does not actually require such a law-like mechanism, only a regular one – if (as already argued[115]) causality does not entail either nomologicality or strict invariance and if the Intentionality of human culture leads us to favor an agency model of causality. Also, (5) is readily instantiated – if informally – by the habitual linguistic, ritual, and related communicative practices of a natural society.[116] Even the very stability and converging structural congruities of a natural language may fairly be taken to instantiate (5). Once again, such a pattern encourages bifurcation.

The single pivot on which the entire issue of the relationship between the natural and human sciences depends – and with it the effective analysis of the whole of the world of human culture – asks whether, in theorizing about the internal structure of the relevant phenomena, in theorizing about "lower-" or more fundamental-level processes underlying the Intentional complexities of our "folk" domains, we have any reason to expect that the Intentional features usually encountered can be eliminated, replaced, or extensionally reduced in improving the explanatory power of the theories developed. The important point to remember is that there is no *a priori* assurance that can be given and that there are very strong empirically pertinent reasons against the expectation. *The Intentional features of molar human behavior may well persist at every level of sub-molar analysis.* If so, then, on an argument already supplied,[117] the human sciences may be obliged to proceed factorially rather than compositionally, top–down rather than bottom–up

– and may do so justifiably, coherently, and without the least loss of explanatory rigor.

The single, decisive barrier against the unity model of science confronts us in the implacable resistance of natural language to sub-linguistic reduction, resistance to the empirical provision of an extensional system of any kind fitting its improvisational power, resistance to its being construed essentially in psychological or infra-organismic terms. Language is the paradigm of human capacities and the *sine qua non* of all the actual work and behavior collected as human culture. The charge that the refusal to subscribe to the unity of science – or to subscribe to at least that much of it as is "methodological" and not yet "ontological"[118] or to that much of it as ensures, by judicious extensional approximation, the best prospects of extracting "covering-law" explanations from the human sciences – is itself a pretense or, worse, an intellectual scandal. There cannot be very much doubt that the question of how the natural and human sciences should be conceived as linked is hardly settled, or hardly settled in unity's favor.

If, however, anyone believes the contrary, let him say what important clue has been neglected. Otherwise, guided by a sense of the rigor proper to the lingual world of culture, we must be hospitable to modes of analysis that the physical sciences contentedly refuse. And, where we admit the thesis here supported, we may construe the irreducible space of Intentional (cultural) structures as the very context of all texts, the context of all histories and all contexts – altogether without benefit of system. Alternatively put, the would-be systems of the physical sciences as well as of the human sciences are context-bound projections of stable structural regularities (thoroughly extensionalized or extensionally regimented) that aspire to a totalizing power they cannot empirically ensure. The irony remains, therefore, that the universality of science is itself an artifact of the historicized informalities of human existence; although the regularities thus projected as context-free or totalized must, on a realist view, be more than mere artifacts. But, then, the artifacts of human culture – preeminently, persons – are, in a most profound sense, the basic realities *of* human culture.

Notes

1   Paul K. Feyerabend, "Mental Events and the Brain," *Journal of Philosophy*, LX (1963). Cf. Paul K. Feyerabend, "Explanation, Reduction, and Empiricism" in Herbert Feigl et al. (eds), *Minnesota Studies in the Philosophy of Science* (Minneapolis: University of Minnesota Press, 1962).

2   Paul K. Feyerabend, "Materialism and the Mind–Body Problem," *Review of Metaphysics*, XVII (1963).

3   Cf. Wilfrid Sellars, "Philosophy and the Scientific Image of Man," *Science, Perception and Reality* (London: Routledge and Kegan Paul, 1963). The two most salient recent attempts along these lines are of course those of Paul M. Churchland, *Scientific Realism and the Plasticity of Mind* (Cambridge: Cambridge University Press, 1979), and "Eliminative Materialism and Propositional Attitudes," *Journal of Philosophy*, LXXVIII (1981); and Stephen P. Stich, *From Folk Psychology to Cognitive Science* (Cambridge, Mass.: MIT Press, 1983). Stich's account attempts to sketch what is required, in a manner that is a most curious version of a strong structuralism (more in the spirit of Chomsky than of Saussure, but one that wishes to make no compromise at all with the semantics of "folk psychology" – roughly equivalent to Sellars's "manifest image"); cf. ch. 8, especially p. 153. Cf. also Bas C. van Fraassen, *The Scientific Image* (Oxford: Clarendon Press, 1980).

4   Churchland, "Eliminative Materialism and Propositional Attitudes," *Journal of Philosophy*, LXXVIII, 76.

5   Noam Chomsky, *Rules and Representations* (New York: Columbia University Press, 1980), p. 87.

6   Noam Chomsky, *Language and Responsibility*, tr. John Viertel (New York: Pantheon, 1979), pp. 140, 139.

7   Ibid., pp. 152–3. Cf. George Lakoff, "On Generative Semantics," in Danny D. Steinberg and Leon J. Jacobovits (eds), *Semantics: An Interdisciplinary Reader* (Cambridge: Cambridge University Press, 1971).

8   At this point in his remarks, he makes the extraordinary confession, "If it proves to be correct [that is, the thesis that grammar is a function of nonlinguistic factors], I would conclude that language is a chaos that is not worth studying . . ." (*Language and Responsibility*, p. 153). Cf. also Noam Chomsky, *Language and Mind*, enlarged edn (New York: Harcourt, Brace, Jovanovich, 1972).

9   Chomsky, *Rules and Representations*, ch. 1.

10   See Chomsky's remark cited in n. 8.

11   See ch. 9, above.

12   If, for instance, the syntactic is really inseparable, modularly, from semantic, pragmatic, experiential contingencies, then Chomsky's theory must be false – but not nativism merely as such. Jerome Bruner has speculated on prelinguistic capacities (socializing capacities) on which the development of the linguistic itself depends; see *Child's Talk* (New York: W. W. Norton, 1983). And Fodor specifically separates his own nativism (explicitly indebted to Chomsky's) from Chomsky's rather extravagant views about linguistic "organs" and concedes the compatibility (under carefully formulated conditions) between "vertical" (modular) faculties and "horizontal" (general, nonmodular) faculties functioning in language, perception, thought and the like; see Jerry A. Fodor, *The Modularity of Mind* (Cambridge, Mass.: MIT Press, 1983), p. 3.

13 See Noam Chomsky, *Knowledge of Language; its Nature, Origin, and Use* (New York: Praeger, 1986), p. 7.

14 See Terry Winograd, *Understanding Natural Language* (New York: Academic Press, 1976), especially the title essay. The accessing problem is very much like the "frame" problem: how do we determine (1) that a given sequence of sounds is rightly construed as meaningful speech; and (2) that that pattern of speech is rightly taken to function in this or that determinate way? (1) is usually treated as a transducing problem, although there are no known physical laws to account for its smooth solution; (2) is the frame (or accessing) problem, which has no known formal or modular solution. Chomsky clearly presupposes a favorable solution of (2), and the force of his own theory depends on such a solution. He nowhere explores it. See Fodor, *The Modularity of Mind*, pp. 112–17.

15 Robert Cummins, *The Nature of Psychological Explanation* (Cambridge, Mass.: MIT Press, 1984), p. 198 n. 1.

16 See Jean Piaget, *Genetic Epistemology*, tr. Eleanor Duckworth (New York: W. W. Norton, 1971).

17 See Massimo Piattelli-Palmarini (ed.), *Language and Learning; The Debate between Jean Piaget and Noam Chomsky* (Cambridge, Mass.: Harvard University Press, 1980).

18 See Jean Piaget, *Structuralism*, tr. Chininah Maschler (New York: Basic Books, 1970), particularly ch. 4.

19 Chomsky, *Knowledge of Language*, pp. 12–13.

20 Ibid., p. 27.

21 Ibid., p. 3.

22 Ibid.

23 Ibid., p. 43.

24 See Piattelli-Palmarini, *Language and Learning*.

25 See ch. 6, above; also Jerry A. Fodor and Zenon Pylyshyn, "How Direct is Visual Perception?," *Cognition*, IX (1981).

26 Fodor, *The Modularity of Mind*, p. 45 (italics added).

27 Ibid., p. 37; see also pt III.

28 Ibid., p. 103.

29 See ch. 6, above.

30 See Fodor, *The Modularity of Mind*, p. 68.

31 Noam Chomsky, *Problems of Knowledge and Freedom* (New York: Random House, 1971), p. 32 (italics added). See also Fred D'Agostino, *Chomsky's System of Ideas* (Oxford: Clarendon Press, 1986), ch. 3.

32 See Nancy Cartwright, *How the Laws on Physics Lie* (Oxford: Clarendon Press, 1983).

33 See ch. 8, above.

34 See Donald Davidson, "Mental Events," *Inquiries into Actions and Events* (Oxford: Clarendon Press, 1980).

35 Louis Hjelmslev, *Prolegomena to a Theory of Language*, rev. edn, tr. Francis J. Whitfield (Madison: University of Wisconsin Press, 1961), p. 9.

Cf. Chomsky, *Language and Responsibility*, pp.140, 152–3.

36 Cummins, *The Nature of Psychological Explanation*, ch. 1.

37 Ibid., p. 1.

38 Ibid., pp. 14–15.

39 Ibid., p. 15, 1; cf. p. 18.

40 Ibid., p. 28; cf. pp. 21, 29.

41 Ibid., p. 31.

42 Ibid., p. 35 (italics added).

43 It may be fairly claimed that Cummins's account is very close to Dennett's homuncularism, under another guise.

44 See Henry M. Hoenigswald, *Language Change and Linguistic Reconstruction* (Chicago: University of Chicago Press, 1960); also Noam Chomsky and Morris Halle, *The Sound Pattern of English* (New York: Harper and Row, 1968).

45 See Fodor, *The Modularity of Mind*, pt III.

46 Ibid., p. 3.

47 See Chomsky, *Rules and Representations*.

48 See Claude Lévi-Strauss, *The Elementary Structures of Kinship*, rev. edn, tr. James Harle Bell, John Richard von Sturmer, and Rodney Needham, ed. Rodney Needham (Boston, Mass.: Beacon Press, 1969).

49 Cf. Cummins, *The Nature of Psychological Explanation*, ch. 3, which (favoring Dennett's approach) rules out the kind of theory Chomsky and Fodor prefer. It is certainly not obvious that theories of the latter sort are incoherent. They could not be theories about computers or about "computers" in the sense Cummins characterizes.

50 There is a very suggestive analogue of this possibility developed, with respect to perceptual and cognitive capacities, in Ulrich Neisser, *Cognition and Reality* (San Francisco: W. H. Freeman, 1976), particularly ch. 4, which introduces the notion of open-ended "schemata." This represents Neisser's rejection of a rigorous Gibsonian psychology – which, in its own way, rejects a computational model of cognition. See J. J. Gibson, *The Sense Considered as Perceptual Systems* (Boston, Mass.: Houghton Mifflin, 1966).

51 See Piattelli-Palmarini, *Language and Learning*.

52 See Jerry A. Fodor, *Representations* (Cambridge, Mass.: MIT Press, 1981).

53 We need not go quite so far, of course, as Rom Harré does in construing the self as a social construction. The issue is empirically open, in any case. See Rom Harré, *Social Being* (Totowa, NJ: Rowman and Littlefield, 1979).

54 Hoenigswald, *Language Change and Linguistic Reconstruction*, p. 88.

55 Ibid., ch. 11.

56 Robert A. Hall, Jr, *Pidgin and Creole Languages* (Ithaca, NY: Cornell University Press, 1966), p. 5.

57 Personal communication. Hoenigswald maintains that the phenomenon is widely known but extremely difficult to document with precision. It is certainly quite reasonable. Further on the phenomenon, though essentially

on the phonological aspects of the matter, see Roman Jakobson, "Über die Phonologischen Sprachbünde" and "Sur la théorie des affinités phonologiques entre les langues," *Selected Writings*, vol. 1 (The Hague: Mouton, 1962). N. S. Trubetzskoy seems to have pioneered this sort of investigation – chiefly in terms of the distinction between *Sprachbünde* (language alliances) and *Sprachfamilien* (language families). There is no very good reason for supposing that the phenomena in question would be (or could be) restricted to the phonological, but pertinent evidence is hard to find. I owe the reference to Jakobson to Hoenigswald.

58 See Fodor, *The Modularity of Mind*, p. 49. See A. M. Liberman et al., "The Perception of the Speech Code," *Psychological Review*, LXXIV (1967), cited by Fodor.

59 Fodor, *The Modularity of Mind*, p. 50 (italics added).

60 Ibid., p. 57 (italics added).

61 This line of argument bears, also, on Fodor's (rather negative) reflections on the so-called "frame" problem, the problem of a "reflexive" view of perception, the problem of "cognitive penetrability," the problem of nonmodular "horizontal" faculties. All we need notice, here, is that these are not closed issues, certainly not closed as a result of Fodor's argument. See for instance ibid., pp. 80–94.

62 See ibid., pp.93–7, especially p. 97; also Eleanor Rosch et al., "Basic Objects in Natural Categories," *Cognitive Psychology*, VIII (1976), cited by Fodor.

63 Chomsky, *Rules and Representations*, p. 6.

64 Stich, *From Folk Psychology to Cognitive Science*, pp. 151, 153, 157–8.

65 Ibid., pp. 153, 158.

66 Ibid., p. 156.

67 Ibid., pp. 158–9 (italics added).

68 Ibid., p. 164. Cf. Zenon Pylyshyn, "Cognitive Representation and the Process–Architecture Distinction," *Behavioral and Brain Sciences*, III (1980), 16, cited by Stich.

69 Emile Durkheim, *The Rules of Sociological Method*, tr. Sarah A. Solovay and John M. Mueller, ed. George E. G. Catlin (New York: Free Press, 1964), p. 103.

70 Ibid., p. 98.

71 Ibid., pp. 103–4n.

72 Ibid., p. 104.

73 Ibid., pp. 97, 101. Cf. also Emile Durkheim, *Suicide*, tr. John A. Spaulding (Glencoe, Ill.: Free Press, 1951), and *Moral Education*, tr. Everett K. Wilson and Herman Schnurer (Glencoe, Ill.: Free Press, 1961).

74 Cf. Emile Durkheim, *The Elementary Forms of the Religious Life*, tr. Joseph Ward Swain (London: Allen and Unwin, 1915); and Steven Lukes, *Emile Durkheim: His Life and Work* (Harmondsworth: Penguin, 1973).

75 H. P. Grice, "Utterer's Meaning and Intentions," *Philosophical Review*, LXXVIII (1969), 151.

76 Cf. for example Paul Ziff, "On H. P. Grice's Account of Meaning,"

Analysis, XXVIII (1967), which was directed primarily against Grice's "Meaning," *Philosophical Review*, LXVI (1957).

77 The point is often missed in standard textbook treatments of Grice's theory: for example in Stephen C. Levinson, *Pragmatics* (Cambridge: Cambridge University Press, 1983), pp. 16–18. Cf. also Stephen R. Schiffer, *Meaning* (Oxford: Clarendon Press, 1972), chs 2, 5; and David K. Lewis, *Convention* (Cambridge, Mass.: Harvard University Press, 1969). Emphasis on convention, as in Lewis, strongly suggests relative fixity, formulability, closure, and the like. Improvisation, possibilities of alternative extension, tacit tolerances, historical existence itself tend to be neglected or minimized. Just how characteristic the suppression of the societal dimension is in Gricean-like accounts of linguistic communication and communicative intent may be readily seen in Dan Sperber and Deirdre Wilson, *Relevance: Communication and Cognition* (Cambridge, Mass.: Harvard University Press, 1986); see particularly chs 1–2.

78 See for example Edmund Husserl, *Cartesian Meditations*, tr. Dorion Cairns (The Hague: Martinus Nijhoff, 1960), Fifth Meditation, and *The Crisis of European Sciences and Transcendental Phenomenology*, tr. David Carr (Evanston, Ill.: Northwestern University Press, 1970). Cf. Paul Ricoeur, *Husserl: An Analysis of his Phenomenology*, tr. Edward G. Ballard and Lester E. Embree (Evanston, Ill.: Northwestern University Press, 1967).

79 On the social insects, see Edward O. Wilson, *The Insect Societies* (Cambridge, Mass.: Harvard University Press, 1971). See also Konrad Lorenz, *Studies in Animal and Human Behavior*, tr. Robert Martin, 2 vols (Cambridge, Mass.: Harvard University Press, 1970)

80 See James J. Gibson, *The Senses Considered as Perceptual Systems* (Boston, Mass.: Houghton Mifflin, 1966) and *The Ecological Approach to Visual Perception* (Boston, Mass.: Houghton Mifflin, 1979).

81 Cf. W. V. Quine, "Natural Kinds," *Ontological Relativity and Other Essays* (New York: Columbia University Press, 1969); and *The Collected Papers of Charles Sanders Peirce*, ed. Charles Hartshorne, Paul Weiss and Arthur W. Burks, 8 vols (Cambridge, Mass.: Harvard University Press, 1931–58), 4.307.

82 This is the sense of "emergent" usefully fixed by Herbert Feigl, *The "Mental" and the "Physical": The Essay and a Postscript* (Minneapolis: University of Minnesota Press, 1967). Feigl's own views do not affect the terminological matter at all.

83 Cf. P. F. Strawson, *Individuals* (London: Methuen, 1959).

84 Cf. for example the concluding remarks in Jean Piaget, "Piaget's Theory," in P. H. Mussen (ed.), *Carmichael's Manual of Child Psychology*, 3rd edn, vol. 1 (New York: John Wiley, 1970), and repr. in Bärbel Inhelder and Harold H. Chipman (eds), *Piaget and his School: A Reader in Developmental Psychology* (New York: Springer, 1976). See also, J. S. Bruner, "The Ontogenesis of Speech Acts," *Journal of Child Language*, II (1975); L. S. Vygotsky, *Thought and Language*, tr. Eugenia Hanfmann and

Gertrude Vakar (Cambridge, Mass.: MIT Press, 1962); L. S. Vygotsky, *Mind in Society*, ed. Michael Cole et al. (Cambridge, Mass.: Harvard University Press, 1978). Cf. Margaret A. Boden, *Jean Piaget* (New York: Viking, 1979).

85 See Tom Rockmore, *Hegel's Circular Epistemology* (Bloomington: Indiana University Press, 1986), especially ch. 7, for a penetrating account of Hegel's opposition to foundationalism.

86 For a general overview of structuralism, see Joseph Margolis, "'The savage mind totalizes,'" *Man and World*, XVII (1984). See also Bourdieu, *Outline of a Theory of Practice*; Jean-Paul Sartre, *Critique of Dialectical Reason*, tr. Alan Sheridan-Smith, ed. Jonathan Rée (London: New Left Books, 1976).

87 See Claude Lévi-Strauss, *Structural Anthropology*, tr. Claire Jacobson and Brooke Grundfest Schoepf (New York: Basic Books, 1963).

88 See ch. 10, above.

89 See ch. 9, above.

90 See ch. 10, above.

91 John Searle, *Minds, Brains and Science* (Cambridge, Mass.: Harvard University Press, 1984), pp. 26–7.

92 Cf. ibid., pp. 18–22.

93 Ibid., p. 99.

94 For a cognate admission, see Fodor, *The Modularity of Mind*, p. 125: "any psychology must attribute some endogenous structure to the mind. ... And it's hard to see how, in the course of making such endogenous structure, the theory [that does so] could fail to imply some constraints on the class of beliefs that the mind can entertain." Cf. also ibid., p. 238f, n. 42.

95 Ludwig Wittgenstein, *Philosophical Investigatons*, tr. G. E. M. Anscombe (New York: Macmillan, 1953), pt II, p. 226.

96 Ludwig Wittgenstein, *Zettel*, ed. G. E. M. Anscombe and G. H. von Wright, tr.G. E. M. Anscombe (Oxford: Basil Blackwell, 1967), sections 297–8, 301, 302, 306, 308. Cf. Joseph Margolis, *Culture and Cultural Entities* (Dordrecht: D. Reidel, 1984), ch. 7.

97 Fodor is reasonably candid about his mistrust of any reliance on "context" and his inability to overcome its open-ended features – especially in characterizing what he calls "central systems"; see *The Modularity of Mind*, pt IV.

98 Chomsky, *Language and Mind*, pp. 26–7, 170–1.

99 Karl R. Popper, *The Poverty of Historicism*, 2nd edn. (London: Routledge and Kegan Paul, 1960), pp. 152–4; 157–8 (italics added).

100 Pierre Bourdieu, *Outline of a Theory of Practice*, p. 72.

101 Ibid., p. 21.

102 Ibid., pp. 78–9. Cf. also, Pierre Bourdieu and Jean-Claude Passeron, *Reproduction in Education, Society and Culture*, tr. Richard Nice (London: Sage Publications, 1977).

103 The remark is cited in Cole et al., *Mind in Society*, p. 126, and is apparently

taken from Edward E. Berg, "L. S. Vygotsky's Theory of the Social and Historical Origins of Consciousness" (Ph.D. Dissertation, University of Wisconsin, 1970). Cf. also the citation in Cole, *Mind in Society*, from A. N. Leontiev's, A. R. Luria's, and B. M. Teplov's preface to the *Development of Higher Psychological Functions* (from which the Cole volume is primarily drawn), p. 139 n. 3.

104   M. M. Bakhtin, "Discourse in the Novel," in Michael Holquist (ed.) *The Dialogic Imagination; Four Essays by M. M. Bakhtin*, tr. Caryl Emerson and Michael Holquist (Austin: University of Texas Press, 1981), p. 272. Cf. also, V. N. Vološinov, *Freudianism: A Marxist Critique*, tr. I. R. Titunik, ed. I. R. Titunik and Neal H. Bruss (New York: Academic Press, 1976), including Appendix II.

105   See Hans-Georg Gadamer, "The Heritage of Hegel," *Reason in the Age of Science*, tr. Frederick G. Lawrence (Cambridge, Mass.: MIT Press, 1981), p. 61.

106   Husain Sarkar has effectively shown in a most detailed way the paradoxical results of failing to distinguish between the rationality of individual scientists and the rationality of scientific groups. In particular, Sarkar shows how the preference for the best theory or theories favored by a single method (at a time or even over time), generalized aggregatively for rational individuals, will tend to reduce or oppose the proliferation of competing theories – which historically prudent sciences require. So what contributes to the rationality of individual scientists appears to defeat the rationality of the (collective) scientific community itself, which must be able to entertain the reliable proliferation of methods of working (as well as the proliferation of particular theories or research programs). Historical or praxical shifts in the orientation of small groups within the bounds of a somewhat Wittgensteinian view of social existence would, in real-time terms, actually tend to thwart the utopian objective of vindicating once and for all the uncompletable (but supposedly rational) preference for any single method (that would of course lead us inexorably to the paradox). See Husain Sarkar, *A Theory of Method* (Berkeley, Calif.: University of California Press, 1983), pp. 178–98.

107   This is perhaps a fair approximation of Anthony Giddens's useful notion of the "double hermeneutic"; see Anthony Giddens, *New Rules of Sociological Method* (New York: Basic Books, 1976), p. 162.

108   See Joseph Margolis, "Thinking about Thinking," *Grazer Philosophische Studien*, XXVII (1986).

109   Jon Elster, "Marxism and Individualism," in M. Descal and O. Gruengard (eds), *Knowledge and Politics: Case Studies on the Relationship between Epistemology and Political Philosophy* (Boulder, Col.: Westview Press, forthcoming). I have seen the essay in manuscript only. Elster notes in passing that the term "methodological individualism" was coined by Joseph Schumpeter, *Das Wesen und der Hauptinhalt der theoretischen Nationalökonomie* (Leipzig: Duncker and Humblot, 1908), pp. 88ff.

110 Karl Marx, *Grundrisse*, tr. Martin Nicolaus (Harmondsworth: Penguin, 1973), p. 651, cited by Elster. Elster's reading may be reasonably disputed. See also Jon Elster, *Making Sense of Marx* (Cambridge: Cambridge University Press, 1985), particularly "Explanation and Dialectics" and "Philosophical Anthropology"; and the perceptive discussion in Gyorgy Markus, *Language and Production* (Dordrecht: D. Reidel, 1986), p. II, section 1.

111 Personal communication.

112 See Ludwik Fleck, *Genesis and Development of a Scientific Fact*, ed. Thaddeus J. Trenn and Robert K. Merton, tr. Fred Bradley and Thaddeus J. Trenn (Chicago: University of Chicago Press, 1979). There is a very recent attempt by Mary Douglas to revive sociological functionalism, with care and skill, in *How Institutions Think* (Syracuse, NY: Syracuse University Press, 1986), particularly ch. 3.

113 Jon Elster, *Explaining Technical Change: A Case Study in the Philosophy of Change* (Cambridge: Cambridge University Press, 1983), p. 57. Elster adds (p. 70), "We explain an action intentionally (or *understand* it), ... when we are able to specify the future state it was intended to bring about."

114 See Arthur Stinchcombe, *Constructing Social Theories* (New York: Harcourt, Brace and World, 1968), cited by Elster.

115 See ch. 8, above.

116 See G. A. Cohen, "Functional Explanation, Consequence Explanation, and Marxism," *Inquiry* XXV (1982); and Robert K. Merton, *Social Theory and Social Structure* (Glencoe, Ill.: Free Press, 1957).

117 See ch. 5, above.

118 This is the dubious formula Grünbaum favors, as in his strong attack on the hermeneutic tradition. See Adolf Grünbaum, Introduction to *The Foundations of Psychoanalysis: A Philosophical Critique* (Berkeley, Calif.: University of California Press, 1984). The formula is "dubious" simply because there is no way to support a methodology in science that does not make some claim about its felicitous grasp of the real structures of the domain it means to explore.

Index